Lecture Notes in Networks and Systems　　1041

The series "Lecture Notes in Networks and Systems" publishes the latest developments in Networks and Systems—quickly, informally and with high quality. Original research reported in proceedings and post-proceedings represents the core of LNNS.

Volumes published in LNNS embrace all aspects and subfields of, as well as new challenges in, Networks and Systems.

The series contains proceedings and edited volumes in systems and networks, spanning the areas of Cyber-Physical Systems, Autonomous Systems, Sensor Networks, Control Systems, Energy Systems, Automotive Systems, Biological Systems, Vehicular Networking and Connected Vehicles, Aerospace Systems, Automation, Manufacturing, Smart Grids, Nonlinear Systems, Power Systems, Robotics, Social Systems, Economic Systems and other. Of particular value to both the contributors and the readership are the short publication timeframe and the world-wide distribution and exposure which enable both a wide and rapid dissemination of research output.

The series covers the theory, applications, and perspectives on the state of the art and future developments relevant to systems and networks, decision making, control, complex processes and related areas, as embedded in the fields of interdisciplinary and applied sciences, engineering, computer science, physics, economics, social, and life sciences, as well as the paradigms and methodologies behind them.

Indexed by SCOPUS, INSPEC, WTI Frankfurt eG, zbMATH, SCImago.

All books published in the series are submitted for consideration in Web of Science.

For proposals from Asia please contact Aninda Bose (aninda.bose@springer.com).

Marcelo Zambrano Vizuete ·
Miguel Botto-Tobar · Sonia Casillas ·
Carina Gonzalez · Carlos Sánchez ·
Gabriel Gomes · Benjamin Durakovic
Editors

Innovation and Research – Smart Technologies & Systems

Proceedings of the CI3 2023, Volume 2

 Springer

Editors
Marcelo Zambrano Vizuete
Instituto Tecnológico Superior Rumiñah
Sangolquí, Ecuador

Miguel Botto-Tobar ⓘD
Eindhoven University of Technology
Eindhoven, The Netherlands

Sonia Casillas
Universidad de Salamanca
Salamanca, Spain

Carina Gonzalez
Universidad de La Laguna
Tenerife, Spain

Carlos Sánchez
Universidad de Zaragoza
Zaragoza, Spain

Gabriel Gomes
Universidad Estatal de Campinas
Campinas, Brazil

Benjamin Durakovic ⓘD
International University of Sarajevo
Sarajevo, Bosnia and Herzegovina

ISSN 2367-3370   ISSN 2367-3389 (electronic)
Lecture Notes in Networks and Systems
ISBN 978-3-031-63436-9   ISBN 978-3-031-63437-6 (eBook)
https://doi.org/10.1007/978-3-031-63437-6

This Springer imprint is published by the registered company Springer Nature Switzerland AG
The registered company address is: Gewerbestrasse 11, 6330 Cham, Switzerland

If disposing of this product, please recycle the paper.

Preface

The 4th edition of the International Research and Innovation Congress – Smart Technologies and Systems, CI3 2023, took place from August 30 to September 1, 2023, at the facilities of the Instituto Tecnológico Universitario Rumiñahui, located in the city of Sangolquĺ, Pichincha, Ecuador.

The conference was organized by the Red de Investigación, Innovación y Transferencia de Tecnología RIT2, made up of the most relevant university institutes in Ecuador, among which are ITCA, BOLIVARIANO, ARGOS, VIDA NUEVA, ESPÍRITU SANTO, SUDAMERICANO, ISMAC, SAN ISIDRO, ARTES GRÁFICAS, ORIENTE, HUMANE, SUCRE, CENTRAL TÉCNICO, POLICÍA NACIONAL and RUMIÑAHUI.

Additionally, the event is sponsored by the Secretaría de Educación Superior, Ciencia, Tecnología e Innovación SENESCYT, Labortorio de Comunicación Visual de la Universidad Estatal de Campinas—Brazil, Universidad Ana G. Méndez—Puerto Rico, Centro de Investigaciones Psicopedagógicas y Sociológicas—Cuba, Instituto Superior de Diseño de la Universidad de La Habana—Cuba, GDEON and the Corporación Ecuatoriana para el Desarrollo de la Investigación y la Academia—CEDIA.

The main objective of CI3 2023 is to generate a space for dissemination and collaboration, where academia, industry and government can share their ideas, experiences and results of their projects and research.

"Research as a pillar of higher education and business improvement" is the motto of the Conference and suggests how research, innovation and academia must coincide with the productive sector to leverage social and economic development.

CI3 2023 had 145 papers submitted, of which 52 were accepted for publication and presentation. To guarantee the quality of the publications, the event has a staff of more than 70 experts, from different countries such as Spain, Argentina, Chile, Mexico, Peru, Brazil, Ecuador, among others, who carry out an exhaustive review of each proposal sent.

Likewise, during the event a series of keynote conferences were held, given by both national and international experts, allowing attendees to get in touch with the latest trends and technological advances around the world. Keynote speakers included: Ph.D. Carina González, University of La Laguna, Spain; Ph.D. Gabriel Gómez, State University of Campinas, Brazil; Ph.D. Carlos Sanchez, University of Zaragoza, Spain; Ph.D. Juan Minango, Instituto Tecnológico Superior Rumiñahui; Dr. Iván Cherrez, Universidad Espíritu Santo, Ecuador; and Ph.D. Nela Paustizaca, Escuela Politécnica del Litoral, Ecuador.

The content of this proceeding is related to the following topics:

- Smart Cities
- Innovation and Development
- Applied Technologies
- Economics and Management
- ICT for Educations.

Organization

General Chairs

Marcelo Zambrano Vizuete   Instituto Tecnológico Universitario Rumiñahui, Ecuador

Organizing Committee

Marcelo Zambrano Vizuete	Instituto Tecnológico Universitario Rumiñahui, Ecuador
Miguel Botto-Tobar	Eindhoven University of Technology, The Netherlands
Sonia Casillas	Universidad de Salamanca, Spain
Carina Gonzales	Universidad de La Laguna, Spain
Carlos Sánchez	Universidad de Zaragoza, Spain
Gabriel Gomes	Universidad Estatal de Campinas, Brazil
Benjamin Durakovic	International University of Sarajevo, Bosnia and Herzegovina

Steering Committee

Marcelo Zambrano	Instituto Tecnológico Superior Rumiñahui, Ecuador
Miguel Botto	Eindhoven University of Technology, The Netherlands
Karla Ayala	Instituto Tecnológico Superior Rumiñahui, Ecuador
Xavier Duque	Instituto Tecnológico Superior Rumiñahui, Ecuador
Ernesto Huerta	Instituto Tecnológico Superior Rumiñahui, Ecuador
Carmita Suárez	Instituto Tecnológico Superior Rumiñahui, Ecuador
Maritza Salazar	Instituto Superior Tecnológico Espíritu Santo, Ecuador
Wilfrido Robalino	Instituto Superior Tecnológico Vida Nueva, Ecuador

Roberto Tolozano	Instituto Superior Tecnológico Bolivariano, Ecuador
Carlos Pérez	Instituto Tecnológico Sudamericano, Ecuador
Alicia Soto	Instituto Superior Tecnológico ITCA, Ecuador
Lorena Ávila	Instituto Superior Tecnológico Policía Nacional, Ecuador
Denis Calvache	Instituto Superior Tecnológico ISMAC, Ecuador
Efraín Paredes	Instituto Superior Tecnológico IGAD, Ecuador
Paola Vasconez	Instituto Superior Tecnológico Oriente, Ecuador
Sandra El Khori	Instituto Superior Tecnológico San Isidro, Ecuador
Santiago Illescas	Instituto Superior Tecnológico Sucre, Ecuador
Jorge Calderón	Instituto Superior Tecnológico Argos, Ecuador
Trotsky Corella	Instituto Superior Tecnológico Central Técnico, Ecuador
Javier García	Instituto Superior Tecnológico Humane, Ecuador
Sandra Salazar	Instituto Superior Tecnológico American College, Ecuador
David Flores	Instituto Superior Tecnológico Cordillera, Ecuador
Marcelo Aguilera	Instituto Superior Tecnológico del Azuay, Ecuador

Program Committee

Carina González	Universidad de La Laguna, Spain
Carlos Sánchez	Universidad de Zaragoza, Spain
Sergey Balandin	Universidad de Helsinki, Finland
Silva Ganesh Malla	SKR Engineering College, India
Alberto Ochoa	Universidad Autónoma Ciudad de Juárez, Mexico
Antonio Orizaga	Universidad de Guadalajara, Mexico
Rocío Maciel	Universidad de Guadalajara, Mexico
Víctor Larios	Universidad de Guadalajara, Mexico
Yorberth Montes	Universidad de Zulia, Venezuela
Juan Minango	Universidad Estatal de Campinas, Brazil
Francisco Pérez	Universidad Politécnica de Valencia, Spain
Víctor Garrido	Universidad Politécnica de Valencia, Spain
Alberto García	Universidad Politécnica de Valencia, Spain
Javier Hingant	Universidad Politécnica de Valencia, Spain
Javier Prado	Universidad Técnica Federico Santa María, Chile
Marco Heredia	Universidad Politécnica de Madrid, Spain

Diego Paredes	Universidad de Zaragoza, Spain
Angela Cadena	Universitat de Valencia, Spain
Sonia Casillas	Universidad de Salamanca, Spain
Marco Cabezas	Universidad de Salamanca, Spain
Francesc Wilhelmi	Centre Tecnològic de Telecomunicacions de Catalunya, Spain
Santiago Vidal	Universidad Nacional Del Centro de la Provincia de Buenos Aires, Argentina
Adailton Antônio Galiza Nunes	Universidad Estatal de Campinas, Brazil
Leandro Bezerra De Lima	Universidad Federal de Mato Grosso Do Sul, Brazil
Andrea Carolina Flores	Embraer Defensa y Seguridad, Brazil
Elmer Levano Huamacto	Universidad Federal Do Parana, Brazil
Ari Lazzarotti	Universidad Federal de Goiás, Brazil
Ellen Synthia Fernandes de Oliveira	Universidad Federal de Goiás, Brazil
Fernanda Cruvinel Pimentel	Universidad Federal de Goiás, Brazil
Yuzo Lanoi	Universidad Estatal de Campinas, Brazil
Gabriel Gómez	Universidad Estatal de Campinas, Brazil
Pablo Minango	Universidad Estatal de Campinas, Brazil
Benjamin Durakovic	International University of Sarajevo, Bosnia and Herzegovina
Jhonny Barrera	Universidad Nacional de La Plata, Argentina
Daniel Ripalda	Universidad Nacional de La Plata, Argentina
Alonso Estrada Cuzcano	Universidad Nacional Mayor de San Marcos, Peru
Marcelo Zambrano	Instituto Tecnológico Superior Rumiñahui, Ecuador
Ana Zambrano	Escuela Politécnica Nacional, Ecuador
Angel Jaramillo	Universidad de las Américas, Ecuador
Ángela Cadena	Universidad de Guayaquil, Ecuador
Oscar Leon	Universidad de Guayaquil, Ecuador
Wladimir Paredes	Consejo de Aseguramiento de la Calidad de la Educación Superior, Ecuador
Miguel Angel Zúñiga	Universidad de Cuenca, Ecuador
Alex Santamaría	Universidad Laica Eloy Alfaro, Ecuador
Fabián Sáenz	Universidad de las Fuerzas Armadas ESPE, Ecuador
Darwin Aguilar	Universidad de las Fuerzas Armadas ESPE, Ecuador
Sergio Montes	Universidad de Las Fuerzas Armadas ESPE, Ecuador
Jhenny Cayambe	Pontificia Universidad Católica del Ecuador Sede Ibarra, Ecuador

Santiago Ushiña	Likatelec, Ecuador
Raisa Bernal	Instituto Tecnológico Superior Rumiñahui, Ecuador
Aníbal Altamirano	Universidad de las Fuerzas Armadas ESPE, Ecuador
Moisés Toapanta	Universidad Católica Santiago de Guayaquil, Ecuador
Ricardo Rosero	Instituto Tecnológico Superior Rumiñahui, Ecuador
Danny De La Cruz	Universidad de las Fuerzas Armadas ESPE, Ecuador
Johana Tobar Quevedo	Universidad de las Fuerzas Armadas ESPE, Ecuador
Jhony Caucha	Universidad Nacional de Tumbes, Peru
Mariana Lima Bandeira	Universidad Andina Simón Bolívar, Ecuador
Norma Molina	Universidad Israel, Ecuador
Elfio Pérez	Universidad Indoamérica, Ecuador
Cristian Tasiguano	Instituto Superior San Antonio, Ecuador
Lizbeth Suarez Morales	Instituto Superior Vida Nueva, Ecuador
Carlos Ruiz	Instituto Superior Vida Nueva, Ecuador
Wilfrido Robalino	Instituto Superior Vida Nueva, Ecuador
William Venegas	Escuela Politécnica Nacional, Ecuador
César Ayabaca	Escuela Politécnica Nacional, Ecuador
Dario Morales	Instituto Superior Vida Nueva, Ecuador
Pamela Villareal	Instituto Superior Vida Nueva, Ecuador
Darwin Tituaña	Instituto Superior Vida Nueva, Ecuador
Ismenia Araujo	Instituto Superior José Chiriboga Grijalva, Ecuador
Luz Marina Rodríguez	Instituto Superior José Chiriboga Grijalva, Ecuador
Luis Alzate	Instituto Superior Bolivariano, Ecuador
Noemi Delgado	Instituto Superior Bolivariano, Ecuador
Marítza Salazar	Instituto Superior Espíritu Santo, Ecuador
Martha Fernández	Instituto Superior Espíritu Santo, Ecuador
Néstor Xavier Maya Izurieta	Instituto Superior Central Técnico, Ecuador
Michael Enrique Carrión Garzón	Instituto Superior Central Técnico, Ecuador
Nela Pastuizaca	Escuela Politécnica del Litoral, Ecuador
Marcelo Flores	Universidad Politécnica Salesiana, Ecuador
Diego Vizuete	Instituto Superior Sucre, Ecuador
Edgar Maya	Universidad Técnica del Norte, Ecuador
Mauricio Domínguez	Universidad Técnica del Norte, Ecuador
Fabián Cuzme	Universidad Técnica del Norte, Ecuador

Contents

Applied Technologies

Mechanical Testing and Durability Evaluation of 3D Printed Magnetic
Closures for Adaptive Fashion 3
    Daniel Ochoa Vallejo, Irlanda Aroca, Paulina Lara, and Angélica Paguay

Study on the Need for Pre-hospital Personnel Within the Office
of Sis-Ecu911 to Mitigate the Emergency 15
    Richard Santiago Cobos Lazo, María del Cisne Cuenca Soto,
    and Pablo Gerónimo Morocho Ochoa

Methodology for Declaration of Conformity Under ISO/IEC 17025
Associating Confidence Levels and Risk Analysis 30
    Carlos Velásquez, Daniela Juiña, Francisco Iturra, Byron Silva,
    and Diego Barona

How Women ICT Specialists Helped Ecuadorian Companies Thrive
During COVID-19 43
    Cynthia L. Román-Bermeo and Segundo F. Vilema-Escudero

Assessment and Selection of Fuel Models in Areas with High
Susceptibility to Wildfires in the Metropolitan District of Quito 54
    Juan Gabriel Mollocana Lara and Johanna Beatriz Paredes Obando

Data Science: Machine Learning and Multivariate Analysis in Learning
Styles 69
    Diego Máiquez, Diego Pabón, Mariela Cóndor, Gonzalo Rodríguez,
    Mauricio Farinango, and Ana Oyasa

Musculoskeletal Disorder Due to Exposure to Data Display Screen 82
    Mónica Monserrath Chorlango García,
    Patricia Lisbeth Esparza Almeida,
    Byron Sebastián Trujillo Montenegro,
    Hugo Jonathan Narváez Jaramillo, Edison Robinson Rodríguez Yar,
    and Kevin Andrés Rivera Vaca

Determination of Fire Affected Areas with BLEVE in Fuel Service
Stations (FSS) 98
    M. Córdova, M. Romero, I. Gavilanez, and A. Robalino

Information Security Management in Higher Education Institutions
in Compliance with the Organic Law for the Protection of Personal Data 110
   Karen Estacio

Food Loss as a Global Problem Identifiable in the Production Chain: The
Case of Oversupply in Agro-Products in Azuay-Ecuador 122
   Rafael Maldonado Yépez, Marco Antonio Gómez Parra,
   Richard Antonio Martínez Villegas, and Diana Sánchez Cabrera

Information Technologies: A Tool for Accident Prevention
in Gerontological Patients 131
   Toapanta Kevin, Novoa Geremy, Solórzano Josselyn,
   Quilumbaquin Mayra, and Vallejo Alejandra

Digital Competency Enhancement in Personnel Training and Development:
A Literature Review of Current Trends and Challenges 144
   Luis Fernando Taruchain-Pozo and Fátima Avilés-Castillo

Modeling Variability in the Readings of an 8-channel Color Sensor and Its
Uncertainty Estimation 155
   Francisco Espín, Eduardo Manzano, Carlos Velásquez,
   Consuelo Chasi, and Paola Andrade

Home Automation System for People with Visual Impairment Controlled
by Neural Frequencies 167
   Cristian Pozo-Carrillo, Jaime Michilena-Calderón,
   Fabián Cuzme-Rodríguez, Carlos Vásquez-Ayala,
   Henry Farinango-Endara, and Stefany Flores-Armas

Application of CNC Systems for the Manufacture of Electronic Boards 178
   Edwin Machay, Patricio Cruz, and Darwin Tituaña

Cuencabike Mobile Application for the Promoting of Bicycle Tourism
in the City of Cuenca, Ecuador 190
   Pablo Crespo López and Marco Guamán Buestán

Comparative Analysis of MH Versus LED Prototype Luminaires: Case
Study for Lighting Public Areas 203
   Gustavo Moreno, Diana Peralta, Jaime Molina, Carolina Chasi,
   and Javier Martínez-Gómez

Control System for Electrical Conductivity in Hydroponic Cultivation
of Lettuce 216
   Paul S. Espinoza, Marco V. Pilco, Vicente Reinaldo Cango Aguirre,
   and E. Fabian Rivera

Building a WhatsApp Chatbot with Node.js: File Processing and AI
Response Generation 229
   Juan Minango, Marcelo Zambrano, Wladimir Paredes, and Karla Ayala

Economics and Management

Software and Methods Applied in the Standardization of the Grinding
Process for Automobile Engines as a Business Solution 243
   Ana Álvarez-Sánchez and Alexis Suárez del Villar-Labastida

Productivity Strategy in the Peruvian Pharmaceutical Industry 251
   *Roli David Rodríguez-Palomino, Omar Bullón-Solís,
   Fernando Alexis Nolazco-Labajos, and Sheylah Hoppe-Coronel*

Efficiency as a Competitive Factor in MSMEs to Generate Profitability 260
   *Verónica Patricia Arévalo Bonilla, Paúl Rodríguez Muñoz,
   Marco Verdezoto Carrillo, Sergio Carrera Guerrero, and Alain Quintana*

ICT for Educations

The Impact of Virtual Reality on Reading Comprehension in Rural Areas 275
   Jorge Beltrán, Carolina Basantes, and Maritza Quinzo

ARTRI: A Gamified Solution for the Motor Stimulation of Older Adults
with Osteoarthritis of the Hands 287
   *Galo Patricio Hurtado Crespo, Ana C. Umaquinga-Criollo,
   Anddy Sebastián Silva Chabla, and Nelson David Cárdenas Peñaranda*

Criteria for Creative Pedagogical Practices in Writing an Essay on National
Reality Using the Logical Framework Approach 301
   *Lizzie Pazmiño-Guevara, Asdrúbal Ayala-Mendoza,
   Jorge Álvarez-Tello, and Marco Duque-Romero*

Collaborative Learning in Higher Education in Distance Mode 314
   Karina Salom é Ayala Jaramillo and Luis Fermando Ávila-Ascanio

Psychosocial Risks in the Work Environment of the ITCA Higher
Technological Institute 324
   *Patricia Lisbeth Esparza Almeida, Marlon Fabricio Hidalgo Méndez,
   Mónica Monserrath Chorlango García,
   Byron Sebastián Trujillo Montenegro,
   Juan Carlos Jaramillo Galárraga, and Luis Shayan Maigua Morales*

Low-Cost Didactic Models for the Teaching-Learning of Automated
Systems in Academic Environments 339
  Marco V. Pilco, Jerry Medina Schaffry, Paul S. Espinoza,
  Liliana M. Calero, and E. Fabian Rivera

Organizational Commitment and Satisfaction in University Teaching
Work: A Comparative Study 350
  Aleixandre Brian Duche-Pérez, Cintya Yadira Vera-Revilla,
  Olger Albino Gutiérrez-Aguilar, Fanny Miyahira Paredes-Quispe,
  Milena Ketty Jaime-Zavala, Brizaida Guadalupe Andía-Gonzales,
  and Javier Roberto Ramírez-Borja

Virtual Classroom Leveling Model Based on Self-regulation for the Higher
Technological Institute "San Isidro" 366
  Ángel Humberto Guapisaca Vargas and Verónica Gabriela Venegas Riera

Author Index 379

Applied Technologies

Mechanical Testing and Durability Evaluation of 3D Printed Magnetic Closures for Adaptive Fashion

Daniel Ochoa Vallejo[1,3] , Irlanda Aroca[2,3] , Paulina Lara[2,3(✉)] ,
and Angélica Paguay[2,3]

[1] Instituto Superior Tecnológico Ocho de Noviembre, Piñas, Ecuador
drochoa@institutos.gob.ec
[2] Instituto Superior Universitario Carlos Cisneros, Riobamba, Ecuador
lplarapadila@gmail.com
[3] Red de Investigación Sostenible y Sustentable en Indumentaria, Ambato, Ecuador

Abstract. This study focuses on the development and evaluation of 3D printed magnetic fasteners or buttons for adaptive fashion, with the goal of improving the dressing experience for people with disabilities. The research employs stereolithography (SLA), a 3D printing method, to create these fasteners or buttons, and compares their mechanical performance and durability to traditional fasteners. A factorial experimental design was used and samples were prepared with various types of fabric commonly used in adaptive clothing. Mechanical performance was evaluated using a tensile testing machine and custom designed devices to measure magnetic separation force. Durability was evaluated under simulated dry and wet conditions by measuring stitch button strength. The study underestimates the potential of these fasteners to improve the wearability and comfort of adaptable garments. The results of this research will serve as a basis for future innovations in adaptive clothing design and contribute to the continuous improvement of adaptive fashion products.

Keywords: Adaptive fashion · 3D printing · Magnetic closures · Mechanical testing · Durability evaluation

1 Introduction

Adaptive fashion is an emerging field within the fashion industry that focuses on designing and creating clothing specifically tailored to the needs of individuals with disabilities, seniors, or those with limited mobility. The primary goal of adaptive fashion is to provide these individuals with comfortable, stylish, and functional clothing options that are easy to put on and remove [1]. The importance of adaptive fashion cannot be overstated, as it can significantly improve the quality of life for many individuals, enabling them to express their personal style and feel confident in their appearance, while also simplifying the dressing process and fostering independence [2]. One of the key challenges in designing adaptive clothing is the development of innovative fastening solutions that cater to the unique needs of the target population. Traditional fastenings such as buttons, zippers, and snaps can be difficult to manipulate for individuals with limited dexterity or fine motor skills and may not provide the necessary ease of use or comfort [3]. Consequently, there is a growing need for novel fastening mechanisms that can be easily used by individuals with diverse abilities and requirements while maintaining durability and aesthetic appeal.

3D printing technology has experienced rapid advancements in recent years and has found applications in a wide range of fields, including fashion [4, 5]. Additive manufacturing techniques, such as Stereolithography (SLA) and Fused Deposition Modeling (FDM), allow for the rapid prototyping and production of custom-designed components, making them particularly well-suited for the development of innovative fastening solutions in adaptive fashion [6]. By harnessing the potential of 3D printing, designers and engineers can develop adaptive fashion solutions that not only meet the functional requirements of users but also reflect contemporary fashion trends and aesthetics [7].

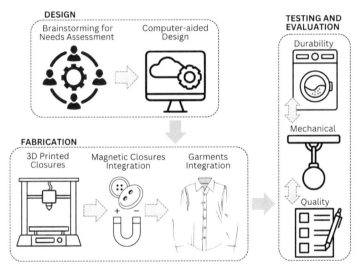

Fig. 1. Roadmap for creating customized clothing for individuals with disabilities through the utilization of 3D printing technology for Adaptive Fashion.

Figure 1 presents a roadmap for the design and fabrication of tailored clothing for people with disabilities using disruptive technologies such as 3D printing. The roadmap can be divided into four main stages: 1) Needs assessment, 2) Design, 3) Fabrication, and 4) Evaluation and feedback.

- Needs assessment: This initial stage involves identifying the specific requirements and preferences of individuals with disabilities. This process may include interviews, focus groups, or surveys to gather information on factors such as comfort, ease of use, and personal style preferences. This stage is crucial for understanding the unique challenges faced by the target population and ensuring that the final product addresses their specific needs [8].

- Design: In this stage, designers use the information gathered during the needs assessment to create clothing prototypes that incorporate innovative fastening solutions, such as 3D printed magnetic closures/buttons. Designers may employ computer-aided design (CAD) tools to create digital models of the garment components, including the fastening mechanisms, and use simulation software to assess their performance [9].
- Fabrication: Once the design has been finalized, the 3D printing process begins. Additive manufacturing techniques, such as Stereolithography (SLA) and Fused Deposition Modeling (FDM), are employed to create the custom-designed fastening components. These components are then integrated into the garments, either during the production process or as a post-production modification [10].
- Evaluation and feedback: After the garments have been produced, they are tested by individuals with disabilities to assess their fit, comfort, and ease of use. Feedback is collected from users to determine the effectiveness of the 3D printed fastening solutions and identify any areas for improvement. This feedback loop allows for continuous refinement of the design and fabrication process, ensuring that the final product meets the needs and preferences of the target population [11].

The objective of this study is to evaluate the mechanical performance and durability of 3D printed magnetic closures/buttons designed for adaptive garments and compare them to traditional fastenings. The study employed a factorial experimental design to assess the mechanical performance of samples using a tensile testing machine and custom-designed fixtures, measuring the magnetic separation force. Durability was evaluated through simulated dry and moist conditions, measuring the stitch button force. The study emphasized the potential of these closures to enhance the ease of use and comfort of clothing for individuals with disabilities. By assessing the performance of these closures under various conditions of materials and assembly used in adaptive clothing, and by measuring the magnetic separation force, we obtained valuable insights that will inform future innovations in adaptive clothing design.

2 Methodology

2.1 Conceptual Design of Magnetic Buttons

The design phase is crucial in applications about adaptive fashion. A key consideration is the target demographic for these magnetic buttons – adults with disabilities, which according to the World Health Organization, represent about 15% of the world's population [5]. As the prevalence of disability is expected to increase due to population ageing, chronic diseases, and improvements in disability measurement methodologies, the design of accessible, user-friendly buttons as closures becomes even more significant (Fig. 1).

3 3D Printing and Assembly of Magnetic Buttons

During the detailed design stage, we employed a design-for-3D-printing methodology using Tinkercad, a user-friendly and versatile free CAD software (Fig. 2). This critical design phase took into account several key factors, including size, shape, magnet

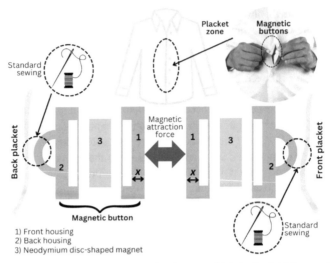

Fig. 2. The conceptual design of the magnetic closure, which is the initial stage in the design process. Magnetic buttons have been chosen for the closure mechanism. The variable 'x' denotes the thickness, which serves as a factor in the experimental design.

incorporation, compatibility with various garments, and the durability of the closure. The housings were manufactured using an FORM 3 (Formlabs, Boston MA), a 3D printing that utilizes SLA photocuring technology. The material used was the standard photoresin of Formlabs. The layer thickness for printing was set at 50 μm (μm). Following the printing of the housings, post-processing was carried out. This involved cleaning with isopropyl alcohol, drying with compressed air, and post-curing for 20 min using UV light. After these steps, the material support from the printing was removed. Finally, a N35 neodymium disc-shaped magnet with a diameter of 10 mm was enclosed between the front and back housings to complete the magnetic buttons (Fig. 3).

4   Magnetic Separation Force Experimental Procedure

Understanding the force required to separate magnetic buttons is crucial, particularly for individuals with disabilities. This force directly impacts the ease with which these individuals can remove their clothing. In this experiment, we measured this force and compared it with the force required to remove typical metal press buttons fasteners used in clothing (Fig. 4b).

A comprehensive experimental design was conducted using a full factorial design with two factors, each having three levels, and three replicates. The factors under consideration were the Front Housing Thickness (FHT) and Button Diameter (BD). The levels for FHT were set at 2 mm, 3 mm, and 4 mm, while for BD, the levels were 10 mm, 12 mm, and 15 mm. The response variable in this experiment was the Separation Force (SF) between the magnetic buttons (Table 1).

The system used to measure the SF was a mechatronic system, composed of a linear actuator and a force sensor with a load capacity of 100 N (FG-3006 Nidec-Shimpo)

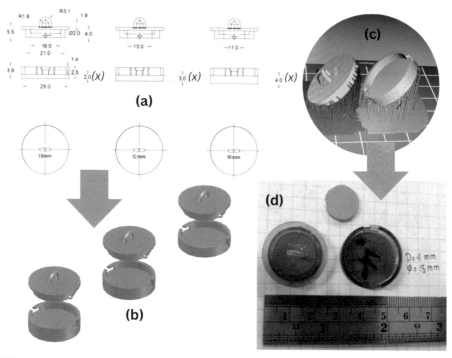

Fig. 3. (a) 2D CAD design of housings for magnetic buttons, (b) 3D models of housings, (c) Study of printability, and (d) 3D printed housings and preliminary assembly of magnetic buttons for closures.

(Fig. 4a). The testing speed was set at 5 mm/min, and all tests were conducted at a consistent temperature of 20 °C. This design allowed for a thorough investigation of the impact of FHT and BD on the SF of the magnetic buttons, providing valuable insights for further development and refinement of the product.

Table 1. Factors used to evaluate the proof of concept.

Factors	Symbol	Units	Levels
Front housing thickness	FHT	mm	2,3, 4
Button diameter	BD	mm	10, 12, 15

5 Button Stitch Separation Force Experimental Procedure

Comprehension of the force required to separate buttons stitched onto clothing is equally crucial, especially for individuals with disabilities. This force directly impacts the ease with which these individuals can remove their clothing. Moreover, an excessive force

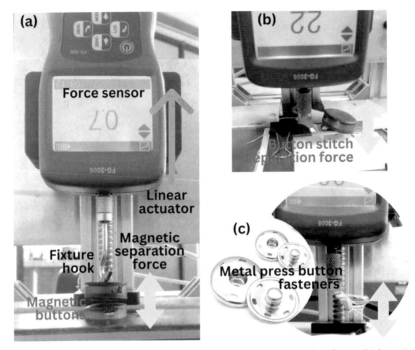

Fig. 4. Experimental procedure to measure (a) the magnetic separation force, (b) button stitch separation force, and (c) separation force of metal press button fasteners in the control system.

applied to separate the magnetic buttons could provoke the failure of the button stitch, further complicating the process of removing clothing. In this experiment, we measured this force using a cotton thread for the stitch, under both dry and moist conditions to simulate washing conditions.

The experimental design was similar to the one used for measuring the Separation Force (SF) of magnetic buttons. The factors under consideration were the Front Housing Thickness (FHT) and Button Diameter (BD). The levels for FHT were set at 2 mm, 3 mm, and 4 mm, while for BD, the levels were 10 mm, 12 mm, and 15 mm. The response variable in this experiment was the Separation Force (SF) between the buttons stitched with cotton thread (Fig. 4c). In this procedure, we did not use replicates, focusing instead on the effect of the cotton thread's condition (dry and moist) on the SF. Initially, we used dry cotton for the stitch. Then, we moistened the cotton thread with water to simulate washing conditions. This design allowed for a thorough investigation of the impact of the thread's condition on the SF of the stitched buttons, providing valuable insights for further development and refinement of the product.

6   Results and Discussion

7   3D Printing and Assembly of Magnetic Buttons

The detailed design stage was essential in the successful fabrication of the 3D printed magnetic buttons. The design process was meticulous, with a strong emphasis on several key factors, including size, shape, magnet incorporation, compatibility with various garments, and the durability of the closure. The use of SLA photocuring technology, was helpful in achieving our objectives. This technology allowed us to manufacture the housings with precision and consistency, which would have been challenging with other types of additive manufacturing technologies. The importance of using SLA 3D enabled us to create customized shapes of buttons, which is crucial for the functionality and aesthetic appeal of the magnetic closures. Other additive manufacturing technologies might struggle to achieve the same level of customization and precision.

However, the assembly process presented a significant challenge due to the properties of the N35 neodymium magnets. Despite their strength, these magnets are notably fragile, necessitating careful handling during assembly to prevent damage. Despite this, the magnets were successfully enclosed between the front and back housings, completing the magnetic buttons.

Creating a straightforward mathematical model to depict the relationship between the magnetic force and the properties of the 3D printed material was a complex endeavor, but necessary in the production of garment accessories. The behavior of the 3D printed photopolymer under different conditions, and its interaction with the magnetic force, is not easily predictable and warrants further investigation. Looking ahead, future studies could explore the use of different types of resins and their impact on the performance of the magnetic buttons. Additionally, refining the design and assembly process to minimize potential damage to the neodymium magnets could be a focus of future research. The development of a more accurate mathematical model to predict the behavior of the magnetic buttons under various conditions would also be a valuable contribution to this field of study. This study serves as a foundation for future investigations and advancements in the field of adaptive fashion. It opens up possibilities for greatly improving the way individuals with disabilities interact with their clothing.

The force necessary to remove typical metal press button fasteners used in clothing was measured three times, yielding an average of 13.5 N with a standard deviation of 0.87 N. This finding indicates that a considerable amount of force is required to manipulate these fasteners, which could pose a significant challenge for individuals with disabilities. Furthermore, it's important to note that many individuals with disabilities may not have the ability to use both hands to manage the closures on their clothing. This underscores the need for more accessible and user-friendly fastening solutions in the field of adaptive fashion.

8   Magnetic Separation Force Experimental Procedure

The force required to separate magnetic buttons is a critical factor, particularly for individuals with disabilities. This force directly influences the ease with which these individuals can remove the closures of their clothing. In this experiment, we measured this force and compared it with the force required to remove typical metal press buttons fasteners used in clothing. The results of the factorial analysis, as shown in Fig. 5, revealed significant understandings. According to the Pareto Chart of standardized effects, simple factors BD, SF, and the interaction BD*SF are significant. The range of the separation force for the magnetic closure varied between 0.8 and 2 N. This range provides valuable information for clothing designers working on adaptive fashion for people with disabilities, allowing them to design adequate closures depending on the type of clothing and the age of the individuals.

The interaction between BD and SF shows a relationship between the shape and dimensions of the housings, which could be studied with more advanced statistical models. This interaction suggests that both the size and shape of the button, as well as the force applied, play a significant role in the ease of separation. The residual plots were also analyzed to verify the assumptions of normality, independence, and equal variance. These plots further validate the reliability of our results and the robustness of our experimental design.

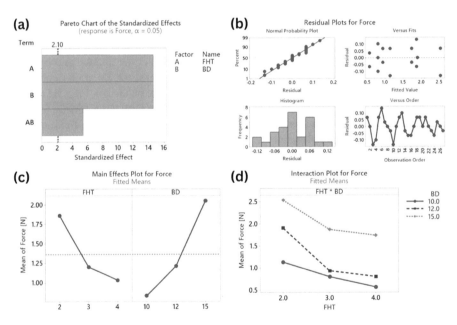

Fig. 5. Factorial analysis of magnetic separation force. (a) Pareto chart for significant effects, (b) Residual plots, (c) Factorial plots for Front Housing Thickness (FHT) and Button Diameter (BD), and (d) Interaction of FHT and BD.

9   Button Stitch Separation Force Experimental Procedure

The force required to separate buttons stitched onto clothing directly impacts the ease with which these individuals can remove their clothing. Moreover, an excessive force applied to separate the magnetic buttons could provoke the failure of the button stitch, further complicating the process of removing clothing. In this experiment, we measured this force using a cotton thread for the stitch, under both dry and moist conditions to simulate washing conditions. Our results showed that the force to break the button stitch for dry conditions varied between 7.6 and 14.6 N, and for moist or humid conditions, it varied between 9.4 and 19.8 N. These results indicate that the magnetic separation force is less than the force to break the stitch in dry conditions, which is a positive outcome for the usability of these magnetic closures.

However, the results under moist conditions suggest that further investigation is needed, especially under washing conditions. The increase in separation force under moist conditions could potentially pose challenges for individuals with disabilities in removing their clothing after washing. Future studies could explore this aspect in more depth, considering other experimental factors that could influence the separation force under washing conditions (Fig. 6).

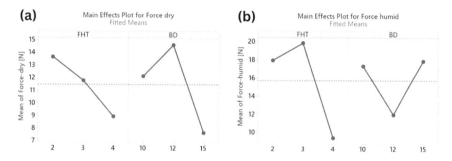

Fig. 6. Factorial analysis of the stitch button separation force under (a) dry conditions and (b) moist conditions.

10 General Discussion

The advent of 3D printing technology in the fashion industry has ushered in a new era of innovation, customization, and sustainability. As highlighted by Wu et al., 3D printing is not just a tool but a transformative force, especially in the realm of smart clothing, where it offers unparalleled opportunities for embedding sensors and electronics seamlessly into garments [12]. This convergence of technology and fashion not only enhances the functionality of clothing but also introduces a new dimension of interactivity and adaptability. The use of 3D printed magnetic buttons, as explored in this study, exemplifies the potential of this technology. Traditional fastenings, such as buttons and zippers, have long been a challenge for individuals with disabilities. The development of 3D printed magnetic closures offers a solution that is not only functional but also aesthetically pleasing and easy to use. This innovation, when viewed in the context of adaptive fashion, underscores the importance of 3D printing in creating garments that are tailored to the unique needs of every individual.

Chakraborty and Biswas delved deeper into the potential of 3D printing in the textile and fashion industry, emphasizing the use of polymer-fiber composites [13]. These materials, when 3D printed, can achieve unique textures and properties, allowing for the creation of garments that are both functional and fashionable. The ability to produce such innovative materials and structures is particularly relevant when considering the design and functionality of magnetic buttons, offering a level of customization and performance that is hard to achieve through traditional methods. Khajavi's exploration of additive manufacturing's role in promoting sustainable business models in the clothing industry is particularly pertinent [14]. The use of 3D printed magnetic closures, for instance, can reduce waste by ensuring precise production, eliminating the need for excess materials. Moreover, the on-demand nature of 3D printing means that garments can be produced as and when needed, reducing overproduction and inventory costs.

However, the broader adoption of 3D printing and its innovations, like magnetic closures, requires a shift in the fashion industry's mindset. Palomo-Lovinski and Hahn's study revealed a spectrum of attitudes towards sustainable practices in the fashion design industry [15]. While the potential benefits of 3D printing and sustainable practices are evident, there remains a need for education and collaboration to ensure widespread adoption. In summary, 3D printing technology, especially in the context of innovations like magnetic closures, holds immense promise for the fashion industry. It offers a unique blend of customization, functionality, and sustainability, paving the way for a future where fashion is not just about aesthetics but also about inclusivity and environmental responsibility.

11 Conclusion

This research has made significant progress in the field of adaptive fashion, particularly in the development and evaluation of 3D printed magnetic closures for adaptive garments. The study underscores the potential of these closures to enhance the ease of use and comfort of clothing for individuals with disabilities. By assessing the performance of these closures with various conditions of materials and assembly used in adaptive clothing and measuring the magnetic separation force, we have obtained valuable insights that will inform future innovations in adaptive clothing design.

The integration of 3D printing technology in the fashion industry, as highlighted in our general discussion, is not just a novel approach but a transformative one. It offers a unique blend of customization, functionality, and sustainability, which is pivotal in the realm of adaptive fashion. The use of magnetic closures, in particular, exemplifies the innovative solutions that 3D printing can bring to address long-standing challenges in the fashion industry.

Furthermore, this research offers a comprehensive understanding of how stitched magnetic buttons function under different conditions. This knowledge is crucial for the ongoing improvement of adaptive fashion products, with the aim of enhancing the clothing experience for individuals with disabilities. The insights gained from this study serve as a foundation for future investigations and advancements in the field of adaptive fashion.

Future Work: As we move forward, our research will expand to evaluate other types of garments, including outerwear, formal wear, and sportswear, to understand how 3D printed magnetic closures can be integrated across a broader spectrum of clothing. Additionally, we plan to explore new experimental configurations of 3D printing, delving into different materials, printing techniques, and post-processing methods to optimize the performance and aesthetics of the closures. There's also potential in collaborating with designers and individuals with disabilities to co-design garments that truly cater to their needs. By doing so, we aim to push the boundaries of what's possible in adaptive fashion, ensuring that clothing is not just functional but also a reflection of one's personal style and identity.

References

1. Kabel, A., Dimka, J., McBee-Black, K.: Clothing-related barriers experienced by people with mobility disabilities and impairments. Appl. Ergon. **59**, 165–169 (2017). https://doi.org/10.1016/j.apergo.2016.08.036
2. Chang, W.-M., Zhao, Y.-X., Guo, R.-P., Wang, Q., Gu, X.-D.: Design and study of clothing structure for people with limb disabilities. J. Fiber Bioeng. Inform. **2**(1), 62–67 (2009). https://doi.org/10.3993/jfbi06200910
3. "Anthony" Chen, X., Kim, J., Mankoff, J., Grossman, T., Coros, S., Hudson, S.E.: Reprise: a design tool for specifying, generating, and customizing 3D printable adaptations on everyday objects. In: Proceedings of the 29th Annual Symposium on User Interface Software and Technology, pp. 29–39. ACM, Tokyo Japan (2016). https://doi.org/10.1145/2984511.2984512
4. Gibson, I., Rosen, D., Stucker, B., Khorasani, M.: Additive Manufacturing Technologies, 3rd edn. Springer, Cham, Switzerland (2021)
5. World Health Organization and World Bank. World report on disability 2011.. https://apps.who.int/iris/handle/10665/44575 (2011). Accessed 05 Jun 2023
6. Ahmad, S., Miskon, S., Alabdan, R., Tlili, I.: Towards sustainable textile and apparel industry: exploring the role of business intelligence systems in the era of industry 4.0. Sustainability **12**(7), 2632 (2020). https://doi.org/10.3390/su12072632
7. Cumming, D.: A case study engaging design for textile upcycling. J. Text. Des. Res. Pract. **4**(2), 113–128 (2017). https://doi.org/10.1080/20511787.2016.1272797
8. Ghoreishi, M., Happonen, A.: The case of fabric and textile industry: the emerging role of digitalization, internet-of-things and industry 4.0 for circularity. In: Yang, X.-S., Sherratt, S., Dey, N., Joshi, A. (eds.) Proceedings of Sixth International Congress on Information and Communication Technology: ICICT 2021, London, Volume 3, pp. 189–200. Springer, Singapore (2022). https://doi.org/10.1007/978-981-16-1781-2_18
9. Jevšnik, S.: 3D virtual prototyping of garments: approaches, developments and challenges. J. Fiber Bioeng. Inform. **10**(1), 51–63 (2017). https://doi.org/10.3993/jfbim00253
10. Rathore, B.: Textile industry 4.0 transformation for sustainable development: prediction in manufacturing & proposed hybrid sustainable practices. Eduzone: Int. Peer Reviewed/Refereed Acad. Multi. J. **11**(01), 223–241, (2022). https://doi.org/10.56614/eip rmj.v11i1.229
11. Wilson, J.: Handbook of Textile Design: Principles, Processes and Practice, Repr. Woodhead, Cambridge (2004)
12. Wu, S., et al.: 3D printing technology for smart clothing: a topic review. Materials **15**(20), 7391 (2022). https://doi.org/10.3390/ma15207391
13. Chakraborty, S., Biswas, M.C.: 3D printing technology of polymer-fiber composites in textile and fashion industry: a potential roadmap of concept to consumer. Compos. Struct. **248**, 112562 (2020). https://doi.org/10.1016/j.compstruct.2020.112562
14. Khajavi, S.H.: Additive manufacturing in the clothing industry: towards sustainable new business models. Appl. Sci. **11**(19), 8994 (2021). https://doi.org/10.3390/app11198994

15. Palomo-Lovinski, N., Hahn, K.: Fashion design industry impressions of current sustainable practices. Fashion Pract. **6**(1), 87–106 (2015). https://doi.org/10.2752/175693814X13916967 094911

Study on the Need for Pre-hospital Personnel Within the Office of Sis-Ecu911 to Mitigate the Emergency

Richard Santiago Cobos Lazo[1]([⊠]) [iD], María del Cisne Cuenca Soto[1] [iD], and Pablo Gerónimo Morocho Ochoa[2] [iD]

[1] American College University Institute, Cuenca, Ecuador
{investigacion,maria.cuenca}@americancollege.edu.ec
[2] Cañar Firefighters, Cuenca, Ecuador
subjefaturacbcc@gmail.com

Abstract. The presence of prehospital personnel within the emergency system office is essential to ensure an efficient and coordinated response to emergency situations. Their training, experience and communication skills play a crucial role in assessing emergencies, allocating resources, and making informed decisions, helping to save lives and provide appropriate care to those in need. The objective of this work is to determine the need for prehospital personnel within the SIS-ECU911 office through the design of the operator profile to mitigate the emergency. The methodology has a quantitative descriptive approach, with a convenience sample of 20 Cañar Canton Firefighters, the collection instrument is through a questionnaire on the importance of a paramedic, Excel tabulation to determine the frequency of responses that tend to the importance of prehospital personnel. As a result, it is observed that 81% mention that the paramedic in the office acts as a crucial link between the callers, the emergency services in the territory, and other support services, such as the police, fire brigades, among others. Your ability to communicate clearly and accurately is essential to ensure a coordinated response and appropriate care. In addition, a professional that the radio operator must comply with and that he must have to implement a priori in the SIS-ECU 911 is designed as a proposal. Additionally, a Professional is proposed that details the basic aspects that the radio operator must comply with to provide a service effectively.

Keywords: Radio operator · Paramedic · Pre-hospital

1 Introduction

Pre-hospital personnel play a crucial role within the emergency system, due to their presence in the office or emergency coordination center it is essential to guarantee a rapid and efficient response to emergency situations.

At the Ecuadorian level, the emergency system is the Integrated Security Service Ecu 911, the institutions that solve the emergency responses of the population and are

M. Z. Vizuete et al. (Eds.): CI3 2023, LNNS 1041, pp. 15–29, 2024.
https://doi.org/10.1007/978-3-031-63437-6_2

articulated to this system are: Traffic and Mobility Management, Municipal Services, Military Services and Armed Forces, Citizen Security and National Police of Ecuador, Sanitary Management and Ministry of Health, Red Cross, IESS, Accident Management and Fire Brigades and Risk Management [1].

In some emergency systems, specific roles are assigned to different professionals based on their training and experience. Dispatchers are typically responsible for receiving and triaging emergency calls, while paramedics provide direct medical care on the ground. This division of duties may be based on considerations of efficiency and specialization.

The lack of paramedical dispatchers could be related to the availability of trained personnel in the area, and it is possible that the number of qualified paramedics to work as dispatchers in the SIS ECU 911 is not sufficient to cover all the operational needs of the system [1].

Emergency systems often have limited resources and must establish priorities based on the needs and demand of the population. It is possible that, in the case of ECU 911, resources may have been directed primarily towards recruiting and training medical and paramedical personnel to provide direct care in the field, rather than focusing on paramedical dispatchers. Due to the aforementioned, the following question arises: Why is it important to incorporate paramedical personnel into the SIS-ECU 911 office?

The importance of this project focuses on the contribution to the correct management of the information that enters the emergency rooms to reach the reduction of the mortality rate of the victims and patients.

By having a professional with a paramedic profile in the SIS-ECU 911 dispatch rooms, their knowledge and experience will allow them to adequately assess the situation and allocate the necessary resources efficiently, optimizing the response time to emergencies. In addition, you can provide basic first aid instructions and basic life support over the phone, which can make all the difference in the survival and well-being of patients [2].

Emergency Communication. It refers to the process of exchanging critical and relevant information during emergency or disaster situations. It is used to coordinate and aid, make informed decisions, and alert the public to events that pose an imminent risk to safety or well-being. Its main objective is to guarantee the safety of people and minimize damage in crisis situations. It may involve disseminating warnings, instructions, and advice on how to act in an emergency, as well as coordinating rescue and response efforts [3].

A good emergency radio operator requires practice, training, exercise, and a fundamental attitude, the same that is necessary to understand how knowledge should be applied. This is not just a radio amateur connecting a set and an antenna and transmitting. It is important to know the most efficient way to communicate a message adequately and effectively, how to operate in an emergency network, how to behave in critical or delicate situations, and how to handle sensitive information. Radio operators must know and accept their own limitations, as well as take advantage of their abilities [3].

Emergency Console Operator. There is a professional in charge of managing radio communications in an emergency control center. His main function is to receive, transmit and coordinate information related to emergencies and incidents that are taking place.

This may include coordinating the location of equipment, updating the situation in real time, and sharing critical information [4].

The emergency console operator maintains accurate and detailed records of all communications and actions taken during the incident. This is important for situation monitoring, post reporting and analysis of emergency responses.

Importance of a Paramedic as a Radio Operator. A radio paramedic is an emergency medical care professional who has skills in both medical care and emergency communications management. This role combines the responsibilities of a paramedic with the duties of an emergency console radio operator. A radio operator paramedic is usually found in emergency control centers, medical dispatch services or mobile health care units [5].

The main advantages of having a paramedic as a radio operator are the following:

Efficient coordination: A radio paramedic can coordinate and dispatch medical resources effectively in emergency situations. By having a solid understanding of emergency medical care, they can quickly assess the severity of a situation and allocate the necessary resources appropriately. This ensures that people who need urgent medical care receive help in a timely manner [6].

Clear and precise communication: Communication in emergency situations is crucial for the safety and proper care of patients. A radio paramedic has the ability to communicate clearly and accurately with medical response teams on the ground. They can provide vital information about the situation and provide up-to-date medical instructions to paramedics and other health professionals at the scene of the emergency [7].

Initial assessment and triage: As a paramedic, the radio operator can use his medical knowledge to perform an initial assessment of emergency calls and determine the severity of situations. This allows them to prioritize responses and allocate medical resources appropriately. In mass emergency situations, the radio paramedic can triage patients and allocate resources based on the severity of injuries or illnesses. [8].

Access to relevant medical information: As a paramedic, the radio operator has a solid understanding of medical procedures, care protocols, and best practices in emergency situations. This enables them to provide evidence-based medical advice and guidance to first responders on the ground. They can also access relevant medical information, such as medical history or allergies, to pass on to healthcare teams prior to their arrival at the scene of the emergency [8].

Stress management and support: Radio operator paramedics are trained to handle high-pressure and stressful situations. They can provide emotional and calming support to callers in times of crisis, which can help reduce panic and provide a sense of security. In addition, their medical knowledge allows them to provide basic first aid or CPR instructions before professional help arrives [8].

The Risk of a Communication Emergency. The risk of a communication emergency refers to the possibility of failures or interruptions in communication systems during emergency situations. These disruptions can have serious consequences and make it difficult to coordinate the response and share critical information. Some of the risks associated with a communication emergency are:

- Loss of connectivity: During an emergency, there may be interruptions in communication networks, either due to infrastructure damage, network overload or technical failures. This may result in the inability to transmit or receive important information in real time [9].
- Network congestion: In emergency situations, the number of people trying to use communication systems can increase dramatically. This can overload networks and make it difficult to transmit crucial messages. Network congestion can make communications slow or even impossible [9].
- Power outages: Emergencies, such as natural disasters or catastrophic events, can cause widespread power outages. This can affect communication systems that rely on electricity, such as telephone networks, radios, and emergency notification systems.
- Lack of redundancy: In some cases, communication systems may not have sufficient backup or redundancy measures to deal with an emergency. This means that if a system fails, there will be no alternatives available to maintain communications, increasing the risk of a communication emergency [9].
- Language or Cultural Barriers: During an emergency, there may be additional challenges to effective communication due to language or cultural barriers. Without translation services or staff trained in intercultural communication, important information may not adequately reach all affected people [9].

Basic Emergency Operator Protocol. It focuses on the evaluation, coordination, and appropriate response to emergency situations. Emergency operators are trained to handle stressful calls and make quick and accurate decisions to provide the help you need. The steps of the protocol are detailed below.

1. Call reception: The emergency operator receives the incoming emergency call. It can be through a telephone line, radio communication system or other communication platform [10].
2. Obtain basic information: The emergency operator obtains basic information from the caller, such as their name, location and contact number. This is important to be able to send help and communicate in case the call is dropped.
3. Emergency Assessment: The emergency operator listens carefully to the caller and asks specific questions to assess the nature and severity of the emergency. These questions may include details about the incident, the exact location, the number of people involved, and any additional risks [10, 11].
4. Prioritization of the response: Based on the information provided, the emergency operator determines the priority of the response. Emergencies are classified into different levels of severity to allocate the right resources and ensure a quick and efficient response [10, 11].
5. Coordination of aid: The emergency operator contacts and coordinates the necessary resources to respond to the emergency. This may include dispatching units of police, fire, emergency medical services or any other specialized team depending on the nature of the incident [12].
6. Provide instructions: In some cases, the emergency operator can provide basic first aid instructions or actions to take until help arrives. This may include cardiopulmonary resuscitation (CPR) techniques, bleeding control, safety maneuvers, etc.

7. Maintain communication: During the entire call, the emergency operator maintains communication with the caller. Provides updates on the progress of the response and can request additional information if necessary [10, 11].
8. Ending the call: Once the response has been coordinated and the necessary instructions have been provided, the emergency operator ends the call. However, it is important that the operator remain available for future communications or to provide additional assistance as needed [13, 14].

The Profesiogram. This is an administrative tool that makes it possible to specifically define the job position, the specific training and training needs of the applicant, as well as determine risk factors to which the employee may be exposed while carrying out their work and, lastly, the job description. It could include, if necessary, exclusions or restrictions on applicants depending on the level of risk declared [15, 16].

2 Materials and Methods

2.1 Methodological Design

Approach: Quantitative.
Type of study: Descriptive.

2.2 Area of Study

Personnel that are part of the dispatch console of the SIS ECU 911 Azogues radio operator and first responders of the Cañar Canton Fire Department.

2.3 Study Population

It was made up of a total of 20 firefighters. 6 of the SIS ECU 911 Azogues. 14 first responder firefighters from the Cañar Canton Fire Department.

2.4 Inclusion Criteria

- First responders to medical emergencies (paramedics).
- Radio operator of the SIS ECU 911 Azogues fire department console.
- Volunteer personnel from the Cañar Fire Department (CBCC).
- Personal emergency vehicle operators.
- First responders (Firefighters).

2.5 Exclusion Criteria

- Administrative staff of the Cañar Fire Department (CBCC).
- Administrative staff of the SIS ECU 911 Azogues.
- Trainee staff at Bomberos Cañar (CBCC) and SIS ECU 911 Azogues.

2.6 Methods, Techniques, and Instruments

Method: Deductive.
Technique: Surveys.
Instruments: Questionnaire.

2.7 Variables

a. Sociodemographics.

- Gender
- Age
- Profession
- Occupation

b. Independent Variable.

- Capacitation

c. Dependent Variable.

- Level of knowledge

2.8 Proposal Intervention

Two trainings were carried out where the project process is indicated to the participating population, they are requested to fill out the informed consent and the authorization for the application of the survey and training is presented.

Training one focused on the theoretical part on the issues of liaison communication between the radio operator and the first responders in attention to extra-hospital emergencies.

Training two was based on radio communication bases, frequencies and radio transmitters based on pre-hospital protocols with drills.

In the drills, scenarios of the radio operator with a paramedic profile and a profile in the health management area were placed, in order to compare the rapid response between these profiles.

Presentation of the profile to the authorities so that they can incorporate it into their institutions.

2.9 Tabulation Plan and Statistical Analysis

For the tabulation, a satisfaction questionnaire of five questions was used for its tabulation and analysis of results in Excel.

3   Results

When carrying out the drills and to verify with the literature of a radio operator, it is evident that the response and decision-making in an emergency is more prepared by a paramedic. These characteristics were observed in the exchange of information when dealing with an emergency with the person who needs help, it must be remembered that in these situations professionalism and the quality of the exact information that must be issued to the other institutions must be shown so that they can provide the best support service.

By implementing a paramedic as a dispatcher, response time and, above all, patient care are improved, seeking to maintain or improve the condition until the first responders arrive. In this project, a satisfaction questionnaire was carried out on the importance of the paramedical profile, where the following was obtained:

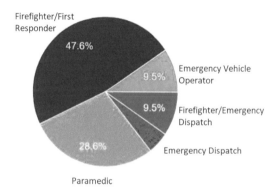

Fig. 1. Characterization of the occupation of Cañar firefighters and SIS ECU 911

In Fig. 1, a characterization was carried out on the occupation of the population where it is obtained that of 100%, 45% have a role of firefighter / First responder, with 28.8% the profile of paramedic and 9, 5% are operators of emergency units who work within the relief institution as first responders and firefighters who carry out emergency dispatch work in the SIS ECU-911. As the population that has been surveyed can be seen, it is linked to intervention actions in emergencies in first response.

Figure 2 shows that 57.1%, which is the majority, give a favorable response to the option where it indicates that the presence of a paramedic is very necessary in a health console of the SIS ECU 9-11, continuing with the analysis there is 23.8% where they answer that it would be necessary, and in a lower percentage it generates 14.3% where they mention that the presence of a paramedic on the console would not be necessary, therefore it is evident that the presence of this professional In dispatch, it is important for him to fulfill the role of radio operator.

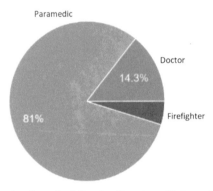

Fig. 3. Results of the question from the following list who will channel a pre-hospital emergency as a radio operator

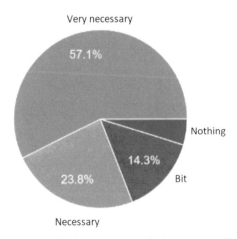

Fig. 2. Result of the response if it is necessary to deploy a paramedic as a radio operator.

Figure 3 shows that 81% of the surveyed population answers that; To mitigate a pre-hospital emergency with accurate knowledge, the radio operator should be a paramedical professional, with 14.3% saying that a doctor should be the professional on the console.

As part of the results obtained, the job description specifically designed for the emergency radio operator is shown, in which the main requirement indicates that they must be a Paramedic or an emergency doctor, this as a result of identifying the need

Table 1. Radio professional format - operator (part 1)

	Recommended professiongram chart for Radio Operator of the Dispatch Room of Sis Ecu 911		
No. 1	**Actualization No.1**	**Issue date:** 2022-11	
JOB PROFILE (For more information, refer to the descriptive letter of the position)	**Tasks Descriptions**	Problem identification, information gathering, oral and written expression, organization of information, judgment of decisions.	
	Environment in which the activity will be carried out	SIS-ECU 911 Situational Call Room	
	Knowledge and Experience	**Studies:** Higher Technological Minimum in paramedicine activities Specific knowledge: Reading and writing, Pre-hospital care, Updated trauma and disaster management course, patient stabilization, Accident Rescue course, Pre-hospital Ambulance driving course, Communication course (Radio operator), psychology in emergencies.	
	Required Competences	**Organizational**	**Required Level**
		Strategic understanding Organizational Commitment Achievement orientation Higher service orientation Teamwork emergency psychology	1 2 2 1 2
	Specific Competences	**Necessary to perform the position**	**Required Level**
		Attention to detail of emergency telephone calls Effective communication Flexibility Orientation to improvement in critical situations Pressure tolerance	2 3 2 3 2

for initial support in a way telematics or by telephone, the job model is shown below (Tables 1, 2 and 3):

Table 2. Radio-operator professional format (part 2)

Musculoskele-tal		X	EVALUACIONES			
Cardiovascular		x	Valora-tions	Pre-occupational exams	Annual Periodic Exams	Exit exam
Dermatological						
Neurological				Type IV blood count	Type IV blood count	NA
Anti-ergo-nomic	Prolonged or maintained postures	x	Labora-tories			
	Extreme and antigravity postures					
	Repetitive movements	x				
	Strength requirements					
	Drawing of the workstation	x	Func-tionals	Audiometry	Audiometry	Oc-cupa-tional Med-icine
Chemicals	particulate matter			Optometry	Optometry	
				Occupational Medicine	Occupational Medicine	
	Smoke			Electrocardiogram in people over 45 years of age	Electrocardiogram in people over 45 years of age	
	Vapors		Neuro-psycho-logical	Neurological (balance tests: Romberg, barany, Unterberg, Babinski Weil, dix-Hallpike)	Neurological (balance tests: Romberg, barany, Unterberg, Babinski Weil, dix-Hallpike)	N/A
	Liquids					
	Gases					

Table 3. Radio-operator professional format (part 3)

			POTENTIAL WORK RESTRICTION CONDITIONS, AND/OR UNFITNESS	
Physical	Noise	x		
	Vibration	x	Neurological Conditions:	Diseases that hinder or alter motor, sensory and cognitive activity. Vertigo or some other pathology that generates balance disturbances. Seizure disorders and/or epilepsy, sleep disturbances, hypersomnia, narcolepsy. syncope
	Radiation			
	Illumination	X		
	Temperatures	x	Breathing Conditions	Chronic or acute bronchopulmonary disease at the time of the psychophysical aptitude test. Bronchial asthma with severe crises up to a period of 2 years prior to the psychophysical aptitude test or that requires the constant use of bronchodilators.
Biological	**Vegetal Material**			
	Microorganisms and their toxins			
	Arthropods: crustaceans, arachnids, and insects		Musculoskeletal conditions:	Musculoskeletal pathologies with limitation or permanent restriction of movement of the neck, spine, upper and lower extremities. Osteoarthritis, rheumatoid arthritis, severe disc disease, acute or chronic untreated low back pain. Total or partial limb amputations. Chronic neuralgia and myopathies
	Vertebrates: urine, saliva, and hair.			
	Invertebrate animals, parasites, protozoa, worms, and snakes			
Mechanical	handling of sharp, sharp, and blunt elements		Visual conditions:	Refractive error that requires permanent optical correction and is not corrected, 20/30 binocular vision required with or without correction. Changes in color vision. Alterations that affect stereopsis.
	moving mechanisms	x		
	Handling of pressure equipment	x		
	Object projection		Hearing Conditions:	No alterations in conversational bands. Severe hearing loss, anacrusis, and chronic otitis, or other ear disorders that can cause balance disturbances.
	falling objects	x		
	fall from heights	x		

(continued)

Table 3. (*continued*)

Electrical	High tension		Psychological Conditions:	Acute or chronic psychiatric disorder (depression, psychosis, neurosis, etc.), History and/or evidence of alcohol and/or illicit drug abuse. Presence of phobia of heights (acrophobia).
	Medium tension			
	Low tension	X		
Locatives	Static Electricity	x	Gastrointestinal Conditions:	Chronic functional and organic pathology (e.g., gastroduodenal ulcer, ulcerative colitis, diverticulitis). No presence of abdominal hernias.
	Inadequate signage			
	Disorder	x		
	Smooth, wet, uneven, or defective floors		Cardiovascular Conditions:	Uncontrolled arterial hypertension, or one that is difficult to manage and with compromise of the target organ. Coronary heart disease producing recurrent angina and/or a history of acute myocardial infarction, Arrhythmias and/or Valvopathies (e.g., mitral, tricuspid, or aortic stenosis and regurgitation), Clinically significant peripheral vascular insufficiencies, maximum with GII varices. Moderate uncontrolled hyperlipidemia (according to the values and concomitant pathologies, the doctor will define if the alteration is restrictive for work at heights)
	Improper storage			
Public order	Riot			
	Kidnapping	X		
	Terrorism	X	Metabolic Conditions:	Unstable and insulin-dependent diabetes, hypoglycemia, obesity, and extreme weight loss (body mass index from 18 to 29.9), marked vitamin or protein deficiencies.
	Heist			
	Car accident			
Naturals	Earthquakes	X	Dermatological Conditions:	Skin diseases hypersensitive to solar radiation.
	Strong Winds			
	Floods			
	Electric Storms		Hematological Conditions:	Clinically significant hematopoietic dyscrasias (e.g., polycythemia, thrombocytopenia, thrombocytopenic purpura, anemia, hemophilia).
Psychosocial	Working Environment	X		
	Task conditions	X		
	Work organization	X	Other conditions:	Minimum age: 21 years and maximum 45 years at admission. Taking remaining medications that alter the state of consciousness and taking anticoagulant medication. Immediate post-operative (for example: abdominal, hand or lower limb surgeries)
	Time organization	X		

4   Discussion

Emergency dispatchers with a paramedic profile play a fundamental role in emergency response systems in Latin America. These professionals have a unique combination of medical skills and technical knowledge that enables them to provide valuable assistance in critical situations [17].

First, paramedical trained emergency dispatchers can quickly assess the severity of an emergency call and provide telephone instructions to callers to provide basic first aid until medical help arrives. Their knowledge of CPR (cardiopulmonary resuscitation) techniques and trauma management allows them to guide callers on how to administer immediate care and stabilize patients [17].

In addition, these paramedical dispatchers can collect crucial information during the initial call, such as a description of symptoms or patient status, and pass it on to medical response teams on the scene. This efficient and accurate communication helps prepare the necessary medical personnel and resources, which can save lives and optimize emergency medical care [18].

In some Latin American countries, emergency dispatchers with a paramedic profile may also provide continuous telephone medical advice while awaiting the arrival of emergency services. This can be especially useful in remote areas where medical resources are limited or access to specialized services may take time.

However, it is important to highlight that the performance of emergency dispatchers with a paramedic profile is subject to the infrastructure and resources available in each Latin American country. Appropriate training, constant updating of knowledge and the availability of efficient communication equipment are essential elements to guarantee the effectiveness of these professionals in their functions.

According to the review of the literature to make a comparison, it is found that, in Mexico, where it mentions (Official Mexican Standard PROY-NOM-227-SCFI-2017, Sect. 7.2.15 the profile for the position of a telephone agent or dispatcher Within the console, you must have at least a technical degree or equivalent as a paramedic. This regulation also includes Argentina and Colombia, seeking a better service for receiving emergency calls [19].

In Colombia and Chile, emergency dispatchers with a paramedic profile also play a fundamental role in emergency response systems. Although the specific details may vary in each country, in general, these professionals provide medical assistance and coordinate the response in emergency situations [19].

In Colombia, there are organizations such as the Urgencies and Emergencies Regulatory Center (CRUE) and the Medical Emergencies Regulatory Center (CRUM), which are responsible for coordinating and managing emergency calls. These centers often have emergency dispatchers with paramedic training. These professionals assess the seriousness of the situation, provide first aid instructions by telephone, and coordinate the mobilization of the necessary emergency services, such as ambulances or specialized medical personnel.

In Chile, the Emergency Medical Care System (SAMU) is in charge of managing medical emergencies. Paramedical emergency dispatchers, known as EMTs (Emergency Medical Technicians), receive emergency calls, perform an initial assessment, and provide first aid instructions over the phone. They also coordinate the dispatch of medical resources, such as ambulances or medicalized helicopters, depending on the severity and location of the emergency.

In both Colombia and Chile, emergency dispatchers with a paramedic profile must have specific training in pre-hospital care and be up to date on medical protocols and procedures. In addition, they must possess effective communication skills, stress management, and the ability to make quick and accurate decisions [19].

It is important to note that these professionals play an essential role in the emergency medical care chain and contribute to saving lives and providing timely and adequate care to those who need it. Their coordinated work and their ability to assess and respond efficiently to emergencies are essential for the proper functioning of the response systems in Colombia and Chile [19].

5   Conclusions

Through the development of the objectives set out in the study, it is concluded that the review of the literature focused on a radio operator and its importance in communication regarding emergency calls with a paramedical profile generates that these roles should be merged for better times. Response to an emergency, because the decision-making and information delivered to other institutions or SIS ECU911 personnel is accurate, allowing service improvement.

By having pre-hospital care personnel (paramedics) there is a plus for the emergency service, since they can carry out a first intervention by telephone, knowing that in an emergency response time is vital, they can guide whoever needs it. Need in a primary stabilization, at the same time the office would be in charge of sending the necessary team for care, in short, having this type of personnel would benefit society in general.

Another of the objectives is to carry out an analysis, based on simulations where a comparison is made where the profile of the paramedic is applied versus the profile he has, this was applied to 20 people, who are linked in first response, where it was observed that the dispatching paramedic provided information under prehospital care protocols, improving the quality of care and recommended that this profile be part of the SIS ECU911.

Acknowledgements. For the development and intervention of this project thanks to MSc. Sandra Salazar, Director of the American College University Institute and Cañar Fire Department.

References

1. Vásconez, J.J.P., Ortiz, C.A.N., Cordero, M.P.O., León, P.A.P., Orellana, P.C.: Evaluación del reconocimiento de voz entre los servicios de Google y Amazon aplicado al Sistema Integrado de Seguridad ECU 911. Revista Tecnológica-ESPOL **33**(2), 147–158 (2021)
2. Fundación Telefónica. Espacio fundación telefónica (2020). https://espacio.fundaciontelefonica.com/evento/historia-de-las-telecomunicaciones/. Access December 2022
3. QUADERNS DEL CAC. Convergencia Tecnológica y Audiovisual. En Rementol S, editor. Convergencia Tecnológica y Audiovisual. (Catalunya): Consell del I"Audiovisula de Catalunya (2019)

4. International Amateur Radio Unión. guía para Telecomunicaciones de Emergencia. Radio Club Argentino (RCA) ed. (Argentina): IARU (2016)
5. PHTLS J&BL. PHTL (edición) EEUU: Intersistemas (2016)
6. Association AH. BLS Texas: Integrador, LTD, 3210 Innovativa Way (2016)
7. Telecomunicaciones UId. U.I.T Suiza: Union Internacional de Telecomunicaciones (2015)
8. Álvarez Sigüenza JF (. Comunicación web 2.0 Madrid: Comunicación, marketing, community manager (2019)
9. https://thedigitalprojectmanager.com/. 9 ejemplos De Metodología De Un Proyecto, Simplificados. [Online]; 2019. Acceso 28 de SEPTIEMBREde 2020. Available in: https://thedigitalprojectmanager.com/es/metodologias-gestion-proyectos-simplificadas/.
10. sinnaps. https://www.sinnaps.com/ (2019). Access 28 de of September 2020. https://www.sinnaps.com/blog-gestion-proyectos/metodologia-cualitativa.
11. Sierra DPSBSDJPC. LOS SERVICIOS DE EMERGENCIAS MÉDICAS EN EL ECUADOR. Resvista de la Facutad de Ciencias Médicas de la Universidad de Cuenca.
12. KUKLINSKI HP. Mobile Web 2.0. Investigación realizada para el Grupo de Investigación de Interacciones.
13. coordinador médico SEMES. características de la coordinación médica de los sistemas de emergencias. 1st ed. Reyes MR, editor. Malagá; (2016)
14. Gobierno de México. Manual Técnico de la Norma para homologar características y TICs de los Centros de Control, Comando, Cómputo y Comunicaciones. única ed. 9–1–1 C, editor. Ciudad de México; (2016)
15. SEMES.ORG. LIBRO DE COMUNICACIONES - SEMES ESPAÑA (2016)
16. S: A:M: U: R protección civil. Protocolo de activación SAMUR civil Sp, editor. Madrid: (España). Sub-Dirección de Gestión Integral de Riesgo y Protección Civil. Protocolo para la activación de emergencias y recepción de llamadas de emergencias Iztalco Ad, editor. Iztacalco: (México) (2022)
17. Giraldo González, L.N., Peña Díaz, J.P.: Revisión narrativa: paramédico comunitario (2021)
18. ATENCIÓN PRE HOSPITALARIA-PARA REDUCIR LOS TIEMPOS DE RESPUESTA EN UNA EMERGENCIA EN EL CENTRO HISTÓRICO DE LA CIUDAD DE CUENCA (Doctoral dissertation).
19. Garnica González, T.M., Mena González, L.M., Moreno Ríos, J.J.: Estudio comparativo de los modelos de atención prehospitalaria entre Colombia y México (Doctoral dissertation, Universidad CES) (2016)

Methodology for Declaration of Conformity Under ISO/IEC 17025 Associating Confidence Levels and Risk Analysis

Carlos Velásquez[1,2,3(✉)] , Daniela Juiña[1] , Francisco Iturra[1] , Byron Silva[1] , and Diego Barona[1]

[1] Instituto de Investigación Geológico y Energético, Quito, Ecuador
`carlos.velasquez@geoenergia.gob.ec`
[2] Universidad Central del Ecuador, Modalidad en Línea, Quito, Ecuador
[3] Department of Applied Mathematics, University of Alicante, Alicante, Spain

Abstract. Accreditation under the ISO 17025 standard is crucial for quality management in testing and calibration laboratories. The methodologies and systematic approaches used for method validation and uncertainty calculations depend on the nature of the tests and calibrations conducted. The declaration of conformity holds significant importance within the standard. Normative references exist that enable the formulation of this declaration through specific methodologies, which require evaluating uncertainty across various scenarios.

In this study, we present a methodology that aims to establish a higher level of confidence in declarations of conformity by formulating hypotheses and quantifying the associated risks. This proposed approach facilitates a more comprehensive analysis of the declarations issued. The methodology has been successfully implemented in accredited laboratories specializing in both physical testing and chemical testing. Our findings demonstrate that this methodology is applicable to laboratories across various testing domains, irrespective of their respective areas of specialization.

Keywords: Quality management · Laboratories · ISO 17025 · Declaration of conformity · Uncertainty

1 Introduction

Laboratories that provide testing or calibration services seek to generate confidence in their customers, therefore, accreditation is a means by which laboratories demonstrate their technical competence and the validity of the results they issue [1]. In addition, accreditation allows worldwide acceptance of the results generated by accredited laboratories and is associated with a quality management system, which through periodic audits allows the maintenance and improvement of quality, leading to continuous improvement in customer services [2].

For a laboratory to be accredited, an authorized and independent accreditation body must evaluate and recognize that the laboratory has the necessary competencies to perform its activities and complies with a specific standard's requirements [3]. ISO/IEC

© The Author(s), under exclusive license to Springer Nature Switzerland AG 2024
M. Z. Vizuete et al. (Eds.): CI3 2023, LNNS 1041, pp. 30–42, 2024.
https://doi.org/10.1007/978-3-031-63437-6_3

17025:2017 "General requirements for the competence of testing and calibration laboratories" is the current standard with which these types of laboratories must be accredited and which aims to promote quality and confidence in the operation of laboratories [4].

Industry, research, and compliance with laws need reliable measurements on a wide variety of measurands in innumerable matrices, always seeking to maintain a certain level of quality [5]. However, due to the costs associated with facility adequacy, acquisition of testing equipment, personnel training, and technical limitations, testing and calibration laboratories are forced to limit the scope of their testing capabilities to certain types of tests and specific matrices [6]. This results in the existence of different "types" of laboratories. Our study case has a geological-environmental chemical testing laboratory and a physical testing laboratory focused on photometric testing [7].

The main quality parameters of the measurements are the bias, which refers to the difference between the obtained value and the true value and is usually already considered in the reported value, and the uncertainty of the measurement [8]. Uncertainty allows the proper use and interpretation of measurement results, as well as knowing the range in which the true value is expected to be found with a certain degree of confidence [9]. The calculation of uncertainty involves the initial identification of the sources of uncertainty, the estimation of the probability distribution that best fits the data sets for each source of uncertainty, the calculation of the individual uncertainties, and the combination of these to obtain the total uncertainty. This is reported with a probability that indicates the confidence that the true value is within the estimated range, called the "coverage factor." [10].

Although both types of laboratories apply the same principles for uncertainty estimation, in practice they are implemented in different ways. In chemical testing laboratories, the CG 4 EURACHEM/CITAC Guide Quantifying Uncertainty in Analytical Measurement recommends a "top-to-bottom" analysis, as opposed to the classical "bottom-to-top" system suggested by the JCGM 100:2008 Guide to the Expression of Uncertainty in Measurement better known as GUM. However, both methods are equally valid and allow the obtained result to be applied to the declaration of conformity [11].

Among the competencies of accredited testing and calibration laboratories, also known as conformity assessment bodies (CABs) [12], is the ability to issue declarations of conformity regarding the result(s) obtained from a test or calibration when requested by the client. The declaration of conformity allows the client to know if the product or service complies with legal and regulatory requirements and quality standards or to know where the product or service could be improved [13].

When making a declaration of conformity, in which compliance with a result is assessed, and a decision is made whether it passes or fails or is within or out of tolerance, ISO/IEC 17025:2018 mentions that the specification or standard against which the assessment is made and the decision rule must be defined [14], considering the level of risk associated with the decision rule employed, i.e. the risk of incorrect acceptance or rejection and the statistical assumption [15].

There are multiple recommendations for the declaration of conformity (ISO, JCMP) when a declaration of conformity is required during a measurement, i.e. to meet a particular specification and to establish whether it passes or fails there are two possibilities,

Fig. 1 shows cases A and B on a correct or incorrect decision regarding compliance with the specification [16].

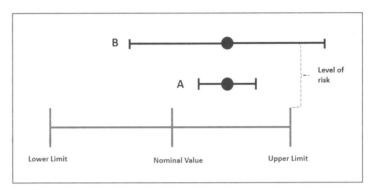

Fig. 1. Measurement decision risk.

Each measurement has its associated uncertainty, Fig. 1, the expanded measurement uncertainty in case A is within the upper and lower limit. Case B has a measurement similar to case A, but its uncertainty is outside the upper limit. At the time of accepting a result, there is a higher risk in case B.

The use of a safety zone helps to decrease the probability of an erroneous decision by taking into account measurement uncertainty. In Fig. 2 the safety zone "w" is defined as the Tolerance/Specification Limit "TL" minus the Acceptance Limit "AL" or "w = TL − AL." So if the measurement result is within the Acceptance Limit "AL," there is conformity between the measurement and the specification.

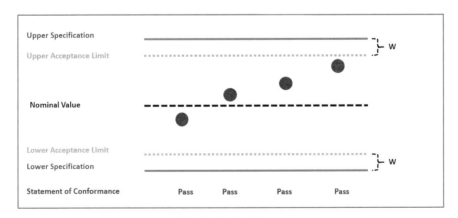

Fig. 2. Safety factor.

When the safety zone is equal to zero, "w = 0", it means that the acceptance is given when the result of a measurement is less than the tolerance limit and is called simple acceptance and implies that there is a shared risk that can reach up to 50% if the measurement value is located exactly in the tolerance limits and the measurement data represents a Normal Distribution. In this particular case, the risk must be assumed by Laboratory and the client.

There are two types of decision rules, the binary decision rule is when the results are limited to pass/fail, and the non-binary decision rule has several terms such as pass, conditional pass, conditional fail, and no pass.

1.1 Binary Statement for a Simple Decision Rule with a Safety Zone "W = 0"

The declaration of conformity will be reported as follows: passes if the measured value is within the acceptance limit or fails if the measured value is outside the acceptance limit (Fig. 3).

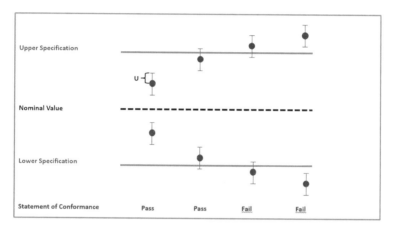

Fig. 3. Representation of a binary statement with simple acceptance.

1.2 Binary Declaration with Safety Zone

The declaration of conformity shall be reported as follows: passes, if the measured result is within the acceptance limit "AL = TL − w," or fails if the measured result is outside the acceptance limit "AL = TL − w" (Fig. 4).

1.3 Non-binary Declaration with Safety Zone

The declaration of conformity shall be reported as follows: passes if the measurement result is within the acceptance limit, "AL = TL − w"; passes conditionally if the measurement result is within the safety zone and is within the tolerance limit, in the range: [TL − w, TL]; do not pass conditionally if the measurement result is outside the tolerance limit, but within the tolerance limit plus the safety zone, in the interval: [TL, TL + w]; or, do not pass if the measurement result is outside the acceptance limit to which the safety zone has been added "TL + w" (Fig. 5).

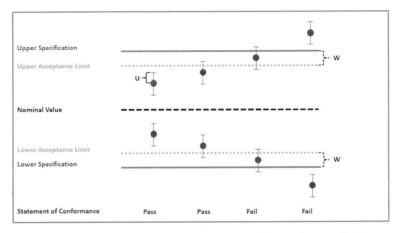

Fig. 4. Representation of a binary statement with security zone "w."

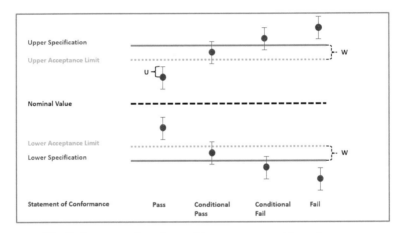

Fig. 5. Representation of a non-binary statement with a safety zone.

The present work proposes a methodology that can use the estimated uncertainty as EURACHEM and GUM and adds a confidence level analysis consistent with the probability distribution of uncertainty to perform a quantification of the risk at the time of making a declaration of conformity and deepen the descriptive cases introduced above, assigning probability values to each of the results obtained by the laboratory. For the case study within the risk analysis in the validation of the method and the decision rule, the 95% confidence level will be used.

2   Methodology

The physical characteristics test used in the experiment of this work determines the luminous efficacy of a visible radiation source, in this case, a public lighting luminaire. Luminous efficacy is defined as the quotient of the luminous flux and the consumed electric power[17]:

$$\varepsilon = \frac{\phi}{P}\left[\frac{lm}{W}\right]$$ (1)

where ϕ is total luminous flux and P is the active power. In the experiment, a LED-type public lighting luminary was used, and tested on a Sensing brand C-type rotating mirror goniophotometer and a YOKOGAWA W310 digital power meter [18]. The test was carried out in a laboratory accredited under the standard NTE INEN ISO/IEC 17025:2018 for total luminous flux evaluation.

The result can be associated with a measurement uncertainty that is obtained under the GUM standard. This uncertainty will have a level of confidence established by the laboratory. Therefore, if a specification is in place, it is possible to make a conformity declaration and mathematically estimate a confidence level to the declaration, and through this, a risk associated with it [19].

The chemical test focused on analysis for the determination of anions in human consumption water. This is carried out using a physicochemical separation technique known as Ion-exchange Chromatography with ion suppression system and conductivity detection, based on the "Standard Methods for the Examination of Water and Wastewater (23rd ed.): 4110B. Ion Chromatography with Chemical Suppression of Eluent Conductivity" [20].

In this test, the components of the liquid sample are separated by the difference in ionic affinities between a stationary phase and a mobile phase. Then the sample components are led to a conductivity detector, where they are identified by comparing their retention times with those of standards. This test was carried out in a laboratory accredited according to the standard NTE INEN ISO/IEC 17025:2018 for the determination of chemical compounds in geological and environmental samples [21]. The aim was to verify whether the water samples complied with the corresponding Ecuadorian regulations.

For the determination of uncertainty in the chemical laboratory, the measurand equation and the set of magnitudes that are directly determined in the measurement are defined, as well as those that are incorporated from external sources. This gives rise to the following mathematical model:

$$C = Xf_d\left[\frac{mg}{L}\right]$$ (2)

where C is the concentration, X is the equipment reading, and f_d is the dilution factor. To quantify the combined standard uncertainty, what is established by EURACHEM is applied, taking into account the mathematical model. And finally, the expanded uncertainty is obtained by multiplying the combined standard uncertainty by a coverage factor k of 2 (which provides a confidence level of around 95% in a normal probability distribution).

The test results are compared with the applicable Ecuadorian regulations, the "Unified Secondary Legislation, Environment" (TULSMA) for waters to determine the conformity of the test result.

3   Results and Discussion

The declaration of conformity under a confidence level obeys the probabilistic interpretation of the value, its uncertainty, and its specification on the probability distribution that is being used; therefore, it is important, to first define the probability distribution with which the analysis, will be performed.

After obtaining the standard uncertainty of the measurand according to the internal processes of the laboratories, it is necessary to check if there is a dominant uncertainty since this will define the probability distribution to be used. Once the confidence level of the expanded uncertainty estimate has been defined, it is necessary to take into account that statistical consistency determines that the confidence level of uncertainty should be the confidence level of the calibration of standards, of the validation of their methods, and environmental conditions and their assurance of the validity of results. Therefore, a good indicator for the confidence level of the conformity statement is the confidence level used by the laboratory. However, as it is an independent analysis, different confidence levels can be established for the declaration of conformity as long as they are justified.

To determine whether the result meets or does not meet objectively, it is possible to assign a hypothesis approach, where given the probability function $F(X) = P(X \leq x)F(X) = P(X \leq x)$, a confidence level NC is associated and consequently a value $\alpha = 1 - NC$. In this probability distribution, the given probability, its measure of central tendency, and its dispersion value will be calculated to meet the specification. If by comparing the probability, with the α value, we can accept the hypothesis that the result complies, otherwise, we reject this hypothesis. Figure 6 shows a simplified flow chart for performing this analysis.

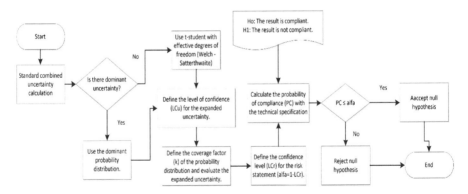

Fig. 6. General proposal for risk analysis in the declaration of conformity at a confidence level.

Based on the ISO/IEC 17025 standard for declarations of conformity, the decision rule and the technical specification must be defined. Thus, for each physical and chemical test, the following results are obtained:

3.1   Results of the Declaration of Conformity at a Given Confidence Level for Tests Performed in a Chemical Laboratory

The following results were executed with a normal probability distribution (dominant uncertainty reported by the laboratories) and the corresponding analyzes were carried out (Fig. 7):

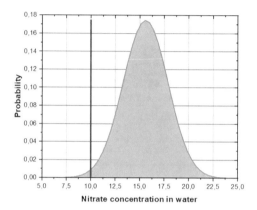

Fig. 7. Probability distribution of nitrate concentration values in the water of sample 1 with a maximum limit.

The value declared by the laboratory for the measurement of nitrate concentration in water is 15.57 mg/L ± 4.57 mg/L for sample 1. The maximum limit of nitrate concentration in water is 10 mg/L according to the TULSMA standard, and the criterion for acceptance of conformity is the declared value with 100% of its combined uncertainty. According to these criteria, the measurement of nitrate concentration in the water of the analyzed sample is not compliant at a confidence level of 95% (Fig. 8).

The value declared by the laboratory for the measurement of nitrate concentration in the water of sample 2 is 7.38 mg/L ± 2.16 mg/L. The maximum limit for nitrate in water is 10 mg/L according to the TULSMA standard, and the conformity acceptance criterion is the declared value with 100% of its combined uncertainty. According to these criteria, the measurement of nitrate concentration in the water of the analyzed sample is not compliant at a 95% confidence level. One interpretation of this result could be that the sum of the result to the maximum value being 9.54 mg/L is less than 10 mg/L, and could pass; however, after the probability calculation, this is correct only for confidence levels equal to or less than 88.7% (Fig. 9).

The value declared by the laboratory for the measurement of sulfate concentration in the water of sample 3 is 223.9 mg/L + 53.9 mg/L. The maximum limit of sulfate in water is 250 mg/L according to the TULSMA standard, and the criterion for acceptance of conformity is the declared value with 100% of its expanded uncertainty. According to these criteria, the measurement of nitrate concentration in water of a sample of the analyzed sample is not compliant at a confidence level of 95%. As in the previous case, this result is acceptable for confidence levels equal to or less than 68.6%.

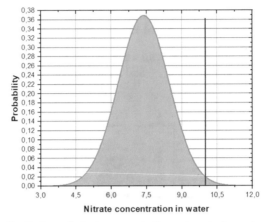

Fig. 8. Probability distribution of nitrate concentration values in the water of sample 2 with a maximum limit.

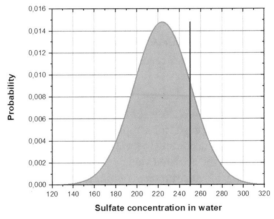

Fig. 9. Probability distribution of the values of sulfate concentration in the water of sample 3 with a maximum limit.

3.2   Results of the Declaration of Conformity at a Given Confidence Level for Tests Performed in a Physical Laboratory

The value declared by the laboratory for the efficacy measurement of a 75W LED luminaire is 138.08 lm/W + 7.78 lm/W. The minimum efficacy limit that the luminaire must have according to the manufacturer's technical specification is greater than or equal to 130 lm/W, and the conformity acceptance criterion is the declared value with 100% of its uncertainty. According to these criteria, the efficacy measurement of the analyzed LED luminaire is not compliant at a confidence level of 95%. In the critical case of the lower limit of the result, 130.3 lm/W is obtained, which is greater than 130 lm/W, and could pass; however, after the probability calculation, this is correct only for confidence levels less than or equal to 85% (Figs. 10 and 11).

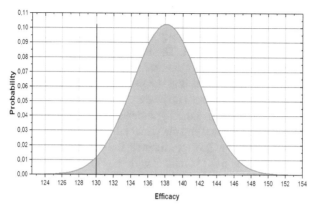

Fig. 10. Probability distribution of the efficacy values of a 75 W LED luminaire when having a minimum limit.

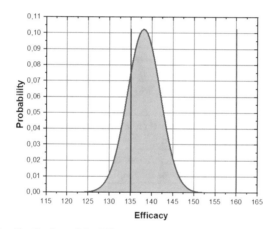

Fig. 11. Probability distribution of the Efficacy values of a 75 W LED luminaire when having a range.

For the previous case with a different technical specification, the luminaire must have an efficacy within the range of $> = 135$ lm/W and < 160 lm/W and with the acceptance criterion of conformity of the declared value with 100% of its uncertainty. According to these criteria, the efficacy measurement of the analyzed LED luminaire is not compliant under a confidence level of 95%.

3.3 Statistical Analysis of Risk in the Declaration of Conformity at a Confidence Level

With the declarations of conformity submitted, the laboratories must perform the statistical analysis of the risk associated with the decision rule of having issued, an incorrect acceptance or rejection. The results of each laboratory are subjected to an evaluation of the probability of occurrence of the limits within which the sample measurements should be, according to the technical specifications used in each case. For this, the hypothesis approach is used, in which the probability of conformity against the risk factor α is evaluated at a certain confidence level. The null hypothesis (Ho) states that the probability of conformity should be lower than the risk factor at a confidence level. While the alternative hypothesis (Ha) states that the probability of non-conformity should be higher than the risk factor at a confidence level.

The probability assessment and the hypothesis approach applied to the results of each test performed by the physical and chemical laboratories, and analyzed at different confidence levels, confirm that at 95% confidence, the declarations of conformity issued are correct. Table 1 shows the risk associated with the decision rule that the laboratories assume in making an erroneous declaration of conformity.

Table 1. Results of the statistical analysis of risk in laboratory declarations of conformity.

Laboratory	No. Sample	Limits	Risk
Chemical	Sample 1	Equal to 10 mg/L	88,9% risk of declaring compliant when it is not
	Sample 2	Equal to 10 mg/L	11,3% risk of declaring compliant when it is not
	Sample 3	Equal to 250 mg/L	31,4% risk of declaring compliant when it is not
Physical	75W Luminaire	Greater than or equal to 130 lm/W	15% risk of declaring compliant when it is not
	75W Luminaire	Greater than or equal to 135 and less than 160 lm/W	35% risk of declaring compliant when it is not

According to the above, the evaluation of the risk associated with the decision rule used in the declaration of conformity under a given level of confidence, it is possible to perform it with the same scheme, regardless of the nature of each test and the own errors generated during the taking of measurements.

Moreover, with this statistical analysis, laboratories can be sure that the declarations of conformity they have made are correct at a given level of confidence and knowing the risk they assume for having made a declaration.

4 Conclusion

The proposed methodology complies with the applicable technical standards for testing and calibration laboratories (ISO/IEC 17025:2017 "General requirements for the competence of testing and calibration laboratories"). Furthermore, regardless of the type of laboratory, it offers an alternative for issuing conformity declarations based on robust statistical tools that allow an understanding of the associated risk.

Issuing a conformity declaration with certainty for a conformity assessment body is paramount, for which, for applying this proposal one must start by accurately calculating the uncertainty associated with each test. This involves the probability distribution of the measured values and the confidence level at which the results will be presented. The conformity declaration is made with a probabilistic analysis of the risk associated with the decision rule employed by each laboratory, and whose results are evaluated at different confidence levels; thus, each laboratory is fully aware of the risk it assumes in its conformity declaration. Moreover, it allows for delivering a reliable result to the client on whether a product or service meets the quality parameters required or not.

From the cases of the studied laboratories, it is evident that the application of statistical tools is necessary to prevent laboratories from issuing a compliance certificate for a test item that is out of specification, or vice versa. This is due to potential legal, reputational, and other implications that could arise. It was also demonstrated that by applying the tools described in this work, the laboratories were able to satisfactorily address each of the issues in the conformity declarations studied, understanding the level of risk associated with each of their decisions. In the particular cases applied in the present work, results of 88.9%, 11.3%, 31.4% risk of declaring compliance when it is not were obtained for the 3 samples from the chemical laboratory. In the case of the physical laboratory, risks of 15% and 35% of declaring compliance when they did not comply were obtained for each sample tested.

Finally, the methodology can be perfectly complemented with the standardized techniques of conformity declaration known and implemented by most laboratories, contributing to the quantification of the confidence level in the declaration, whether binary, non-binary, with or without safety bands.

References

1. Braga, F., Panteghini, M.: The utility of measurement uncertainty in medical laboratories. Clin. Chem. Lab Med. **58**, 1407–1413 (2020)
2. Zima, T.: Accreditation of medical laboratories-system, process, Benefits for Labs. J. Med. Biochem. **36**, 231–237 (2017)
3. Ghernaout, D., Aichouni, M., Alghamdi, A.: Overlapping ISO/IEC 17025:2017 into big data: a review and perspectives. Int. J. Sci. Qual. Anal. **4**, 83–92 (2017)
4. INEN/ISO: Requisitos generales para la competencia de los laboratorios de ensayo y calibración (ISO/IEC 17025:2017, IDT) (2018)
5. JCGM: Vocabulario Internacional de Metrología – Conceptos fundamentales y generales, y términos asociados (VIM). Int. Organ. Stand. Geneva ISBN. 3ª Edición, 104 (2012)
6. Ram, A., Tiwari, S.K., Pandey, H.K., Chaurasia, A.K., Singh, S., Singh, Y.V.: Groundwater quality assessment using water quality index (WQI) under GIS framework. Appl. Water Sci. **11**, 46 (2021)
7. Sönmez, K.B., Kılınç, T.O., Yüksel, İ.A., Ön Aktan, S.: Inter-laboratory comparison on the calibration of measurements photometric and radiometric sensors. 10004 (2019)
8. Ma, L.-L., Wang, Y.-Y., Yang, Z.-H., Huang, D., Weng, H., Zeng, X.-T.: Methodological quality (risk of bias) assessment tools for primary and secondary medical studies: what are they and which is better? Mil. Med. Res. **7**, 7 (2020)
9. Araguillin, R., Toapanta, A., Juiña, D., Silva, B.: Design and characterization of a wireless illuminance meter with IoT-based systems for smart lighting applications. In: Botto-Tobar, M., Zambrano Vizuete, M., Diaz Cadena, A., Vizuete, A.Z. (eds.) Latest Advances in Electrical Engineering, and Electronics, pp. 129–140. Springer International Publishing, Cham (2022)

10. GUM: Evaluation of measurement data — Guide to the expression of uncertainty in measurement. Int. Organ. Stand. Geneva ISBN. 50, 134 (2008)
11. Eurolab: Cuantificación de la Incertidumbre en Medidas Analíticas. Eurachem/Citac. 2–27 (2012)
12. Singh, B., Ranjan, M.R., Chauhan, A., Verma, V.K., Jindal, T.: Accreditation of conformity assessment bodies in the field of environmental testing in India. Mapan 37, 261–268 (2022). https://doi.org/10.1007/s12647-022-00563-4
13. Stajkovic, S., Vasilev, D., Dimitrijevic, M., Karabasil, N.: Uncertainty of measurement and conformity assessment. IOP Conf. Ser. Earth Environ. Sci. 854, 12093 (2021)
14. Allocca, A., Panizo, M.M.: Modelo de acreditación de laboratorios de ensayos basado en la norma ISO/IEC 17025:2017 y el ciclo de Deming: Assay laboratory accreditation model based on the ISO/IEC 17025: 2017 standard and the Deming cycle. Tekhné. 24, 20 (2021)
15. Islek, D., Yukseloglu, E.: Accreditation of forensic science laboratories in Turkey in the scope of TS EN ISO/IEC 17025:2017 standard. Med. Sci. | Int. Med. J. 7, 962-966 (2018)
16. ILAC: ILAC-G8:09/2019 Guidelines on Decision Rules and Statements of Conformity (2019)
17. Pekur, D.V., Sorokin, V.M., Nikolaenko, Y.E.: Features of wall-mounted luminaires with different types of light sources. Electrica 21, 32–40 (2020)
18. Velásquez, C., Castro, M.A., Rodríguez, F., Espín, F., Falconi, N.: Optimization of the Calibration Interval of a Luminous Flux Measurement System in HID and SSL Lamps Using a Gray Model Approximation. ETCM 2021 – 5th Ecuador Tech. Chapters Meet. 1–7 (2021)
19. EPPO STANDARD: PM 7/98 (5) Specific requirements for laboratories preparing accreditation for a plant pest diagnostic activity. EPPO Bull. 51, 468–498 (2021)
20. Diana Buitrón, O., Barona, D., Iturra, F., Johana León, F.: Validación Del Método Para La Determinación De Oro Por Ensayo Al Fuego Combinado Con Espectrometría De Absorción Atómica En Muestras Geológicas Mineras Metalúrgicas. infoANALÍTICA 9(1), 119–136 (2021). https://doi.org/10.26807/ia.v9i1.188
21. Zhang, W., Hu, Z.: Recent advances in sample preparation methods for elemental and isotopic analysis of geological samples. Spectrochim. Acta Part B At. Spectrosc. 160, 105690 (2019)

How Women ICT Specialists Helped Ecuadorian Companies Thrive During COVID-19

Cynthia L. Román-Bermeo[1]([✉]) [iD] and Segundo F. Vilema-Escudero[2] [iD]

[1] Instituto Superior Tecnológico ARGOS, Daule, Guayas, Ecuador
c_roman@tecnologicoargos.edu.ec

[2] Universidad ECOTEC, Km 13 1/2 Vía Samborondón, Samborondón, Guayas, Ecuador
svilema@ecotec.edu.ec

Abstract. This study investigates the role of women Information and Communication Technology (ICT) specialists in the performance of Ecuadorian companies during the COVID-19 pandemic. Using regression analysis, the study explores the relationship between the participation of women ICT specialists and total sales, controlling for company size, industry sector, use of ICT in different business areas, and geographical location. The results reveal an unexpectedly significant, negative relationship between the number of women ICT specialists and total sales. Larger companies and those utilizing ICT in their supply chains were found to have an advantage during the pandemic. The geographical location did not significantly impact company performance, suggesting that the effects of the pandemic were felt across different regions. The study also highlights the persistent barriers faced by women in the ICT sector, including gender stereotypes, bias, discrimination, and a lack of access to education, training, and mentorship. These findings underscore the importance of women ICT specialists in bolstering company performance during crises and call for further research and efforts to promote gender diversity and equality in the ICT sector. The study contributes to the understanding of gender dynamics in digital transformation and the resilience of businesses in the face of global crises.

Keywords: ICT · Women · Ecuador

1 Introduction

The COVID-19 pandemic has had a profound impact on companies worldwide, affecting their operations, financial performance, and workforce[1]. The pandemic has caused a global economic downturn, with many companies experiencing reduced demand for their products and services, leading to job losses, reduced hours, and financial difficulties [2]. The pandemic has also accelerated the digital transformation of many companies, creating new opportunities for growth and innovation, but also posing challenges in terms of technology, cybersecurity, and data privacy [3]. The pandemic has affected the workforce of many companies, with many employees working from home or facing reduced hours, and women have been particularly affected, taking on additional caregiving responsibilities due to the closure of schools and daycare centers [4]. The pandemic has also

M. Z. Vizuete et al. (Eds.): CI3 2023, LNNS 1041, pp. 43–53, 2024.
https://doi.org/10.1007/978-3-031-63437-6_4

disrupted global supply chains, affecting the availability of raw materials, components, and finished products, and this has affected companies in various industries, such as manufacturing, retail, and hospitality [5]. Governments have implemented policies to support companies during the pandemic, such as financial assistance, tax relief, and regulatory flexibility, but the effectiveness of these policies has varied across countries and industries [6]. Companies that have been able to adapt to the new challenges posed by the pandemic have been more successful in maintaining their business performance. The COVID-19 pandemic has incited a global business crisis, raising significant challenges and opportunities[7].

The COVID-19 pandemic has had a significant impact on the workforce, with women being disproportionately affected [8]. Women are more likely to be employed in sectors that have been hard hit by the pandemic, such as tourism and hospitality, and they are also more likely to be working in informal or precarious jobs [2]. The pandemic has also led to an increase in caregiving responsibilities for women, making it more difficult for them to participate in the workforce and leading to a decline in their employment rates [9]. Against this backdrop, the importance of women ICT specialists in companies has become even more evident. Women ICT specialists are in high demand, as businesses increasingly rely on technology to operate during the pandemic [10]. They bring a unique set of skills and perspectives to the table, including the ability to design and develop technology that is inclusive and accessible, bridge the digital divide, and promote gender equality in the workplace [11]. By leveraging the skills and expertise of women ICT specialists, companies can position themselves for success in the post-pandemic world. Telecommuting has offered women flexibility in work schedules and location, facilitating a better balance between professional and personal responsibilities [12]. However, the actual benefits depend on individual circumstances and company policies.

Women ICT specialists in Latin America played a vital role in helping businesses to prosper during the COVID-19 pandemic. Their skills and expertise were essential for the adoption of new technologies and the transition to remote work [13]. In Mexico, women ICT specialists were involved in a wide range of activities, such as developing and implementing e-commerce platforms, creating online learning tools, and providing cybersecurity services. Their work helped Mexican businesses to adapt to the new digital landscape and remain competitive [14]. In Colombia, women ICT specialists were instrumental in supporting businesses during the pandemic. They were involved in the development of contact tracing apps, the implementation of telehealth services, and the creation of digital marketing campaigns. Their work helped Colombian businesses to stay connected with their customers and employees, even when they were forced to operate remotely [15]. In Argentina, women ICT specialists made significant contributions to the country's economy during the pandemic. They were involved in developing new technologies, such as artificial intelligence and blockchain, which helped businesses improve their operations. They also played a key role in providing workers with digital skills training, which helped create new jobs and boost the economy [16]. In Ecuador, teleworking has emerged as a strategic response to the COVID-19 pandemic, helping women companies to continue with their operations and workers to better balance their work and personal lives. Women ICT specialists were responsible for setting up and maintaining IT infrastructure, providing technical support to employees, and developing new digital solutions. As a result of their efforts, many Ecuadorian companies were able to continue operating during the pandemic and even thrive [5]. Therefore, the objective of this study is to analyze the relationship between the labor participation of women specialized in ICTs and business performance during COVID-19 in the context of Ecuadorian companies.

2 Methods

The COVID-19 pandemic has had a significant impact on the workforce, with women being disproportionately affected [10]. Women are more likely to be employed in sectors that have been hard hit by the pandemic, such as tourism and hospitality, and they are also more likely to be working in informal or precarious jobs [17]. The pandemic has also led to an increase in caregiving responsibilities for women, making it more difficult for them to participate in the workforce and leading to a decline in their employment rates [18]. Against this backdrop, the importance of women ICT specialists in companies has become even more evident. Women ICT specialists are in high demand, as businesses increasingly rely on technology to operate during the pandemic [19]. They bring a unique set of skills and perspectives to the table, including the ability to design and develop technology that is inclusive and accessible, bridge the digital divide, and promote gender equality in the workplace [20].

Studies have shown that companies with more gender-diverse workforces tend to be more innovative, productive, and profitable. Linear regression and econometric models can be used to analyze the relationship between the representation of women ICT specialists in a company's workforce and its business performance, controlling for other factors such as company size, industry, and location [21]. Public policies and institutional support can play a critical role in promoting the participation of women ICT specialists in the workforce. In countries like Morocco and Jordan, governments have implemented policies to support women entrepreneurs and promote gender diversity in the ICT sector [22].

Telecommuting has offered women flexibility in work schedules and location, facilitating a better balance between professional and personal responsibilities [23]. However, the actual benefits depend on individual circumstances and company policies. In Ecuador, teleworking has emerged as a strategic response to the COVID-19 pandemic, helping women companies to continue with their operations and workers to better balance their work and personal lives [5].

Overall, the participation of women ICT specialists in a company's workforce is critical to its success during COVID-19. Companies that prioritize gender diversity and support the advancement of women in the ICT sector are more likely to succeed in the post-pandemic world [24]. By leveraging the skills and expertise of women ICT specialists, companies can position themselves for success in the digital age. The COVID-19 pandemic has highlighted the importance of women in the workforce and the need for gender equality in the ICT sector [25]. Governments and institutions can play a critical role in promoting gender diversity and supporting women entrepreneurs in the ICT sector. By doing so, they can help to ensure that women are not left behind in the digital age and that companies can thrive in the post-pandemic world [26].

Table 1. Table captions should be placed above the tables.

Variable	Description	Unit
INCOME	Total sales income	Millions of dollars
WOMEN_ICT	Participation of women specialists in ICTs	Workers
MEDIUM	Median company size	= 1 yes = 0 not
BIG	Large company size	= 1 yes = 0 not
SALES	Use of ICT in services and sales support	= 1 if you use = 0 does not use
SUPPLY	Use of ICT in the supply chain	= 1 if you use = 0 does not use
ECOMMERCE	The company has internet sales	= 1 yes = 0 not
MANUFACTURE	Manufacture sector	= 1 yes = 0 not
TRADE	Trade sector	= 1 yes = 0 not
SERVICES	Service sector	= 1 yes = 0 not
COASTAL	Coastal Region	= 1 yes = 0 not
HIGHLANDS	Highlands Region	= 1 yes = 0 not
EAST	East Region	= 1 yes = 0 not

For the present study, the following question is posed: How Women ICT Specialists Helped Ecuadorian Companies Thrive during COVID-19? In addition, the following hypothesis: The participation of women ICT specialists has significantly improved total sales in Ecuadorian companies during the COVID-19 pandemic.

The participation of women ICT specialists in a company's workforce is critical to its success during COVID-19. Women ICT specialists are in high demand, as businesses increasingly rely on technology to operate during the pandemic. They bring a unique set of skills and perspectives to the table, including the ability to design and develop technology that is inclusive and accessible, bridge the digital divide, and promote gender equality in the workplace. Studies have shown that companies with more gender-diverse workforces tend to be more innovative, productive, and profitable. Public policies and institutional support can play a critical role in promoting the participation of women ICT specialists in the workforce. In countries like Morocco and Jordan, governments have implemented policies to support women entrepreneurs and promote gender diversity in the ICT sector.

The COVID-19 pandemic has highlighted the importance of women in the workforce and the need for gender equality in the ICT sector. Governments and institutions can play a critical role in promoting gender diversity and supporting women entrepreneurs in the ICT sector. By doing so, they can help to ensure that women are not left behind in the digital age and that companies can thrive in the post-pandemic world. Telecommuting has offered women flexibility in work schedules and location, facilitating a better balance between professional and personal responsibilities. However, the actual benefits depend on individual circumstances and company policies. In Ecuador, teleworking has emerged as a strategic response to the COVID-19 pandemic, helping women companies to continue with their operations and workers to better balance their work and personal lives.

Studies have shown that women ICT specialists bring specific skills to companies during COVID-19. For example, they can help to develop and implement digital solutions that support remote work and collaboration, improve the security of company data and systems, promote digital literacy and adoption among employees, and create a more inclusive and diverse workplace. By leveraging the skills and expertise of women ICT specialists, companies can position themselves for success in the digital age. However, women continue to face barriers to entry and advancement in the ICT sector, including gender stereotypes, bias, and discrimination, as well as a lack of access to education, training, and mentorship. Therefore, it is important to promote gender diversity and support the advancement of women in the ICT sector to ensure that companies can thrive in the post-pandemic world.

To analyze the relationship between the proportion of workers with internet access and internet sales in companies in Ecuador, the 2021 Business Structural Survey (ENE-SEM) is used, which is a statistical survey carried out by the National Institute of Statistics and Censuses. (INEC) of Ecuador between the years 2020 and 2021, the most relevant period during the pandemic. The survey generates statistical information on the structure and evolution of micro, small, and medium-sized companies in Ecuador. The survey calculates economic aggregates such as production and employment, among other indicators (INEC, 2023). A sample of 396 companies categorized into a) medium-sized company A with a sales volume between 1 and 2 million dollars and a staff employed between 50 and 99 employees, b) medium-sized company B with a sales volume between 2 and 5 is considered. Million dollars and employed personnel between 100 and 199 employees and c) a large company with a sales volume greater than 5 million dollars and employed personnel greater than 200 employees. Table 1 shows the study variables used to test the hypothesis and the following linear regression equation is proposed. The following shows the linear equation to use:

$$INCOME_i = \beta_0 + \beta_1 WOMEN\_ICT_i + \beta_j \sum_{n}^{j=2} X_i + \varepsilon$$

where, $INCOME_i$ is the dependent variable (Total sales income of company i). $WOMEN\_ICT_i$ is the independent variable (Participation of women specialists in ICTs company i). $\sum_{n}^{j=2} X_i$ is a set of characteristic variables of firm i. β_0, β_1, β_j are the regression coefficients that represent the relationship between the variables. Y ε is the error term, which captures the variation not explained by the independent variables.

3   Results

Table 2 shows the descriptive statistics of the study variables. The average income of these companies is 80 million dollars, with a standard deviation of 3.36 million dollars. This relatively small standard deviation suggests that the incomes of the companies are closely clustered around the mean, indicating a small spread in income levels. The variable WOMEN_ICT, likely representing the number of women specialists in Information and Communication Technologies (ICTs) in a company, has an average of approximately 3 but a high standard deviation of 7.88. This wide dispersion in the data indicates a significant variation in the number of women ICT specialists across different companies. The binary variables MEDIUM and BIG likely represent the size of the companies. The data suggests that a majority of the companies are big (85.4%), with a smaller proportion being medium-sized (14.6%). The standard deviations for these binary variables suggest a binary distribution, indicating that the variables are categorical. The variables SALES, SUPPLY, and ECOMMERCE likely represent whether the companies use ICT in services and sales support, in the supply chain, and engage in e-commerce, respectively. A significant majority of companies appear to use ICT in these areas, as indicated by the high mean values. Geographically, the companies are primarily located in the highlands (57.6%) and coastal areas (41.9%), with a small percentage in the east (0.5%), as represented by the binary variables COASTAL, HIGHLANDS, and EAST. In terms of industry, represented by the binary variables MANUFACTURE, TRADE, and SERVICES, the companies are fairly evenly distributed across manufacturing (23%), trade (33.6%), and services (40.4%).

   Table 3 shows the results of 6 online regression models, the WOMEN_ICT variable, representing the participation of women ICT specialists, is significant across all models, indicating a substantial impact on the dependent variable. However, the negative coefficients suggest an inverse relationship, indicating that as the number of women ICT specialists increases, the dependent variable decreases. The variables MANUFACTURE, TRADE, and SERVICES, representing different industry sectors, are included in Models 3 and 6. In Model 3, none of these sectors demonstrate a significant impact on the dependent variable. However, in Model 6, the services sector shows a significant negative impact, suggesting that companies in the services sector may perform differently compared to those in other sectors. The variable BIG, likely indicating large-sized companies, is significant in Models 2 and 6. This suggests that being a large company positively influences the dependent variable. The geographical variables COASTAL and HIGHLANDS, included in Models 4 and 6, do not demonstrate a significant impact on the dependent variable, suggesting that the geographical location of the companies may not be a significant factor in this context. The variables SUPPLY, SALES, and ECOMMERCE, representing the use of ICT in different areas of the companies, are included in Models 5 and 6. In Model 5, only the SUPPLY variable shows a significant positive

Table 2. Descriptive Statistics

Variable	Mean	Std. Dev.
WOMEN_ICT	2.92	7.88
MEDIUM	0.15	0.35
BIG	0.85	0.35
SALES	0.91	0.29
SUPPLY	0.89	0.31
ECOMMERCE	0.76	0.43
COASTAL	0.42	0.49
HIGHLANDS	0.57	0.49
EAST	0.005	0.07
MANUFACTURE	0.23	0.42
TRADE	0.33	0.47
SERVICES	0.40	0.49

impact on the dependent variable. In Model 6, the SUPPLY variable remains significant, but the SALES and ECOMMERCE variables do not show a significant impact. The constant term, representing the baseline value of the dependent variable when all other variables are zero, is significant in all models except Model 4. The R-squared values are relatively low across all models, suggesting that these models explain only a small portion of the variance in the dependent variable.

Table 3. Linear regressions

VARIABLES	Model 1	Model 2	Model 3	Model 4	Model 5	Model 6
WOMEN_ICT	−2.85***	−1.91***	−2.36***	−2.95***	−2.58***	−1.35***
	(1.068)	(0.583)	(0.805)	(0.992)	(0.906)	(0.420)
MANUFACTURE			13.33			−8.16
			(63.211)			(34.213)
TRADE			−38.22			−20.11
			(45.258)			(23.085)
SERVICES			−41.42			−42.83*
			(41.990)			(25.852)

(*continued*)

Table 3. (*continued*)

VARIABLES	Model 1	Model 2	Model 3	Model 4	Model 5	Model 6
BIG		85.59***				89.95***
		(18.246)				(21.391)
MEDIUM		−				−
COASTAL				−164		−172
				(218.9)		(176.7)
HIGHLANDS				−187		−154
				(219.3)		(178.3)
EAST				−		−
SUPPLY					52.77**	32.78*
					(25.456)	(17.452)
SALES					−24.88	−15.88
					(34.815)	(26.273)
ECOMMERCE					−45.79	−40.68
					(44.633)	(37.283)
Constant	491.65***	11.85***	77.25*	230.90	60.30**	208.18
	(14.73)	(3.006)	(42.009)	(218.7)	(26.158)	(183.9)
Observations	396	396	396	396	396	396
R-squared	0.007	0.040	0.014	0.011	0.015	0.052

Robust standard errors in parentheses.
***$p < 0.01$, **$p < 0.05$, *$p < 0.1$

4 Discussion

The regression analysis results presented in the table align with the hypothesis of this study, which posits that the participation of women ICT specialists significantly improved total sales in Ecuadorian companies during the COVID-19 pandemic. The WOMEN_ICT variable, representing the participation of women ICT specialists, is significant across all models, suggesting that the presence of women ICT specialists has a significant impact on the dependent variable, which in this case is total sales. However, the negative coefficients indicate that as the number of women ICT specialists increases, the total sales decrease. This unexpected result may be due to other factors not included in the model or could be an area for further investigation. The variable BIG, indicating large-sized companies, is significant in Models 2 and 6, suggesting that larger companies may have been better positioned to maintain or increase sales during the pandemic. This could be due to a variety of factors, such as having more resources to pivot to remote work or being more likely to have an established online presence before the pandemic. The variables SUPPLY, SALES, and ECOMMERCE, representing the use of ICT in different areas of the companies, are included in Models 5 and 6. In Model 5, only the SUPPLY variable shows a significant positive impact on the dependent variable, suggesting that the use of ICT in the supply chain may have been particularly beneficial for companies during the pandemic. The geographical variables COASTAL and HIGHLANDS, included in Models 4 and 6, do not demonstrate a significant impact on the dependent variable, suggesting that the geographical location of the companies may not have been a significant factor in their sales performance during the pandemic. These findings underscore the importance of women ICT specialists in helping companies navigate the challenges of the pandemic. However, they also highlight the need for further research to better understand the barriers that women face in the ICT sector and how these can be overcome to promote gender diversity and equality in the post-pandemic world.

5   Conclusions

The findings of this study underscore the pivotal role of women ICT specialists in bolstering the resilience and performance of Ecuadorian companies during the COVID-19 pandemic. Despite the unexpected negative relationship between the number of women ICT specialists and total sales, the significance of the WOMEN_ICT variable across all models suggests that their participation is a crucial factor in company performance. The results also highlight the advantage of larger companies during the pandemic, as indicated by the significant positive impact of the BIG variable in Models 2 and 6. This suggests that larger companies, potentially due to their greater resources and established online presence, were better equipped to navigate the challenges of the pandemic. The use of ICT in supply chains, represented by the SUPPLY variable, was found to have a significant positive impact on company performance in Model 5. This underscores the importance of digital transformation in business operations, particularly in times of crisis. However, the geographical location of companies, represented by the COASTAL and HIGHLANDS variables, did not significantly impact company performance. This suggests that the effects of the pandemic transcended geographical boundaries, affecting companies across different regions similarly. Despite the crucial role of women ICT specialists, the study also brings to light the persistent barriers they face in the sector, including gender stereotypes, bias, and discrimination, as well as a lack of access to education, training, and mentorship. Addressing these barriers is essential to promoting gender diversity and equality in the ICT sector, and by extension, enhancing the resilience and performance of companies in the post-pandemic world. In conclusion, this study emphasizes the importance of women ICT specialists in the business landscape during the COVID-19 pandemic. It calls for further research to delve deeper into the dynamics of gender and digital transformation in business, and for concerted efforts to dismantle the barriers impeding women's full participation in the ICT sector.

References

1. Phogaat, V.: Impact of Lockdown and Government Policies on Indian Stock Market Particularly in Pharmaceutical Industry (2021)
2. Vindhya, P.J., Varma, R.R.: The impact of Covid19 in tourism sector. Indian J. Appl. Res. (2021). https://doi.org/10.36106/ijar/4313909

3. Laxe, F.G., Piña, J.F.A., Lago, S., Sánchez, P.: Impacto económico del COVID19 en una economía regional. El caso del confinamiento para Galicia [Economic impact of COVID19 in a regional economy. The case of the lock-down in Galicia] (2020)
4. Gorzelany, M.: COVID-19: business innovation challenges. Sustainability **13**, 11439 (2021). https://doi.org/10.3390/su132011439
5. Alvarez, E., Hernandez-Villafuerte, K., Saldarriaga, C., Zuccardi, J.: COVID-19 in latin america: impacts and policy responses. World Dev. **137**, 105217 (2021). https://doi.org/10.1016/j.worlddev.2020.105217
6. Liu, Y., Lee, J.M., Lee, C.: The challenges and opportunities of a global health crisis: the management and business implications of COVID-19 from an Asian perspective. Asian Bus. Manag. **19**, 277–297 (2020). https://doi.org/10.1057/s41291-020-00119-x
7. Meyer, B.H., Prescott, B., Sheng, X.S.: The impact of the COVID-19 pandemic on business expectations. Int. J. Forecast. **38**, 529–544 (2022). https://doi.org/10.1016/j.ijforecast.2021.02.009
8. Gilli, K., Lettner, N., Guettel, W.: The future of leadership: new digital skills or old analog virtues? J Bus Strategy ahead-of-print: (2023). https://doi.org/10.1108/JBS-06-2022-0093
9. Zhao, N., Hong, J., Lau, K.H.: Impact of supply chain digitalization on supply chain resilience and performance: a multi-mediation model. Int. J. Prod. Econ. **259**, 108817 (2023). https://doi.org/10.1016/j.ijpe.2023.108817
10. Athanasiadou, C., Theriou, G.: Telework: systematic literature review and future research agenda. Heliyon **7**, e08165 (2021). https://doi.org/10.1016/j.heliyon.2021.e08165
11. Franconi, A., Naumowicz, K.: Remote work during covid-19 pandemic and the right to disconnect – implications for women´s incorporation in the digital world of work. Z Problematyki Prawa Pracy i Polityki Socjalnej **19**, 1–20 (2021). https://doi.org/10.31261/zpppips.2021.19.09
12. Bouziri, H., Smith, D.R.M., Descatha, A., Dab, W., Jean, K.: Working from home in the time of covid-19: how to best preserve occupational health? Occup. Environ. Med. **77**(7), 509–510 (2020). https://doi.org/10.1136/oemed-2020-106599
13. Sandoval, J., Idrovo-Carlier, S., Duque-Oliva, E.J.: Remote work, work stress, and work-life during pandemic times: a latin america situation. Int. J. Environ. Res. Public Health **18**, 7069 (2021). https://doi.org/10.3390/ijerph18137069
14. Domínguez, J.G., Chen, A., McMurtrey, M., Cohernour, E.C., Gabriel, M.: ICT competencies in eight Mayan-speaking communities of Mexico: preliminary findings. J. Int. Technol. Inf. Manag. **28**(1), 90–116 (2019). https://doi.org/10.58729/1941-6679.1389
15. Della, M., Jaramillo-Gutiérrez, A., Henao-Rosero, A.: Universities and digital skills' development in Colombia. In: Visvizi, A., Troisi, O., Grimaldi, M. (eds.) Research and Innovation Forum 2022, pp. 385–394. Springer International Publishing, Cham (2023)
16. Egana, P., Bustelo, M., Ripani, L., Soler, N., Viollaz, M.: Automation in latin america: are women at higher risk of losing their jobs? Technol. Forecast. Soc. Change **175**, 121333 (2022). https://doi.org/10.1016/j.techfore.2021.121333
17. Hilbrecht, M., Shaw, S.M., Johnson, L.C., Andrey, J.: 'I'm home for the kids': contradictory implications for work-life balance of teleworking mothers. Gend. Work Organ. **15**, 454–476 (2008). https://doi.org/10.1111/j.1468-0432.2008.00413.x
18. Craig, L., Powell, A.: Dual-earner parents' work-family time: the effects of atypical work patterns and non-parental childcare. J. Popul. Res. **29**, 229–247 (2012). https://doi.org/10.1007/s12546-012-9086-5
19. Heiden, M., Widar, L., Wiitavaara, B., Boman, E.: Telework in academia: associations with health and well-being among staff. High. Educ. **81**, 707–722 (2021). https://doi.org/10.1007/s10734-020-00569-4

20. Ton, D., et al.: Teleworking during COVID-19 in the Netherlands: understanding behaviour, attitudes, and future intentions of train travellers. Transp Res Part Policy Pract **159**, 55–73 (2022). https://doi.org/10.1016/j.tra.2022.03.019
21. Constantinescu, L., Nichita, E.-M.: Bucharest University of Economic Studies, Romania, Păunescu M, Bucharest University of Economic Studies, Romania (2022) INVESTIGATING THE IMPACT OF TELEWORKING MODEL ON WORK PERFORMANCE AND WORK CONDITIONS IN A ROMANIAN FAMILY BUSINESS. https://doi.org/10.24818/IMC/2021/04.05
22. Thaher, L.M., Radieah, N.M., Wan Norhaniza, W.H.: Factors affecting women micro and small-sized enterprises' success: a case study in Jordan. J. Asian Finance Econ. Bus. **8**, 727–739 (2021). https://doi.org/10.13106/JAFEB.2021.VOL8.NO5.0727
23. Yossef, M.W., Ahmed, M.N.A., Ragheb, M.A.S.: Business environment and their readiness to implement the teleworking: a field study on the application of the egyptian private commercial banks. OALib **07**, 1–18 (2020). https://doi.org/10.4236/oalib.1106578
24. Benavides, F.G., et al.: O futuro do trabalho após a COVID-19: o papel incerto do teletrabalho no domicílio. Rev Bras Saúde Ocupacional **46**, e31 (2021). https://doi.org/10.1590/2317-6369000037820
25. Verma, S., Gustafsson, A.: Investigating the emerging COVID-19 research trends in the field of business and management: a bibliometric analysis approach. J. Bus. Res. **118**, 253–261 (2020). https://doi.org/10.1016/j.jbusres.2020.06.057
26. Molleví, G., Álvarez, J., Nicolas, R.: Sustainable, technological, and innovative challenges post Covid-19 in health, economy, and education sectors. Technol. Forecast. Soc. Change **190**, 122424 (2023). https://doi.org/10.1016/j.techfore.2023.122424

Assessment and Selection of Fuel Models in Areas with High Susceptibility to Wildfires in the Metropolitan District of Quito

Juan Gabriel Mollocana Lara[1]([✉]) and Johanna Beatriz Paredes Obando[2]

[1] Grupo de Investigación Ambiental en el Desarrollo Sustentable GIADES, Carrera de Ingeniería Ambiental, Universidad Politécnica Salesiana, Quito 170702, Ecuador
jmollocana@ups.edu.ec
[2] Carrera de Ingeniería Ambiental, Universidad Politécnica Salesiana, Quito 170702, Ecuador

Abstract. Fuel models are a crucial component of modeling fire behavior in wildfires, and their appropriate selection is essential for developing effective management strategies. This research presents and evaluates a simple methodology for selecting fuel models based on comparing the main features of standardized models with open-access geographic information and dead fuel loads sampled on field. For evaluating fuel models selection FlamMap simulation of recent wildfires that occurred in three high susceptibility areas within the Metropolitan District of Quito: Casitagua, Ilalo, and Guagua Pichincha volcanoes were carried out. The final extension of simulated and real wildfires was compare using Cohen's kappa coefficient, obtaining values of 0.53, 0.35, and 0.47, respectively. These results indicated a moderate to acceptable relationship between simulations and actual fires. Additionally, simulations of the effects of fire barriers to estimate the reduction in the final wildfire extension were carried out, obtaining reductions of 22%, 47%, and 37%, respectively. In conclusion, this methodology provides a useful approach to selecting appropriate fuel models for effective wildfire management strategies in highly susceptible areas.

Keywords: FlamMap · Wildfire simulation · Fuel models

1 Introduction

Wildfires are phenomena that occur without control on combustible materials from terrestrial ecosystems [1]. They can be of natural or anthropic origin and affect a large number of hectares of vegetation [2]. Its behavior depends on the combination of several factors such as orientation, slope, and wind direction [3, 4].

There are several wildfire simulation models capable of emulating the possible behavior of a real event, obtaining data and analyzing scenarios with extreme conditions [5]. Several of these fire behavior models require fuel models as input data that group together the characteristics of the different fuels present in the area [6]. To determine the precise fuel models of an area, it is possible to choose either to use standardized models [7] or to generate custom models [8]. The standardized fuel models require using the features

© The Author(s), under exclusive license to Springer Nature Switzerland AG 2024
M. Z. Vizuete et al. (Eds.): CI3 2023, LNNS 1041, pp. 54–68, 2024.
https://doi.org/10.1007/978-3-031-63437-6_5

of a predefined model that are similar to those in the study area; these models are chosen based on the type of vegetation and the rate of fire spread expected [9]. On the other hand, the personalized fuel models require a large amount of field information and the application of statistical methods of data clustering.

There are several wildfire simulation programs that require fuel models compatible with the Rothermel's fire spread models (i.e. BehavePlus, Farsite, FlamMap). These programs use information on fuels, topography, and meteorological data [10] in order to have outputs such as the extension of the fire, major paths, arrival times, length and intensity of the flame, among others [11]. One of the most widely used software today is the FlamMap software. [12] assesses flame intensity, speed, and length using remote sensing estimates of fuel load and moisture and maps the probability and severity of wildfires by predicting the final extent of an event. Lastly, [13] studies the behavior of a wildfire occurred at Atacazo Hill in Ecuador using the classification of the Scott and Burgan fuel models, to obtain the possible final extension of the wildfire and simulates mitigation measures to reduce the impact of the event.

To validate the results of the wildfire simulation, the Sorensen and Cohen's kappa coefficients can be used [14, 15]. In addition, the reliability of the results increases if the simulations are compared with several fires of different sizes, that cross various types of fuels and that occur in different humidity scenarios [13]. The error made in the simulations largely depends on the definition of the fuel models that best describe the characteristics of the vegetation in an area. Therefore, a precise selection of the fuel models has a great influence on the applicability of fire management strategies developed from the simulations [16].

This research aims to propose and assess a methodology for choosing standardized fuel models that utilize field information collected an easy manner. In addition, wildfires of different sizes and types of vegetation that occurred in three high susceptibility areas of the Metropolitan District of Quito will be simulated, these are: Casitagua, Ilalo and Guagua Pichincha volcanoes. The results of the simulations serve to assess the proposed methodology by comparing the extension of the wildfires between the ones obtained from satellite images and those simulated through software. To measure the similarity of the fire extensions Cohen's kappa coefficient will be calculated. Finally, the effectiveness of the use of fire barriers is assessed to reduce the impact of a wildfire in the study areas.

2 Materials and Methods

To achieve the objectives of the investigation and obtain reliable results, three case studies of wildfires of different extensions that occurred on zones with a variety of fuels are chosen. Analysis of topographic information from the Military Geographic Institute of Ecuador (IGM), land use and cover maps from the Ministry of Agriculture, Livestock and Fisheries of Ecuador (MAGAP), field information on dead fuels, meteorological data from the Network of weather stations in Quito (REMMAQ), Sentinel 2 satellite images, and information on roads, highways, and coverage surfaces from the *open-streetmap* platform are performed. It is necessary to count with a landscape file that contains the information of five rasters: fuel model, canopy cover, elevation, orientation, and slope, as input for the FlamMap software. To generate the fuel models raster, the data sampled in the field are related to the Scott & Burgan standardized fuel models; and for the canopy cover raster de observed percentages on canopy cover are classified in five categories. The slope, elevation and orientation rasters are made from the contour lines of the IGM topographic maps. Simulations of each wildfire are made in FlamMap using the generated landscape (LCP) file, meteorological data, satellite images and information from *openstreetmap*. The results of the simulations are validated by comparing the simulated extensions with the real extensions of each fire, through Cohen's kappa coefficient. Finally, fire simulations were carried out, adding fire barriers with the aim of reducing the consumed area (Fig. 1).

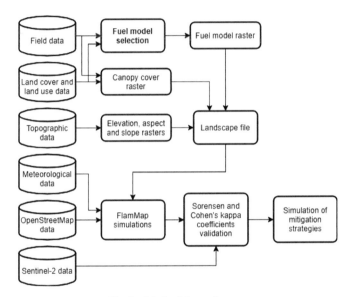

Fig. 1. Methodology diagram

2.1   Study Cases

The study examines three previous wildfires in high susceptibility areas of DMQ. The first fire occurred on January 14, 2020, at Casitagua volcano and lasting three days; the second fire occurred on September 9, 2019 on the slopes of Guagua Pichincha volcano and lasting 2 days; and the third fire occurred on August 4, 2020 at Ilalo volcano and lasting 18 h. The final wildfire extents are determined using Sentinel-2 satellite images: February 11, 2020, for Casitagua, September 17, 2019 for Guagua Pichincha, and August 8, 2020 for Ilaló. Meteorological data are collected from nearby stations (Fig. 2).

2.2   Field Data

A stratified random sampling is developed based on the categories of the MAGAP land use and cover map, obtaining a total of 72 sampling points. At each point, the

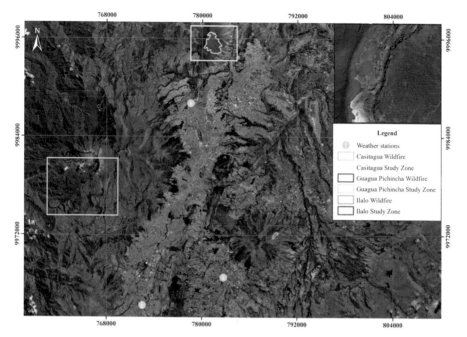

Fig. 2. Study cases

methodology of Brown (1974) was applied to estimate dead fuel loads with delays of 1h, 10h and 100h, and the fuel bed depth. These are the easiest parameters to obtain from a fuel model since their estimation does not involve clearing and drying vegetation in laboratory. The general fire-carrying fuel and the percentage of canopy cover are determined by direct observation. For the percentage of canopy cover, the information is recorded using five categories that represent different percentages of canopy cover according to the requirements of the FlamMap software (Tables 1, 2 and 3), these are:

0. 0%
1. 1–20%
2. 21–50%
3. 51–80%
4. 81–100%

2.3   Fuel Model Selection

For the selection of fuel models, the field information is related to the 40 standardized models of Scott & Burgan. These models are classified according to the general fire-carrying fuel type and its extinction moisture. The general fire-carrying fuel is determined based on the mode of the field observations of each stratum. The percentage of extinction moisture is determined based on the MAATE ombrotype map that classifies the terrain into humid and dry areas. Finally, to choose the fuel model corresponding to each stratum, the Euclidean distance is calculated between the sampled dead fuel loads and the fuel

Table 1. Casitagua field data

Cover and land use stratum	Dead fuel load 1h (t/ac)	Dead fuel load 10h (t/ac)	Dead fuel load 100h (t/ac)	Fuel bed Depth (cm)	Fire-carrying fuel	Canopy cover
Moderately altered dry scrub	0.0596	0.3799	0.2390	30.477	Shurb	3
Heavily altered dry scrub	0.0157	0.0301	0.1863	29.207	Shurb	1
Highly altered herbaceous vegetation	0.0390	0.0340	0.0000	16.837	Grass	1
Eucalyptus	0.1822	12.898	12.931	0.6232	Timber litter	2

bed depth and those reported by Scott & Burgan. The selected model is the one that shows a smaller distance to the field data. Table 1 shows the averages of the sampled parameters and the fuel model that is closest to these data according to the result of the Euclidean distance (Tables 4, 5 and 6).

2.4   Landscape File

The Landscape is one of the main inputs for the simulation in the FlamMap software. This file is generated based on at least five rasters with topographic and fuel information. The fuel model raster contains the fuel code for the selected model (Fig. 3); the vegetation cover raster contains the vegetation cover categories determined through field observations (Fig. 4); the elevation (Fig. 5), slope (Fig. 6), and aspect (Fig. 7) rasters are generated from the contour lines of the IGM topographic sheets. In all the rasters is used a cell size of 20 m and the projected coordinate system WGS 84/UTM zone 17S.

2.5   Weather Data

Data on temperature, relative humidity, precipitation, wind speed and wind direction from the closest weather stations to the study areas are used to generate the weather stream file (.wxs). This file is used by the FlamMap WindNinja tool to create spatially varying wind fields in the study areas.

2.6   Openstreetmap Data

Rocky areas, water bodies, and wide paths behave as fire barriers and greatly influence fire behavior. For this reason, fire barrier maps are made for each study area based on the information from the OpenStreetMap platform.

Table 2. Guagua Pichincha field data

Cover and land use stratum	Dead fuel load 1h (t/ac)	Dead fuel load 10h (t/ac)	Dead fuel load 100h (t/ac)	Fuel bed Depth (cm)	Fire-carrying fuel	Canopy cover
Cultivated pastures	0.0000	0.0000	0.0000	0.0000	Grass	1
Little disturbed herbaceous wasteland	0.1730	0.0000	0.0000	1.66733	Grass	4
Moderately altered humid scrubland	0.0243	0.1847	0.1923	102.281	Shurb	3
Cultivated patures with trees	0.0462	0.5697	0.2862	46.959	Grass	1
Moderately altered herbaceous vegetation	0.0177	0.2517	0.2529	67.459	Grass	1
Severely altered humid scrubland	0.0232	0.2954	0.2772	95.366	Grass	2
Moist herbaceous vegetation moderately altered	0.0375	0.5968	0.2538	103.375	Grass	2
Moderately altered humid forest	0.0951	10.491	0.3112	120.649	Timber-understory	2

2.7 Error Measurements

To validate the results of the generated model, the simulated fire extension is compared with its real extension, using Cohen's kappa coefficient. This coefficient varies between 0 to 1 and allows knowing the concordance between two qualitative variables (Lopez & Pita, 1999) (Table 7).

$$KC = \frac{N \sum_{i=1}^{r} x_{ii} - \sum_{i=1}^{r} (x_{i+} x_{+i})}{N^2 - \sum_{i=1}^{r} (x_{i+} x_{+i})} \tag{1}$$

Table 3. Ilalo field data

Cover and land use stratum	Dead fuel load 1h (t/ac)	Dead fuel load 10h (t/ac)	Dead fuel load 100h (t/ac)	Fuel bed Depth (cm)	Fire-carrying fuel	Canopy cover
Moist scrub moderately altered	0.0099	0.0496	0.2636	2.7197	Shurb	3
Grass	0.0143	0.0240	0.0000	2.3985	Grass	1
Little disturbed moist scrub	0.0659	0.0702	0.1075	4.4280	Shurb	4
Eucalyptus	0.0804	0.9193	0.8992	3.9370	Timber-understory	2

Table 4. Casitagua fuel model selection

Cover and land use stratum	Selected fuel model	Fuel model code
Moderately altered dry scrub	SH4	144
Heavily altered dry scrub	SH4	144
Highly altered herbaceous vegetation	GR4	104
Eucalyptus	TL3	183

Table 5. Guagua Pichincha fuel model selection

Cover and land use stratum	Selected fuel model	Fuel model code
Cultivated pastures	GR1	101
Little disturbed herbaceous wasteland	GR6	106
Moderately altered humid scrubland	SH5	145
Cultivated patures with trees	GR8	108
Moderately altered herbaceous vegetation	GR9	109
Severely altered humid scrubland	SH5	145
Moist herbaceous vegetation moderately altered	GS4	124
Moderately altered humid forest	TU3	163

where:

KC: Cohen's kappa coefficient.
r: number of rows in the error matrix.
N: total number of observations.
x_{ii}: number of observations in row i and column i of the error matrix.

Table 6. Ilalo fuel model selection

Cover and land use stratum	Selected fuel model	Fuel model code
Moist scrub moderately altered	SH4	144
Grass	GR3	103
Little disturbed moist scrub	SH4	144
Eucalyptus	TU2	162

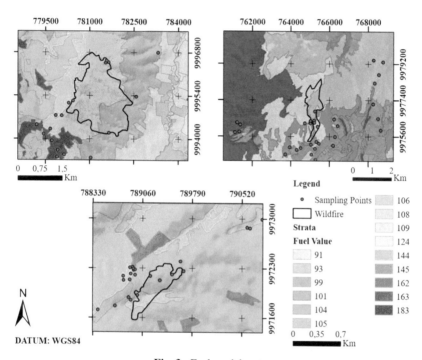

Fig. 3. Fuel model raster

x_{i+}: row i marginal total.
x_{+i}: column i marginal total.

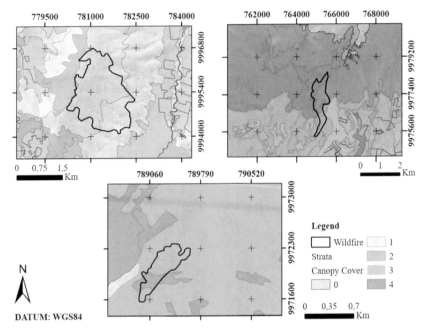

Fig. 4. Canopy cover raster

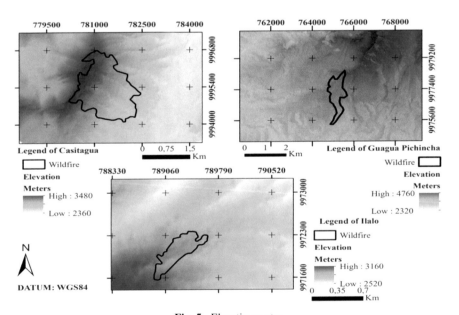

Fig. 5. Elevation raster

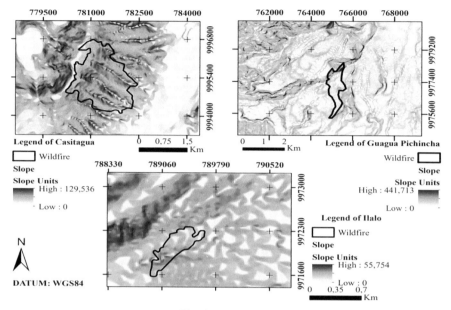

Fig. 6. Slope raster

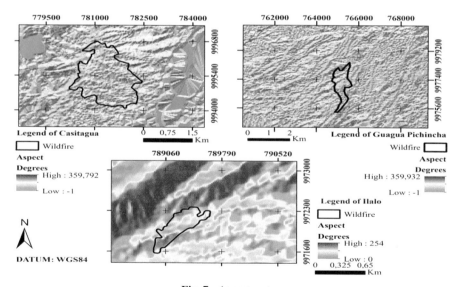

Fig. 7. Aspect raster

Table 7. Cohen's Kappa coefficient meaning

Cohen's Kappa coefficient	Agreement force
0.00	Poor
0.01 – 0.20	Mild
0.21–0.40	Aceptable
0.41–0.60	Moderate
0.61–0.80	Considerable
0.81–1.00	Almost perfect

3 Results

After making the landscape file with the 5 raster maps in the FlamMap software, simulations of arrival times and major paths are carried out using the minimum travel time (MTT) algorithm. Default initial moistures and the Scott-Reinhardt method were used to calculate crown fire. In addition, the meteorological data and fire barriers generated were loaded. to create the simulation and know the possible behavior of the fire in the areas of interest (Fig. 8).

Fig. 8. Arrival time and major paths simulations

The results obtained were validated using Cohen's kappa similarity coefficient; for this it was necessary to calculate the total number of cells burned in the simulation and in the real fire, the number of cells burned in the simulation, but not in the real fire, and the number of cells not burned in the simulation, but in the actual fire (Fig. 9). These data were tabulated in an error matrix and later used with Eq. 1 to calculate the coefficient. The results show fair to moderate force of agreement (Table 8).

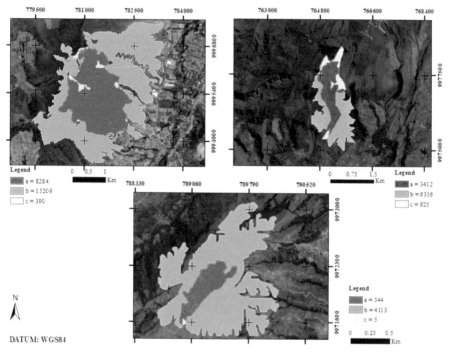

Fig. 9. Cohen's kappa data

Table 8. Cohen's Kappa coefficient results

Study case	Cohen's kappa coefficient
Casitagua	0.53
Ilalo	0.35
Guagua Pichincha	0.47

Once the model was validated, proposals were generated for fire impact mitigation strategies that seek to prevent the spread of fire in large areas through the design of fire barriers. This implies the creation of new barriers and the widening or maintenance of roads that exist in the study areas but do not prevent the passage of fire. The simulations show how a decrease of between 22 and 47% of the consumed area can be achieved (Fig. 10, Table 9).

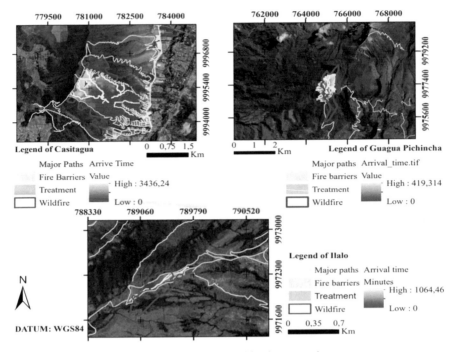

Fig. 10. Simulation of mitigation strategies

Table 9. Mitigation strategies results

Study case	Total hectares consumed in the actual fire	Total hectares consumed in simulations
Casitagua	288	234
Ilalo	60	32
Guagua Pichincha	210	134

4 Discussion

Simulations of three wildfires of different extensions that occurred in the Metropolitan District of Quito in areas with varied vegetation have been carried out. The results of the validation stage show an acceptable to moderate strength of agreement according to the results of the calculation of the Cohen's kappa coefficient. Although the values obtained are lower than those reported in other studies [13, 17], they have the advantage of being the product of the simulations of three different events with extensions, climate, and diverse vegetation.

Among the possible sources of error in the simulations, it can be mentioned that the stratified random sampling had to exclude areas of difficult access due to high slopes, being on private property or the presence of animals that endanger the safety of sampling staff. On the other hand, the land cover and land use map used contains information from the year 2015, which means that it does not consider the regeneration of vegetation from burned areas or recent reforestation plans. In addition, it was not possible to obtain the coordinates of the point of ignition of the wildfires; Therefore, it was decided to assume that they occurred near tourist sites with free circulation or agricultural borders. It can also be mentioned that the fire extinction actions carried out by the DMQ fire department cannot be simulated and make the final extensions of the fires decrease. Finally, precise meteorological values of the study areas were not obtained, since nearby meteorological stations were located at a considerable distance.

5 Conclusions

While specific wildfire simulation data for the DMQ is lacking, it is feasible to make estimations based on land cover and land use maps. These maps facilitate the estimation of fuel characteristics, which, when combined with field data, can lead to improved fuel model approximations. This approach enables simulations in data-scarce areas, yielding valuable insights to mitigate wildfire impacts. Nonetheless, the model's accuracy diminishes for smaller-scale fires.

Finally, the integration of field data and land use information into wildfire simulation models significantly augments their predictive precision. By accounting for dead fuel and canopy data, these models can offer more realistic projections of fire behavior. Moreover, the simulation of fire barriers offers pivotal insights for wildfire management strategies. Pinpointing critical fuel treatment sites along main pathways facilitates the implementation of fire barriers, leading to substantial reductions in wildfire extension. This approach equips fire management agencies with valuable knowledge, empowering them to make well-informed choices and optimize resource allocation for proficient wildfire control and mitigation.

References

1. Davíla, A., Cuesta, R., Villagómez, M., León, F., Vallejo, J.: Atlas de espacios geográficos expuestos a amenazas naturales y antrópicas (2018)
2. Rebolo, G.O., del Río, R.L., Pérez, C.R., Guarinos, J.C., Villamayor, R.H.: Metodología aplicada en los planes de defensa de Aragón para la determinación de áreas de defensa y rodales/parcelas estratégicos en la defensa frente a incendios forestales. Cuadernos de la Sociedad Española de Ciencias Forestales **46**(1), 161–172 (2020)
3. Pausas, J.G.: Incendios forestales: ¿desastre o biodiversidad? (2014)
4. Barros, A.M.G., et al.: Spatiotemporal dynamics of simulated wildfire, forest management, and forest succession in central Oregon, USA. Ecol. Soc. **22**(1), 24 (2017)
5. Bakhshaii, A., Johnson, E.A.: A review of a new generation of wildfire–atmosphere modeling. Can. J. For. Res. **49**(6), 565–574 (2019)
6. Yavuz, M., Sağlam, B., Küçük, Ö., Tüfekçioğlu, A.: Assessing forest fire behavior simulation using FlamMap software and remote sensing techniques in Western Black Sea Region, Turkey. Kastamonu Üniversitesi Orman Fakültesi Dergisi **18**(2), 171–188 (2018). https://doi.org/10.17475/kastorman.459698

7. Scott, J.H., Burgan, R.E.: Standard fire behavior fuel models: A comprehensive set for use with Rothermel's surface fire spread model. *USDA For. Serv. – Gen. Tech. Rep. RMRS-GTR*, no. 153 RMRS-GTR, pp. 1–76 (2005)

8. Elia, M., Lafortezza, R., Lovreglio, R., Sanesi, G.: Developing custom fire behavior fuel models for Mediterranean wildland-urban interfaces in southern Italy. Environ. Manage. **56**(3), 754–764 (2015)

9. Anderson, H.E.: Aids to determining fuel models for estimating fire behavior (1982)

10. Andrews, P.L.: BehavePlus fire modeling system, version 5.0: Variables. Gen. Tech. Rep. RMRS-GTR-213 Revised. Fort Collins, CO Dep. Agric. For. Serv. Rocky Mt. Res. Station. vol. 213, 111 p. (2009)

11. Stratton, R.D.: Assessing the effectiveness of landscape fuel treatments on fire growth and behavior. J. For. **102**(7), 32–40 (2004)

12. Córdova Luspa, B.G.: Estrategias de mitigación y prevención ante incendios forestales en zonas de alta y crítica susceptibilidad, en el Distrito Metropolitano de Quito, utilizando el programa FlamMap (2022)

13. Mollocana Lara, J.G., Álvarez Mendoza, C.I., Jaramillo Coronel, L.J., Mollocana Lara, J.G., Álvarez Mendoza, C.I., Jaramillo Coronel, L.J.: Evaluación de información relacionada con combustibles en el Distrito Metropolitano de quito para el modelado y simulación de incendios forestales, caso de estudio: incendio del cerro atacazo. La Granja **34**(2), 45–62 (2021). https://doi.org/10.17163/lgr.n34.2021.03

14. Kraemer, H.C.: Kappa coefficient. In: Kenett, R.S., Longford, N.T., Piegorsch, W.W., Ruggeri, F. (eds.) Wiley StatsRef: Statistics Reference Online, pp. 1–4. Wiley (2014). https://doi.org/10.1002/9781118445112.stat00365.pub2

15. Conver, J.L., Falk, D.A., Yool, S.R., Parmenter, R.R.: Modeling fire pathways in montane grassland-forest ecotones. Fire Ecol. **14**(1), 17–32 (2018)

16. Botequim, B., Fernandes, P.M., Borges, J.G., González-Ferreiro, E., Guerra-Hernández, J.: Improving silvicultural practices for Mediterranean forests through fire behaviour modelling using LiDAR-derived canopy fuel characteristics. Int. J. Wildl. Fire **28**(11), 823–839 (2019)

17. Jahdi, R., Salis, M., Arabi, M., Arca, B.: Fire modelling to assess spatial patterns of wildfire exposure in Ardabil, NW Iran. Int. Arch. Photogramm. Remote Sens. Spatial Inf. Sci. **XLII-4/W18**, 577–581 (2019)

Data Science: Machine Learning and Multivariate Analysis in Learning Styles

Diego Máiquez[✉] [iD], Diego Pabón [iD], Mariela Cóndor [iD], Gonzalo Rodríguez [iD], Mauricio Farinango [iD], and Ana Oyasa [iD]

Instituto Tecnológico Universitario ISMAC, Tumbaco, Belermo S2-02 Y, Quito 170184, Ecuador
dmaiquez@tecnologicoismac.edu.ec

Abstract. Data science is responsible for the analysis, interpretation and prediction of simple data to generate significant knowledge, it applies to any area that produces data, for example: sales, finance, production, health, education, etc. Data science in education uses machine learning techniques and multivariate analysis. Currently, Educational Data Mining or EDM is spoken, which unites the areas of education, Big data, and data science to improve learning. In this study, data science models are analyzed and compared to identify patterns of behavior, similarity, and anomalous data from students to generate new unknown knowledge and characterize the profile of students according to learning styles such as those proposed by Felder and Silverman in addition to relating these styles to development by competencies. The results show that the most efficient methods are: Clustering and its k-means algorithm with which group characteristics are obtained, decision trees with its ID3 algorithm that through the gain of information a better classification is obtained and the PCA mathematical model or Principal Component Analysis that by its properties of variability analysis and dimension reduction allows to obtain more information from data with noise or outliers. A characterization of the data to be processed is also carried out, classifying it into profile data, class data and test data. These analyzed models will be implemented in a next phase in students of the Software Development career of the ISMAC Institute, which allows obtaining promising results to predict student learning styles and improve said learning process.

Keywords: Data science · Machine learning · Multivariate analysis · Data warehousing · Learning style

1 Introduction

The use of technology produces data, in recent years the growth of this data has been exponential. The question arises, what to do with so much data? Data science is responsible for the analysis, interpretation, and prediction of a set of data to obtain significant knowledge from simple data, this is achieved with the help of information technology. Big Data and/or BI. Data science is present in areas such as: marketing, sales, finance, production, telecommunications, health, education, etc., and any area where data is produced [2, 19, 20].

M. Z. Vizuete et al. (Eds.): CI3 2023, LNNS 1041, pp. 69–81, 2024.
https://doi.org/10.1007/978-3-031-63437-6_6

In the area of education, there are traditional and modern data science techniques to evaluate psychometric and educational psychology data, among which are machine learning techniques and multivariate analysis techniques. Today there is talk of EDM or Educational Data Mining, which is composed of the areas of Big Data, data science and education whose objective is to improve learning, for this it uses computer mathematical methods and techniques such as those mentioned [1, 2, 4, 5].

Machine learning is a data science technique that is responsible for performing non-computationally programmed tasks from a training data set, it contains supervised and unsupervised methods. Among the unsupervised are the clustering models that allow to associate data in groups of similar characteristics, one of the main clustering algorithms is K-means [1–3, 7].

On the other hand, decision trees are supervised machine learning methods that allow discrete data classification tasks and continuous data regression tasks to be carried out, it is ideal for processes where there are several variables of different characteristics, among the most used algorithms is the ID3 that is useful in the analysis of the educational area as mentioned by some studies [1, 9–11].

The multivariate analysis allows to analyze variables of different characteristics simultaneously, among the most used methods are the PCA or Principal Component Analysis that allows to reduce dimensions to obtain the main characteristics of a data set, a method closely related to the factorial analysis that is used in the area of education through psychometrics and educational psychology. [2, 21].

On the other hand, there is the model for learning styles proposed by Felder and Silverman (1988) whose objective is to distinguish the learning style of students, it consists of four dimensions with their classifications of styles and they are: Comprehension with global/sequential styles, processing with active/reflective styles, perception with sensory/intuitive styles, and representation with visual/verbal styles [3, 23, 24].

On the other hand, competency-based learning refers to the approach to problems applied to real-world situations, a study proposes 8 elementary competencies that cover mathematics, native language, foreign language, social, etc. [25, 26]. The development by competences is supported by e-learning technologies to improve the learning process [27].

The lack of knowledge of the student's profile causes problems for said student to learn efficiently, knowing the student's learning styles is useful to improve the aforementioned learning process [3–5, 9]. In this study, the data science techniques, and models applicable to analyze data from the area of education, particularly student processes, were analyzed and the following question was posed: Are predictive data science models adaptable to learning styles?

The decision tree machine learning methods with its ID3 algorithm and clustering with the k-means algorithm and the PCA multivariate analysis method or Principal Component Analysis turned out to be the most efficient for analyzing student process data related to learning styles. These data were classified into profile data, test data and class data according to the characteristics and resources available in the study group of the ISMAC Technological Institute in the Software Development career.

This study corresponds to the first phase of the project "Learning capacities of students in the Software Development career of the face-to-face modality, using Big Data and/or BI tools" (project in progress) from which the results have been obtained, mentioned modeling that will be implemented in the second phase of said project hoping to obtain promising results that provide us with new knowledge of the student data of the sample of the mentioned institution, for this the mathematical-computer model will allow predicting the learning style of each student to later personalize and improve the student's own learning.

2   Data Science in Education

2.1   Data Science

Data science is responsible for the study of data, its interpretation, analysis and inference or prediction, to generate information and then significant knowledge about an area of study from simple data. With the advancement of technology, the data has increased exponentially, this large amount of data today is treated with Big Data technology and analyzed with data science [2, 20].

Data science analyzes any area or sector that produces data, among the main ones are *Finances:* Detection of credit card fraud, customer or product risk analysis, etc. *Sales:* Loss prevention, activity-based recommendations, dynamic pricing, etc. *Manufacture:* Product research, improvement of production processes and product quality, etc. *Cybersecurity and threat intelligence:* Intelligent detection of cyber-attacks, etc. *Medicine*: disease prevention, pharmaceutical research, etc.; and other areas such as telecommunications, government, education etc. [2].

Among the most used techniques in data science are machine learning and multivariate analysis. *Machine Learning* or learning machines are models that take advantage of the data and resources of a computer to predict or decide and make a decision on a computationally unprogrammed task. On the other hand, multivariate analysis allows analyzing several variables of different characteristics simultaneously. [1, 2, 8, 11, 20].

2.2   Machine Learning in Education

Machine learning is a data science technique that handles supervised and unsupervised methods, the supervised ones contain methods and algorithms such as decision trees and the unsupervised ones contain methods and algorithms such as clustering which are related to data analysis in the area of education due to its properties and characteristics [1, 2]. Machine learning, in turn, is related to Educational Data Mining EDM as per its acronym in Spanish, which is one of the main ways to analyze information from the educational area [1].

Educational data mining or EDM is born from the union of data science, Big Data, and the field of education in order to improve educational learning. The EDM has a historical trajectory that goes back to the year 2000 and was extended in formalized EDM conferences that took off in 2008. In 2011, a group of researchers formed the EDM society, with which they proposed a series of more formal definitions of this area [1, 5].

Some of the most important definitions mentioned in the EDM handbook are *Learning analytics. –* It is the measurement, collection, analysis, and reports of data about students/learners and their context, with the purpose of understanding and optimizing learning and the environment where it occurs. *Academic Analytics.* – It is the application of statistical techniques and institutional data mining to produce business intelligence and solutions to universities and administrators [1].

The objective of the EDM is to mine or find a new unknown knowledge, this process is called KDD or Knowledge Discovery of Databases, this process is carried out with the aforementioned machine learning models, which according to its properties and characteristics [1, 3].

The generation of new knowledge is used to solve problems such as the prediction of academic performance, creating intelligent tutors, analyzing behavior patterns, similarities and anomalous ones, quantifying the effectiveness of the teaching-learning process, developing new EDM technological tools, developing algorithms, carry out replication studies in other domains of education, etc., which allows optimizing and improving the learning process [1, 3–5].

To manage the EDM, different technological tools are used, among the most important are the Moodle platform, which allows you to manage tasks, evaluations, grades, attendance, platform use logs, interaction in forums, etc., this information is useful for working in processing models [14]. WEKA software written in java language provides many artificial intelligence functionalities and statistical methods with much support for experimentation [5]. There is also the KEEL software written in open-source java (GPLv3) which solves KDD or knowledge discovery tasks using training techniques, feature selection, discretization, missing data analysis, etc., this software has a particularity, it is focused on research and education [17].

2.3   Clustering

Clustering is an unsupervised machine learning technique that allows grouping or associating data into groups that contain similar characteristics, in such a way that pattern recognition is performed when analyzing the information. Some of the methods for clustering are by partition, by density, hierarchies, etc., and some of the most used algorithms are:

K-means. – This algorithm is fast, robust, and simple, it allows the grouping of data based on the least squared distance of a centroid, it is sensitive to outliers and noise.
Mean-shift. – This algorithm performs a grouping by updating centroids, it is computationally expensive, so it does not work properly with large amounts of data.
DBSCAN. – Data clustering based on non-parametric density to separate high-density and low-density groups into clusters, is resistant to outliers.
Hierarchical. – Technique that creates a hierarchy in the grouping of data according to their similarity with a bottom-up approach that allows to have a more informative organization of the data than other algorithms, information represented in a dendrogram graph [1–3, 7].

2.4   Decision Trees

Supervised model used in machine learning that allows categorizing or classifying a data set according to established rules. It is composed of nodes, edges, in the nodes the variables to be studied are placed, this method consists of a training stage and a test stage where underfitting or overfitting are usually given in training [1, 2, 10]. This model solves classification tasks with discrete data and regression tasks with continuous data, they are efficient for analyzing user behavior and cybersecurity analysis. Decision trees use criteria for their operation, among the main ones are the Gini impurity or Gini index, given by

$$G = \left(\frac{1}{2n^2\mu}\right) \sum_{j=1}^{m} \sum_{k=1}^{m} n_j n_k \left| y_j - y_k \right| \tag{1}$$

and the information gain that arises from information theory and entropy [2, 11]. Given by

$$I(p, n) = \left(\frac{-p}{p+n}\right) log_2\left(\frac{p}{p+n}\right) - \left(\frac{n}{n+p}\right) log_2\left(\frac{n}{p+n}\right) \tag{2}$$

Among the most used decision tree algorithms are: IBK, ID3, j48, C4.5, CART.

IBK. – It is an algorithm that is called lazy type and does not create a single decision tree, each time it finds a new instance it generates a calculation of the relationship with other instances [6].

ID3. – It is based on the use of information gain through the entropy analysis of information theory, it consists of selecting an attribute as the root of the tree and creating a branch with each of the possible values of attribute, is focused for categorical classification, also related in the field of educational prediction [6, 9].

J48. – Induction algorithm that generates rules from subsets of the total data, an optimization is performed according to a calculation of goodness [6, 9].

C4.5. – Algorithm that works in a similar way to ID3, it works with continuous and discrete data sets, it also allows handling incomplete data sets.

CART. – Tree classification and regression algorithm, uses the Gini index and is applied to carry out splitting processes [1, 6, 11, 14].

2.5   Multivariate Analysis in Education

Multivariate analysis is a technique that allows analyzing several variables with different characteristics at the same time. There are two types of multivariate analysis: explanatory or dependency models and descriptive or interdependence models. Descriptive models allow analyzing data from several variables that do not depend on a single variable or interdependence, some of the most used models are factor analysis and PCA or Principal Component Analysis, methods related to the area of education through psychometrics and educational psychology [2, 21].

2.6   Principal Component Analysis or PCA

PCA or Principal Component Analysis is a data analysis technique that allows evaluating several variables of different kinds or that are not initially correlated, where patterns are established through the use of eigenvalues and eigenvectors in conjunction with correlation methods between the data, analyzing the variance, this method also allows a reduction of variables and works as a predictive method [2, 12, 13, 21].

PCA is closely related to factor analysis. The PCA in education allows evaluating variables of various types such as: sociodemographic, socioeconomic, use of formal and informal resources for education, etc., and obtaining efficient results as shown by some studies [12, 13, 21].

2.7   Methods Used in Psychometrics

Psychometrics allows us to evaluate or measure variables, aspects, and characteristics in the area of education, in particular it is also applied in the evaluation of learning styles related to the learning abilities of students [15]. Among the most used methods are the Cronbach's Alpha coefficient to assess the reliability of a measurement scale and the Bartlett test to assess the variance of different population samples [16, 21].

On the other hand, factor analysis allows analyzing correlation in unobserved variables from a set of observed variables, there are two types: Exploratory factor analysis or AFE and confirmatory factor analysis or AFC. On the other hand, the R-squared coefficient of determination allows evaluating the quality of fit of a linear regression model. If the coefficient is close to 1, it will have a greater linearity fit and therefore better quality; otherwise, the closer to 0 the model will be not so efficient [2, 8, 15, 16, 21].

2.8   Data Warehousing

Big Data and/or BI allow the data warehousing process to be carried out, with which different resources of different characteristics can be joined to analyze data, obtain information and be able to generate greater knowledge. These resources are usually internal to an organization or institution such as databases, Excel resources, pdf, etc., and external such as web links, social network forums, etc. [18].

One of the important threads is the ETL or Extraction, Transformation and Load and they are defined as follows: *Extraction.* – These are the tasks of extracting information from different data sources, web resources, etc. *Transformation.* – These are the tasks dedicated to the transformation, cleaning, elimination, or treatment of null, ambiguous, incomplete, anomalous data, etc., according to the study area. For example, in the field of education, once the data has been correctly identified, it is not advisable to delete anomalous data, since precisely this type of data could give us important information about the students [1]. *Load.* – Data loading refers to the process of placing the purified or transformed data in a big data warehouse or big database [6, 7, 18].

The type of data used for the Datawarehouse is structured and unstructured, the largest amount of information on the Internet is unstructured, all this information is processed in distributed type software architectures managed by software tools such as Hadoop, Spark, Power BI, etc. [18]. Power BI is a Microsoft tool that allows ETL processes to be carried out for data processing, it is connected to multiple sources of information such as databases, web resources, Excel files, etc.; on the other hand, it is connected to certain data science tools and to R and Python execution files, these languages being the most widely used for data science today [18, 19].

3   Learning Styles

3.1   Felder and Silverman Model

As proposed by Felder and Silverman (1988), it classifies learning styles into four dimensions: *Understanding.* – Identifies the Global/Sequential concept, where the global student is governed neither by time nor by a calendar, he focuses on the objective that must be met, on the contrary, a sequential student is governed exclusively by time, his performance is presented using small logical steps to move quickly in the process. *Processing.* – Where the students are of the Active/Reflective type, the active ones are identified by manipulating things and working in a team, on the other hand, the reflective learns by thinking about things and likes to work alone. *Perception.* – Within this large group, the Sensing/Intuitive styles are identified, the sensitive student is the one who shows interest in the practical part while the intuitive student predominates in theory-oriented training. *Input.* – Here the Visual/Verbal styles are identified, where the visual style is related to the development of skills through the use of visual resources such as concept maps, statistical graphs, photos, videos, etc., on the other hand, a verbal student it is supported by the use of guided readings, active listening and note-taking, it learns from the discussion [3, 6, 22–24].

3.2   Development by Competencies

The development by competences proposes a learning applied to real world situations, according to a study 8 elementary competences are proposed: Communication in mother tongue, communication in foreign languages, mathematical competence, science, technology and digital (CD), social (CS), learning to learn (CAAP), social and civic competences (CSC), sense of initiative and entrepreneurship (SIEE) and cultural awareness and expressions (CEC); as per their acronyms in Spanish. Competences are related to learning styles, an important relationship occurs between the competence of learning to learn and learning styles, it is important to know what the learning style of each person is to enhance their competence [25, 26].

The development by competences is supported by e-learning technologies to improve the learning process. A competency-based assessment improves and personalizes the teacher's management while allowing a better characterization of the student's profile to be obtained [27].

4   Method

In this study, a methodology of qualitative approach, exploratory, documentary, bibliographic and bibliometric research was used to analyze and compare the main methods, techniques and computer mathematical models that allow processing data related to the area of education. To reach a study object, bibliometric analysis was used with the search equation given in Table 1.

The bibliometric analysis was carried out in the Scopus database where the information was processed by the Vos Viewer software where after performing a concurrency

Table 1. Search equation, bibliometric analysis.

Equation	Number of documents found in Scopus
"Learning capabilities "and "data science"	316

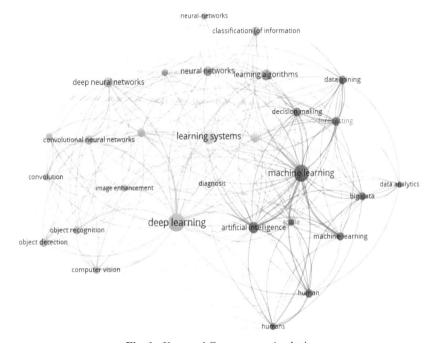

Fig. 1. Keyword Concurrency Analysis

analysis of the keywords in the articles referring to the defined search equation, the results shown were obtained in Fig. 1.

In which the keywords are highlighted: "learning systems," "machine learning", "deep learning" and "artificial intelligence". In addition to VOS viewer, the Bilbliometrix software was used to process the information from the documents found and shows that, in terms of the trends of the topics, it is relevant that they begin to develop since 2018, the concepts remaining constant over time. "machine learning" and "big data", in terms of the topics that have been most relevant during 2022 are "Deep learning", "object detection" and "computer vision", as shown in Fig. 2.

The bibliographic analysis was carried out with searches for articles referring to the proposed theme, meetings and discussions between researchers, where the best methods to use both in the educational area and in the data science area were analyzed, taking into account that the subject of the educational data analysis with data science methods and Big data and/or BI technologies is relatively new [1].

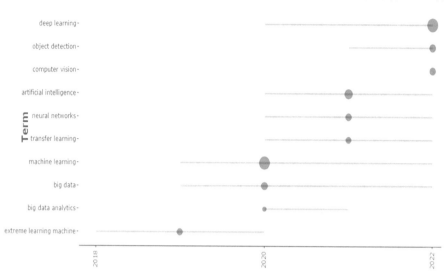

Fig. 2. Topic trends over time

5 Predictive Models of Learning Styles and Competencies

Table 2 shows the results of the analyzed models.

Table 2. Comparison of predictive models before and after completing an educational cycle based on the data to be processed to characterize the student profile.

Model or method	Before finishing an educational cycle	After completing an educational cycle
Clustering	Felder and Silverman test data	Competency assessment data
Decision tree	Felder and Silverman test data	Competency assessment data
Principal Component Analysis (PCA)	Felder and Silverman test data	Competency assessment data

To generate a new unknown knowledge or KDD Knowledge Discovery Databases in the area of education at the ISMAC Technological Institute in the students of the Software Development career, the machine learning and multivariate analysis models described in Table 2 are proposed, as follows:

Decision trees with the ID3 algorithm and its processing based on the gain of information, is ideal for characterizing the student profile from the point of view of learning styles and student competencies, since in each node there is an information maximization update or information gain through entropy minimization.

Clustering with the k-means algorithm to obtain group characteristics is efficient to characterize the student profile from the point of view of learning styles and student competencies to analyze the correlations between the different students and the classifying group they belong to, in the same way, verify the distinction between groups, this from a perspective that does not have predefined rules as it happens in decision trees.

The PCA or Principal Component Analysis is used to obtain a correlation between several variables of different kinds and reduce dimensions, which provides a focus on correlation and variability of the data to be treated, particularly the modeling of learning styles and student competencies with PCA to characterize the student profile is ideal since the model performs the correlation between variables based on variances and gives an approach that allows reducing variables to work with the most important ones. Note that the method with a variance approach does not rule out outliers of the sample or noise, which are important cases to analyze in the area of education.

Fig. 3. Keyword concurrency analysis (word cloud)

Figures 2 and 3 show important results obtained in the bibliometric analysis; the use of machine learning is very present in data science studies. A study mentions the most used methods, among these very present clustering and decision trees [20], which are part of the proposed models, with a focus on education detailing specific algorithms such as ID3 for decision trees and k – means for clustering.

Table 3. Student data types.

Type of data	Description
Profile data	Sociodemographic and socioeconomic information on students
Class data	Historical information of total and partial grades, subjects, attendance, etc., both historical and current
Test data	Information on the student profile evaluated with psychometric tests

Table 3 shows a classification of data to be processed with the machine learning and multivariate analysis models from different sources, both external and internal to the study group (Software Development students), as follows: Own or internal data of the study group. – Profile data and class data. External data of the study group. – Test data. It should be noted that class data is managed today by e-learning platforms such as Moodle [14] where there is very important stored information such as grades, student participation in discussion forums, access levels and test completion time. And tasks, all this information is stored in logs, database tables, web resources, etc., in e-learning platforms [27].

Table 2 proposes the machine learning and multivariate analysis methods to process the data according to the classification in Table 3, with the aim of carrying out a KDD process to obtain new unknown knowledge with the identification of behavior patterns, similarity and anomalous through a predictive process to identify the student learning styles of the students of the Software Development career of the ISMAC Institute before or during their student life within the institution, this will allow to personalize the student's learning and improve their skills and abilities.

In addition, it is proposed to implement this data infrastructure, processing with machine learning models and multivariate analysis and generation of knowledge in the second stage of the project "Learning capacities of students in the Software Development career of the face-to-face modality, using tools of Big Data and/or BI", hoping to obtain promising results that provide us with new knowledge of the student data from the sample of the aforementioned institution, which allows establishing decision-making to improve student learning.

6   Conclusions

The most efficient data science methods for data processing of student processes and the prediction of learning styles in Software Development technology careers are decision trees with the ID3 algorithm, clustering with the k-means algorithm and the analysis of principal components or ACP.

The EDM or Educational Data Mining is a new area that unites data science, Big data technology and education to improve learning.

The bibliometric analysis shows us a notable presence of machine learning techniques within data analysis with Big Data technologies.

The classifications of the data of the study group of the Software Development technology careers of the ISMAC Institute, were defined as: *Profile data.* – That correspond to sociodemographic and socioeconomic information. *Class data.* – Which correspond to historical information from past and current academic periods managed by e-learning platforms such as Moodle and Test data. – Corresponding to the Felder and Silverman learning styles tests.

The definition of a software infrastructure for data warehousing with Power BI tools and Python language has been chosen to implement the different computer mathematical models in current technologies and customize the KDD or knowledge discovery process.

References

1. Romero, C., Ventura, S., Pechenizkiy, M.: Handbook of Educational Data Mining. CRC Press, Boca Ratón (2011)

 2. Sarker, I.H.: Machine Learning: Algorithms, Real-World Applications and Research Directions. SN Comput. Sci. **2**(3), 1–21 (2021)
 3. Costaguta, R.: Identificación de estilos de aprendizaje dominantes en estudiantes de informática. XV Congreso Argentino de Ciencias de la Computación, pp. 331–340 (2009)
 4. Porcel, E., Dapozo, G.: Modelos Predictivos y técnicas de minería de datos para la identificación de factores asociados al rendimiento académico de alumnos universitarios. In: XI Workshop de Investigadores en Ciencias de la Computación, pp. 635–639 (2009)
 5. Ayala, E., López, R.: Modelos predictivos de riesgo académico en carreras de computación con minería de datos educativa. Revista de Educación a Distancia (RED) **21**(66) (2021)
 6. Caman, R., Salguero, A.: Herramientas para detección de estilos de aprendizaje en estudiantes de educación superior. Revista Tecnológica ESPOL – RTE **30**(3), 106–121 (2017)
 7. Ballesteros, A., Sánchez, D.: Minería de datos educativa: Una herramienta para la investigación de patrones de aprendizaje sobre un contexto educativo. Latin-Am. J. Phys. Educ. **7**(4), (2013)
 8. Dabhade, P., Agarwal, R.: Educational data mining for predicting student's academic performace using machine learning algorithms. Materials Today: Proc. **47**(15), 5260–5267 (2021)
 9. Costa, E.B., Fonseca, B., Santana, M.A., de Araújo, F.F., Rego, J.: Evaluating the effectiveness of educational data mining techniques for early prediction of students' academic failure in introductory programming courses. Comput. Hum. Behav. **73**, 247–256 (2017). https://doi.org/10.1016/j.chb.2017.01.047
10. Agarwal, S., Pandey, G.N., Tiwari, M.D.: Data mining in education: data classification and decision tree approach. Int. J. e-Educ., e-Bus., e-Manag. e-Learn. **2**(2), 140 (2012)
11. Patel, H.H., Prajapati, P.: Study and analysis of decision tree based classification algorithms. Int. J. Comput. Sci. Eng. **6**(10), 74–78 (2018)
12. Todhunter, F.: Using principal components analysis to explore competence and confidence in student nurses as users of information and communication technologies. Nurs. Open **2**(2), 72–84 (2015)
13. Dumpit, D.Z., Fernandez, C.J.: Analysis of the use of social media in Higher Education Institutions (HEIs) using the Technology Acceptance Model. Int. J. Educ. Technol. High. Educ. **14**, 5 (2017)
14. Romero, C., Ventura, S., Espejo, P.G., Hervás, C.: Data mining algorithms to classify students. In Educational data mining 2008 (2008)
15. Dubois, J., Galdi, P., Paul, L.K., Adolphs, R.: A distributed brain network predicts general intelligence from resting-state human neuroimaging data. Phil. Trans. R. Soc. B **373**(1756), 20170284 (2018)
16. Freiberg-Hoffmann, A., Fernández-Liporace, M.: Learning styles type indicator: una adaptación para estudiantes medios y universitarios argentinos. Academo (Asunción) **10**(1), 1–12 (2023)
17. Alcalá-Fdez, J., et al.: KEEL: a software tool to assess evolutionary algorithms for data mining problems. Soft. Comput. **13**, 307–318 (2009)
18. Santoso, L.W.: Data warehouse with big data technology for higher education. Procedia Comput. Sci. **124**, 93–99 (2017)
19. Vásquez, R.A.D., Espinoza, J.L.A., Cabrera, M.A.C.: Power bi como herramienta de apoyo a la toma de decisiones. Universidad y Sociedad **14**(S3), 195–207 (2022)
20. Liao, S.-H., Chu, P.-H., Hsiao, P.-Y.: Data mining techniques and applications – A decade review from 2000 to 2011. Expert Syst. Appl. **39**(12), 11303–11311 (2012)
21. Closas, A.H., Arriola, E.A., Kuc Zening, C.I., Amarilla, M.R., Jovanovich, E.C.: Análisis multivariante, conceptos y aplicaciones en Psicología Educativa y Psicometría. Enfoques **25**(1), 65–92 (2013)

22. Figueroa, N., Cataldi, Z., Méndez, P., Rendon Zander, J., Costa, G., Lage, F.J.: Los estilos de aprendizaje y las inteligencias múltiples en cursos iniciales de programación. In VII Workshop de Investigadores en Ciencias de la Computación (2005)
23. Nivela-Cornejo, M.A., Echeverría-Desiderio, S.V., Morillo, R.: Estilos de aprendizaje y rendimiento académico en el contexto universitario. Domino de las Ciencias **5**(1), 70–104 (2019)
24. Roa, K., Martínez, C.: Diseño de un ambiente virtual de aprendizaje soportado en los estilos de aprendizaje. Virtu@ lmente **8**(2), 68–86 (2020)
25. Marcos, B., Alarcón, V., Serrano, N., Cuetos, M., Manzanal, A.: Aplicación de los estilos de aprendizaje según el modelo de Felder y Silverman para el desarrollo de competencias clave en la práctica docente. Tendencias pedagógicas (2020)
26. González, B., Hernández, M., Castrejón, V.: Estilos de aprendizaje para el desarrollo de competencias en estudiantes de la Licenciatura en Enfermería/Learning styles for the development of competences in students of the Degree in Nursing. RIDE Revista Iberoamericana para la Investigación y el Desarrollo Educativo **8**(16), 351–369 (2018). https://doi.org/10.23913/ride.v8i16.345
27. Moodle: https://moodle.org/. Last accessed 5 June 2023

Musculoskeletal Disorder Due to Exposure to Data Display Screen

Mónica Monserrath Chorlango García(✉) ⓘ, Patricia Lisbeth Esparza Almeida ⓘ,
Byron Sebastián Trujillo Montenegro ⓘ, Hugo Jonathan Narváez Jaramillo ⓘ,
Edison Robinson Rodríguez Yar ⓘ, and Kevin Andrés Rivera Vaca ⓘ

Instituto Superior Tecnológico ITCA, Ibarra EC100150, Ecuador
monicachorlango@gmail.com

Abstract. During daily activities, ergonomic conditions are essential due to the health and work performance that depend on them. Ergonomics is based on active breaks and comfort during work, significantly contributing to labor productivity. This research has been carried out with a mixed approach, using the Rapid Office Strain Assessment method to determine the characteristics of the job and the Rapid Upper Limb Assessment method to evaluate individual postures; A survey was also applied using the Nordic Questionnaire for the evaluation of musculoskeletal disorders. The research was carried out on the Higher Technological Institute for Research, Technology, Science, and the Academy staff, with a population of teachers and administrative staff. The personnel are exposed to ergonomic risks, with a percentage of 44% with the Rapid Office Strain Assessment method, with the Rapid Upper Limb Assessment 42.23% of the personnel, and with the Nordic Questionnaire the appearance of discomfort at the neck and shoulder due to the type of equipment and furniture used by the staff. Modifying furniture in the work area (ergonomic chairs, desks), acquiring appropriate instruments (ergonomic mouse, pad), and taking active and passive breaks with all staff have reduced risks and, therefore, musculoskeletal disorders.

Keywords: Musculoskeletal diseases · Ergonomic risk factors · Postural loading · ROSA method · RULA method

1 Introduction

Musculoskeletal disorders (MSD) have been defined as the interaction of an imbalance between the biomechanical demands and the functional capacity of each person [1]. MSD currently represent a significant public health problem due to their high prevalence and incidence and because functional capacity decreases, limiting work activity and impacting the health system and quality of life of the people who suffer from them. [2].

People who spend much time in front of a Data Display Screen are more exposed to MSD, as occurred during the pandemic caused by COVID-19. When physical requirements such as posture, strength, and movement exceed the individual's response capacity, this effort may be related to MSD originating during work [3]. Four theories explain the

M. Z. Vizuete et al. (Eds.): CI3 2023, LNNS 1041, pp. 82–97, 2024.
https://doi.org/10.1007/978-3-031-63437-6_7

mechanism of the appearance of MSD: the theory of the multivariate interaction of genetic, morphological, psychosocial, and biomechanical factors; the differential theory of fatigue due to kinetic and kinematic imbalance; the cumulative theory of repetition loading and the theory of overexertion [4]. MSD are the most common cause of long-lasting severe pain and physical disability. Epidemiological studies in various countries show that MSD occur in various human activities and all economic sectors and imply a high cost to society [5].

MSD are injuries or alterations suffered by body structures such as muscles, joints, tendons, ligaments, nerves, bones, and the circulatory system, caused or aggravated mainly by work and the effects of the environment in which it takes place [6]. Most are cumulative disorders resulting from repeated exposure to heavy loads and postures over a long period. However, MSDs can also be caused by acute trauma, such as fractures caused by work accidents [6].

MSD primarily affect the back, neck, shoulders, and upper extremities, although they can also affect the lower extremities. Some MSD, such as carpal tunnel syndrome, are specific due to their well-defined symptoms. Others are not so since only pain or discomfort is observed, without apparent symptoms of a specific disorder [6].

Some studies indicate that the prevalence of MSD in office positions ranges from 10% to 62%, generally related to the upper extremities, neck, and back [7]. The most common risk factors in this type of position are derived from using the keyboard and mouse. For example, repetitive movements of the fingers, hands, and wrists, holding the forearm and wrist in awkward postures, or elevated contact pressure on the mouse operating wrist. On the other hand, maintaining a sitting posture for a long time, primarily if it is maintained incorrectly, increases muscle fatigue [7].

Physical effort occurs because of fundamental work activity. Intense or repetitive mechanical effort in an environment with inadequate working conditions can affect the musculoskeletal system, which comprises bones, joints, ligaments, tendons, nerves, and vessels [6].

MSD increase in jobs that are intensive in physical activity and offices whose tasks require prolonged permanence in certain work positions. The Instituto Tecnológico Superior Investigación, Tecnología, Ciencia y Academia, ITCA, does not have an ergonomic risk study by Data Display Screen (DDS), which generates the need to carry out investigations to determine and prevent MSD and the appearance of occupational diseases to which personnel are exposed. The activities of many workers are not usually carried out in optimal conditions for their health due to ignorance of ergonomic risk factors such as DDS, forced postures, or repetitive movements; Hence the importance of investigating what are the ergonomic risks due to DDS exposure and their relationship with the MSD of ITCA personnel.

The modern use of the term is due to Murrell and was officially adopted during the creation, in July 1949, of the first ergonomics society, the Ergonomics Research Society, founded by British engineers, physiologists, and psychologists with the aim of fitting work to man [8].

Risks correspond to those that originate when the worker interacts with his job and when work activities present movements, postures, or actions that can cause damage to health [9]. There are characteristics of the work environment capable of generating a series of disorders or injuries, giving rise to risks due to forced postures, risks caused by repetitive movements, health risks caused by vibrations, application of forces, and environmental characteristics in the work environment (lighting, noise, heat, among others, and risks due to musculoskeletal disorders derived from physical load: back pain, hand injuries) [10].

Screens with electronically generated images or DDS represent the most characteristic element of computerized work equipment in the workplace and private life. A workstation may be designed to include, as a minimum, a DDS, and data entry device (usually a keyboard); it may also include space for additional technical equipment, such as multiple displays and recording devices. Input and output data. A few years ago, in the early 1980s, data entry was the most common task for computer users. However, today, these devices are used daily in all professions.

Among the risk factors that MSD can generate are repetitive movements, force, dynamic and static load, posture, precision, visual demand, vibration, and inappropriate cycles of work and rest. It is necessary to have recovery periods that allow sufficient time for the physiological rest of the compromised area. The main risks associated with using equipment with DDS are musculoskeletal discomfort, visual fatigue, and mental fatigue. It must be considered that a worker is considered a computer user with DDS if their effective activity is more than 4 h a day or 20 h a week [11].

The injuries related to the use of DDS are progressive, and the symptoms are different; they worsen according to the different stages, such as pain and fatigue in the wrists, arms, shoulders, or neck that appear during work; it improves during the night and the weekly rest. It usually lasts weeks or months; pain and fatigue start early in the day, persist longer into the night, and can even interrupt sleep. Workers usually take pain pills in this phase can last several months, but they continue to work with pain, fatigue, and weakness, even when rested. The injuries can also interrupt sleep; they do not allow them to do daily tasks at work or home, an affectation that can sometimes last months or years; some people do not fully recover, and others become disabled.

The injuries produced can be in the back (herniated discs, low back pain, sciatica, muscle pain, disc protrusion); neck (pain, muscle spasms, disc injuries); shoulders (tendonitis, periarteritis, bursitis); elbow (tennis elbow, epicondylitis); hands (carpal tunnel syndrome, tendonitis, numbness, strain); legs (sciatica and varicose veins) [12].

For the prevention of MSD, ergonomic risk management allows the identification of factors to which workers are exposed, which allows them to protect and make decisions and the permanence of a better comfortable and safe work environment for better development [13].

This research aims to determine the ergonomic risks due to exposure to DDS and their relationship with the MSD of the ITCA staff to improve the conditions in which the tasks are carried out, promoting the greater well-being of the institute's workers.

2   Methodology

The research is of a non-experimental, descriptive, and mixed approach type. Through observation, the different activities carried out by the ITCA staff were evaluated with the help of videos and photographs, allowing the identification of the ergonomic risk by DDS applying the Rapid Office Strain methods. Assessment (ROSA), Rapid Upper limb Assessment (RULA), and the MSD with the application of the Standardized Nordic questionnaire, which took place at ITCA. The population under study was 45 people belonging to the administrative staff (eleven people), teachers (thirty people), and research teachers (four people). The Human Resources department was asked for the list of the Institution's officials, with information on their position, dependency, and length of employment. The selection criteria were having a job seniority equal to or greater than one year.

The ROSA method calculates the deviation between the characteristics of the evaluated position and those of an office position with ideal characteristics. Scoring diagrams assign a score to each of the elements of the position: chair, screen, keyboard, mouse, and telephone. The method includes a checklist to assess the risks commonly associated with office jobs. The method applies to jobs where the worker sits in a chair before a table and handles a computer with DDS. The most common elements of these workstations (chair, work surface, screen, keyboard, mouse, and other peripherals) are considered in the evaluation. As a result of its application, an assessment of the measured risk and an estimate of the need to act on the position to reduce the level of risk are obtained [7].

To apply the method, the evaluator observed the workstation while the worker carried out his task. Taking photographs of it allowed its subsequent analysis. After the observation, a brief interview was held with the worker to clarify the aspects of the task and the position [14].

Once the necessary data was obtained, the different elements of the job were scored using the scoring diagrams. These diagrams were designed so that if a job element's situation is ideal, it is assigned a score of 1. As the element's situation deviates from the ideal, the score increases linearly to 3. On the other hand, certain specific situations concerning each element increase the score obtained by the element (+1). If the position of the chair's armrests is not adjustable, the score is increased by one point. Additionally, the time that the worker uses each element during the working day increases or decreases the score obtained. Once the scores of the five job elements considered by ROSA have been obtained, partial and final scores are obtained [7].

The value of the ROSA score can range from 1 to 10, with the higher the risk for the person occupying the position being higher. The value 1 indicates that no risk is appreciated. Values between 2 and 4 indicate that the level of risk is low but that some aspects of the position can be improved. Values equal to or greater than 5 indicate that the level of risk is high. From the final ROSA score, 5 Action Levels are proposed for the position. The Action Level establishes whether intervention in the position is necessary and its urgency and can range from level 0, which indicates that action is not necessary, to level 4, corresponding to action on the position is urgent. The priority actions can be established from the partial scores obtained for each element of the position [15].

One of the risk factors most associated with the appearance of MSD is excessive postural load. If inappropriate postures are adopted continuously or repeatedly at work, fatigue is generated, and, in the long run, health problems can be caused. The evaluation of the postural or static load, and its reduction, if necessary, is one of the fundamental measures to be adopted in improving jobs [7].

The RULA method was developed in 1993 by McAtamney and Corlett from the University of Nottingham (Institute for Occupational Ergonomics) to evaluate the exposure of workers to risk factors that cause a high postural load that can cause disorders in the upper limbs of the body. For the evaluation of the risk, the adopted posture, its duration and frequency, and the forces exerted when it is maintained are considered in the method [7].

RULA obtains a score for a particular posture from which a specific Action Level is established. The Action Level indicates whether the position is acceptable or to what extent changes or redesigns are necessary. In short, RULA allows the evaluator to detect possible ergonomic problems derived from an excessive postural load. The RULA method evaluates individual postures and not sets or sequences of postures. Therefore, it is necessary to select those postures that will be evaluated from those adopted by the worker in the position. Those that, a priori, involve a more significant postural load due to their duration, their frequency, or because they present a more significant deviation from the neutral position will be selected [16].

The first step consists of observing the tasks performed by the worker. Several work cycles were observed, and the evaluated postures were determined. If the cycle is long or there are no cycles, evaluations can be carried out regularly. In this case, the time spent by the worker in each position was also considered [7].

The measurements that must be taken on the employee's postures are mainly angular (the angles formed by the different body members concerning specific references). These measurements can be made directly on the worker using protractors, electro goniometers, or any other device that takes angular data. In addition, photographs of the worker adopting the studied posture can be taken, and their angles measured. If photographs are used, it is necessary to take an adequate number of photographs from various viewpoints (front, profile, and detail views). It is crucial to ensure that the angles to be measured are shown in their accurate magnitudes on the images, i.e., that the plane in which the angles are measured is parallel to the camera plane in this case. For this task, the rulers (RULA), the Argonauts' tool for measuring angles on photographs, can be used [7].

The method must be used separately for the right and left sides of the body. Although the expert evaluator can choose a priori the side that faces the most excellent postural load, it is preferable to analyze both sides in case of doubt. Two groups are divided into RULA: Group A covers the upper limbs (arms, forearms, and wrists), and Group B covers the legs, trunk, and neck. A score is assigned to each body part (legs, wrists, arms, and trunk) using the tables associated with the method to assign global values to each A and B group [7].

Measuring the angles formed by the different parts of the operator's body is crucial to assigning scores to the members. The way to measure the angle is determined by the method for each member. Subsequently, the global scores of groups A and B are adjusted based on the type of muscular activity developed and the force applied during the task. Finally, these modified global values are used to calculate the final score. The final value of the RULA method is proportional to the risk associated with performing the task, so high values indicate a higher risk of MSD. The method divides the final scores into action levels so the evaluator can decide after the analysis. The proposed action levels range from 1 to 4, considering that the evaluated position is acceptable [7].

Kuorinka Nordic Questionnaire is a standardized questionnaire to detect and analyze MSD symptoms, applicable in ergonomic or occupational health studies, to determine if the initial symptoms are not yet a disease or have not yet led to consulting a doctor. Its value lies in providing information that allows a proactive estimation of the level of risks and rapid action. The questions are multiple-choice and can be applied in one of two ways. One is self-administered, which means that the respondent responds on their own without the presence of a surveyor [17].

The reliability of the questionnaires is acceptable. Some specific characteristics of the efforts made at work are shown in the frequency of questionnaire responses. This questionnaire collects information about pain, fatigue, or discomfort in different areas of the body. The questionnaire is anonymous; nothing in it can tell which person has answered which form. All information collected is used to investigate possible factors that cause fatigue at work.

3   Results

The results of this study helped to verify the risks that DDS and the MSD give that these positions cause in ITCA workers. Figure 1 presents the results of the application of the ROSA method.

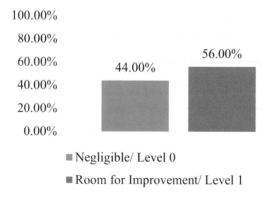

Fig. 1. Evaluation with the application of the ROSA Method.

It can be seen in Fig. 1, that the values obtained with the ROSA Method show 44% of the personnel with a level of 0, which corresponds to a negligible risk that does not necessarily require action. In comparison, 56% of the workers present a level 1 with risk in which elements of the position can be improved, and at levels 2, 3, and 4 with 0%.

Santamaría (2023) [18], has reported that when applying the ROSA Method to people in the accounting area in the workplace with a computer in the office, it was evidenced that 66.6% of the areas evaluated present a very high risk and 33.3% have a high risk this is attributed to the fact that in certain areas there are no ergonomic chairs. The areas with this tool are not adapted to the worker. These results are lower than those obtained in the

ITCA research since, in Santamaría's proposal, the personnel studied were permanently working in offices that were not conditioned to prevent musculoskeletal conditions. In contrast, at ITCA, the offices and administrative staff are duly conditioned. Figure 2 presents the results of the application of the RULA Method.

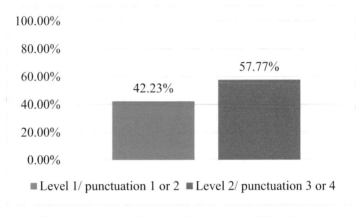

Fig. 2. Evaluation with the application of the RULA Method

As observed in Fig. 2, 42.23% of the personnel is exposed to a risk level 1, which means that the evaluated posture is acceptable; 57.77% corresponds to a risk level 2 in which changes may be required in the posture and 0 at levels 3 and 4. Farez (2022) [19], applying the RULA Method, reports scores of 5 and 7, with risk levels of 3 and 4 considered high, interpreting that there is a need to implement strategies to prevent damage to workers, especially at the level of the hands and wrists due to the frequent use of keyboards and mice. These results were superior to those obtained in applying the RULA Method to ITCA, where there was no evidence of high-risk regarding posture changes.

The results of the application of the Nordic Assignment are presented in Fig. 3.

■ Not presented discomfort   ■ Presented discomfort

Fig. 3. Discomfort detected with the application of the Nordic questionnaire.

When applying the Standardized Nordic questionnaire to the ITCA staff to observe who presented discomfort or not, it was obtained that 62.2% of the staff had presented musculoskeletal discomfort. Aguilera and Ortiz (2022) [20], using the same instrument, demonstrated that there is a 37% prevalence of symptoms of musculoskeletal disorders associated with shoulder pain; 32% on the right wrist; 24% in the right elbow (of which 28% are associated with previous car accidents, with incurrence of fissures and fractures in the upper limbs) and in the lower back a percentage of 21%, of which 7% have a history musculoskeletal disorders unrelated to work tasks. In the ITCA, a high incidence of personnel with musculoskeletal discomfort was observed concerning those in the study by Aguilera and Ortiz [16], which is because the personnel investigated in the ITCA spend more time on activities in front of a computer, concerning the personnel of nursing analyzed by the authors in which the ailments were mainly due to MSD sequelae.

The areas of presence of annoyances are presented in Fig. 4.

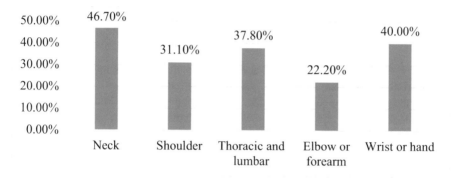

Fig. 4. Areas of presence of discomfort.

Figure 4 shows the data obtained on the inconveniences presented by the ITCA staff. The affected areas were the neck at 46.7%, shoulder at 31.1%, dorsal or lumbar region at 37.8%, elbow or forearm at 22.2%, and wrist or hand at 40%. Vahdatpour, Bozorgi, and Reza (2019) [21], have reported that users who work with computers presented more significant musculoskeletal discomfort in the upper extremities. Pain in the cervical region (77.5%) represented the greatest complaint of total discomfort in different organs of the body, followed by pain in the waist (73.2%), shoulders (64.9%), upper back (62.0%), and dolls (59.9%). In the leg region, the sole and the thigh presented a prevalence of 28.2%, 29.6%, and 32.4%, respectively.

In the study carried out at the ITCA, a predominance of musculoskeletal discomfort in the upper extremities was observed, like that found in the Vahdatpour study, because the users of both studies had poor posture when working in front of computers; however, when choosing the sitting position, few discomforts were reported in the lower limbs.

Figure 5 presents the values found for the temporality of the discomforts presented.

In Fig. 5, 22% of the population have presented discomfort in the neck for a time equal to or greater than six months, followed by 13% who have felt discomfort for less than one month and 11% for three months. Regarding shoulder discomfort, 27% have

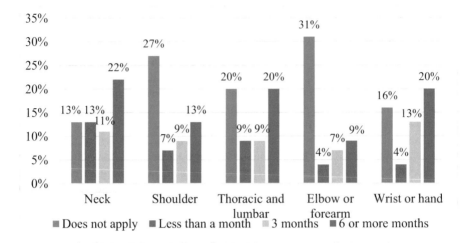

Fig. 5. Nuisance temporality values

not presented discomfort, while 13% have felt discomfort for six months or more. In the dorsal or lumbar area, 20% of the staff have not felt discomfort, and 20% have felt discomfort for a period equal to or greater than six months. In the elbow or forearm area, 31% of the staff did not feel discomfort, while 9% felt discomfort for six or more months. In the wrist or hand, 20% of the staff have felt discomfort for six or more months, and 13% have felt discomfort for three months.

Santamaría (2023) [18], in his research, states that, of 27 people surveyed, 26 responded that they sometimes have foot discomfort, and one worker stated that he often has foot discomfort. Regarding the knees and elbows, 25 workers report that the discomfort occurs rarely, and two workers respond that many times; while in the legs, hands, and wrists, 24 and 23 workers present discomfort at times; therefore, 3 and 4 many times, respectively; the least prevalent are discomfort in the lumbar back, neck and upper back.

The results provided by the ITCA show a higher prevalence in teachers with discomfort in the neck, wrist, or hand and the dorsal or lumbar area for more than six months, which is also presented in the Santamaría [18], study where musculoskeletal discomfort exists. Also, in the hands and wrists, several times reported by the people surveyed, this is due to the prolonged use of computers with inappropriate positions, which leads to the appearance of these discomforts for several months.

Figure 6 presents the results obtained on the changes of postures required by ailments (A) and discomforts (B) presented.

Workers have presented discomfort during the last year in areas such as the neck 42%, shoulder 22%, dorsal or lumbar 33%, elbow or forearms 18, and 36% have had discomfort in the wrist or hand (Fig. 6A). The discomfort is due to the tasks carried out regularly by the surveyed personnel and the inadequate posture in front of the computers. Due to the inconvenience, the staff has had to modify their positions. Of the participating population, 47% were due to neck discomfort, 27% to shoulder discomfort, 38% to

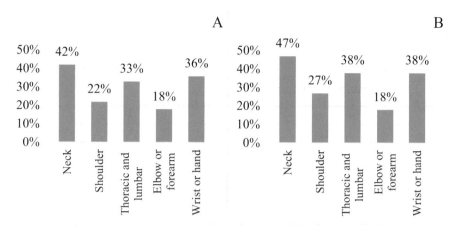

Fig. 6. Changes in postures required by ailments presented (A) and Discomfort presented in the last 12 months (B).

dorsal or lumbar areas, 18% to elbow or forearm discomfort, and 38% to wrist or hand discomfort (Fig. 6B). These changes in posture decrease discomfort in the ITCA staff for a short time because it leads to relaxation of the affected area.

Figure 7 presents the Duration of Discomfort results in the last 12 months.

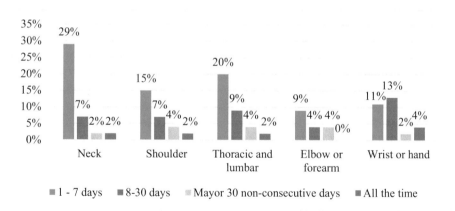

Fig. 7. Duration of the complaints presented in the last 12 months.

The highest percentage of complaints presented by the staff in the ITCA in the last seven days were in the neck and dorsal or lumbar region. From eight to 30 days the discomfort occurred in the wrist or hand, followed by the dorsal or lumbar region. Ailments of more than 30 days do not exceed 4%. If the complaints in the last month are considered, 36% of the staff have had complaints in the neck, followed by 29% in the dorsal or lumbar region and 22% in the shoulders. Knowledge of these results prompted preventive measures to be taken at the institute.

The values obtained for the episodes caused by the MSDs are presented in Fig. 8.

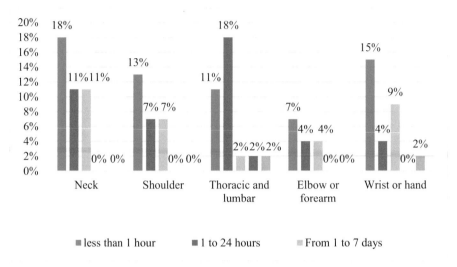

Fig. 8. Duration of MSD episodes.

The duration of the episodes of discomfort in the neck has been for less than an hour in 18%; 22% of the personnel have felt discomfort from one hour to seven days; in the shoulder, 13% have had discomfort for less than an hour and 14% have had more than an hour to seven days; In the dorsal or lumbar zone, 11% felt discomfort for less than an hour and 18% from one to twenty-four hours. In the area of the elbow or forearm, 7% have felt discomfort for less than an hour, and 8% from one hour to 7 days. In the wrist or hand, 15% have felt discomfort for less than an hour, and 9% for one to seven days.

Figure 9 shows the time that the inconvenience has prevented from working.

The values corresponding to the time that staff has been prevented from carrying out their work for one to seven days due to neck discomfort were: 4% due to neck discomfort, 2% due to shoulder discomfort, 4% due to dorsal or lumbar discomfort, and a 2% for wrist or hand discomfort.

Next, Fig. 10 presents the values corresponding to the treatment applied for discomfort.

The personnel studied who presented osteoarticular discomfort required treatment in the last 12 months; according to the affected area, there was a frequency like 15% for neck discomfort, 9% for shoulder, 13% for dorsal or lumbar, 11% for elbow or forearm and 18% due to discomfort in the wrist or hand.

Vera et al., in 2023 [22], state that the users needed medical treatment and the most affected part that has caused hospitalization in 2.4% is the shoulder area: left side 47.7%; right side 31.6%, both shoulders 21.1%; while medical assistance or physiotherapist weighted with 18.2 in the lumbar region followed by 12.3% in the neck region. In the study carried out at the ITCA, the staff needed medical treatment after approximately 12 months of presenting musculoskeletal discomfort. In Vera's study, the users also

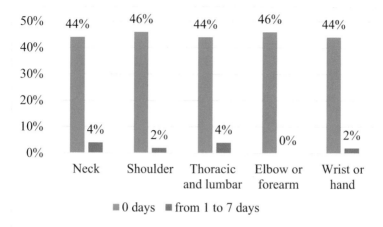

Fig. 9. Time that has prevented you from doing your job.

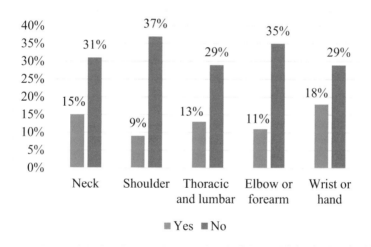

Fig. 10. Frequency of treatment needed for discomfort.

needed hospitalization and physiotherapy because, in this study, the discomfort presented was acute, which required more specialized treatment.

Figure 11 includes the results of the frequency of complaints presented in the last seven days.

The participating population presented discomfort in the last seven days in the neck, 29%; 15% on the shoulder, 22% in the dorsal or lumbar: 9% in the elbow or forearm, and 22% in the wrist or hand.

Vera et al., in 2023 [22], reported that identifying problems in the locomotor system demonstrates the prevalence of musculoskeletal pain symptoms concerning the most affected area of the body. The neck was found at 63.1%, followed by the lumbar region

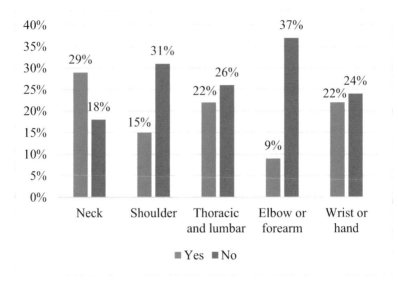

Fig. 11. Frequency of complaints presented in the last 7 days.

at 58.2%, dorsal region at 45.9%, knees at 33.6%, shoulders at 32.8%, and wrist at 27%, but in the last seven days, the area most exposed to discomfort is the lumbar area at with 45.9%, the neck area with 40.6% and the lumbar region 40.4%. The area least exposed to discomfort or pain is the elbow. In the study carried out at ITCA, there was a higher prevalence of discomfort towards the upper limbs due to the activities carried out by the teachers in front of the computers due to the repetitive movements carried out by the professionals. In contrast, in the Vera study [22], there was a higher prevalence in the lumbosacral area since the personnel studied adopted incorrect positions and did not have active breaks to achieve relaxation of the lower limbs muscle area.

In the review of the results obtained in the ITCA, after the application of the ROSA and RULA methods, it is verified that there is a relationship between the levels found since in ROSA, we have a percentage of 44% of personnel and with RULA 42.23%. of the personnel, which means that they are not exposed to a risk that could affect their health. Regarding the need to apply corrective actions in both methods, it was found that actions must be carried out to improve the position and tasks. The Nordic questionnaire showed that 32.60% of the population had no discomfort or pain from the postures. 68% of the staff have many conditions in the neck, shoulder. Corrective measures have been generated based on the risks found, generalizing to personnel, and are shown in Table 1.

Table 1. Corrective actions

Method	Level	Meaning	Corrective action	Responsible
ROSA	0	Negligible	No action required	N/A
RULA	1	Acceptable		
ROSA	1	Room for Improvement	Modify the furniture in the work area (ergonomic chair, desks)	Human talent and occupational health and safety and rectory
			Implement instruments that reduce risk (ergonomic mouse, pad)	
RULA	2	Task changes	Time and movements study	

4 Conclusions

ITCA personnel are exposed to low-risk levels of presenting MSD, which means that immediate job changes should not be considered; however, it can be concluded that, due to the presence of discomfort in the neck, shoulders, and dorsal or lumbar area, it is essential to make changes to some office elements such as screens, keyboards, and mice. In addition, the correct use of ergonomic chairs must be verified, and those that do not have the appropriate characteristics that adapt to the body and have support points should be replaced.

A small part of the staff has discomfort at the neck and shoulder level due to poor posture in the workplace, in many cases prolonged, considering that in the future, the pain may increase, even leading to absences due to illness, highlighting the need for Take active breaks to stretch the muscles. The personnel must be trained concerning good postures in their work activities as preventive measures, generating a corrective measures guide, which aims to reduce or even eliminate the MSD of the personnel.

Depending on the risk levels and the MSD found, the following corrective measures have been proposed: modify the furniture in the work area (ergonomic chair, desks), which allow postures to be improved and thereby reduce discomfort; for which instruments are implemented that reduce the risk (ergonomic mouse, pad), which will prevent the appearance of carpal tunnel, de Quervain's tendinitis and other diseases related to repetitive movements for extended periods; and take active and passive breaks with all the staff, which will prevent muscular contractures from being generated due to the stress of long hours.

This study's results reveal an association between exposure to biomechanical risk factors and the presence of musculoskeletal injuries, indicating which forced work postures mean more significant risk. Therefore, this type of disorder could incapacitate the worker in activities of daily living.

References

1. Mora, A.: Prevalencia y prevención de síntomas de trastornos musculoesqueléticos en teletrabajadores del Departamento de Servicios Generales del Instituto Costarricense de Electricidad en modalidad de teletrabajo como medida de contingencia ante la pandemia por COVID-19. Universidad Nacional de Costa Rica y Escuela de Seguridad Laboral e Higiene Ambiental, Instituto Tecnológico de Costa Rica (2023)
2. Jordán, R.: Trastornos musculoesqueléticos de origen laboral por la exposición en actividades operacionales en los trabajadores de una empresa de aluminios. Universidad Técnica del Norte, Ecuador (2023)
3. Riihimäkii, H.: Sistema musculosqueléticos. Enciclopedia de salud y seguridad en el trabajo 1(1), 6–39 (2001)
4. Kumar, S.: Theories of musculoskeletal injury causation. Ergonomics 44(1), 17–47 (2001). https://doi.org/10.1080/00140130120716
5. Robles, J.B., Iglesias Ortiz, J.: Relación entre posturas ergonómicas inadecuadas y la aparición de trastornos musculo esqueléticos en los trabajadores de las áreas administrativas que utilizan pantalla de visualización de datos, en una empresa de la ciudad de Quito en el año 2015. Revista de Ciencias de Seguridad y Defensa, IV(2), 158–181 (2019)
6. Agencia Europea para la Seguridad y la Salud en el Trabajo: Trastornos musculoesqueléticos. Europa.eu. https://osha.europa.eu/es/themes/musculoskeletal-disorders (2022)
7. Diego-Mas, J.A.: Evaluación de puestos de trabajo de oficinas mediante el método ROSA. Obtenido de Ergonautas. https://www.ergonautas.upv.es/metodos/rosa/rosa-ayuda.php (2015)
8. Mondelo, P., Gregori, E., Barrau, P.: Ergonomía 1 Fundamentos. Edicions UPC, Barcelona (1999)
9. Universidad Nacional de La Plata: I Jornadas Nacionales de TIC e Innovación en el Aula - 2011 UNLP. https://unlp.edu.ar/ensenanza/educacionadistancia/jornadas_2011_eduacion_a_distancia_publicaciones-3676-8676/ (2012)
10. Confederación regional de organizaciones empresariales de Murcia: Prevención y riesgos Ergonómicos (2017)
11. Rueda, M.J., Zambrano, M.: Manual de ergonomía y seguridad. Alpha Editorial (2018)
12. Comisiones Obreras de Asturias: Lesiones musculo esqueléticas de origen laboral. Asturias: Gráficas Careaga (2016)
13. Soares e Silva, J., Batista de Carvalho, A., Carvalho Santos Leite, H., Neves de Oliveira, E.: Reflexiones sobre los riesgos ocupacionales en trabajadores de salud en tiempos pandémicos por COVID 19. Revista Cubana de Enfermería 36(2), 1–11 (2020)
14. Emerson, S., Emerson, K., Fedorczyk, J.: Computer workstation ergonomics: current evidence for evaluation, corrections, and recommendations for remote evaluation. J. Hand Ther. 34(2), 166–178 (2021)
15. Sonne, M., Villalta, D.L., Andrews, D.M.: Development and evaluation of an office ergonomic risk checklist: ROSA–Rapid office strain assessment. Appl. Ergon. 43(1), 98–108 (2012)
16. Shariat, A., Cleland, J.A., Danaee, M., Kargarfard, M., Sangelaji, B., Tamrin, S.B.M.: Effects of stretching exercise training and ergonomic modifications on musculoskeletal discomforts of office workers: a randomized controlled trial. Braz. J. Phys. Ther. 22(2), 144–153 (2018)
17. García, S.R.C., Burbano, E.D.Y., Constante, L.F.F., Álvarez, M.G.A.: Gestión del talento humano: Diagnóstico y sintomatología de trastornos musculoesqueléticos evidenciados a través del Cuestionario Nórdico de Kuorinka. INNOVA Res. J. 6(1), 232–245 (2021)
18. Santamaría Flores, M.K.: Diseño de un programa de control de Riesgos Ergonómicos para los trabajadores de la Cooperativa de Ahorro y Crédito Crediya (2023)

19. Farez Rosas, O.D.: Diseño de un programa de control de riesgo ergonómico para el personal sanitario materno infantil Belly Moran (2022)
20. Mateus Aguilera, E.T., Ortiz, Y.P.: Desordenes Musculo Esqueléticos Relacionados Con Actos Inseguros En El Personal De Cuidado De La Salud De La UT Villa San Francisco (2022)
21. Vahdatpour, B., Bozorgi, M., Reza Taheri, M.: Investigating muscolosketal discomforts and their relation to workplace ergonomic conditions among computer office workers at Alzahra hospital, Isfahan. Iran. Phys. Med. Rehab. Electrodiagnosis 1(2), 52–58 (2019)
22. Vera Márquez, M., Valle Delgado, V., Mazacón Gómez, M., Nuñez Olalla, P.T., Vargas Bedoya, C.V.: Problemas ergonómicos existentes en el puesto de trabajo del personal administrativo académico y dependencia de planta central en el contexto Universitario. Revista Pertinencia Académica 7(1), 1–12 (2023)

Determination of Fire Affected Areas
with BLEVE in Fuel Service Stations (FSS)

M. Córdova$^{(\boxtimes)}$ ⓘ, M. Romero ⓘ, I. Gavilanez ⓘ, and A. Robalino ⓘ

Universidad Nacional de Chimborazo, Av. Antonio José de Sucre 060108, Riobamba, Ecuador
manolo.cordova@unach.edu.ec

Abstract. Accidents due to fuel fires in Service Stations (SS) present a serious threat to human activities and the local economy. In order to improve emergency planning, intervention zones were determined. First, the areas of affectation by thermal irradiation in a fire with Boiling Liquid Expanding Vapor Explosion (BLEVE) were calculated at the Service Station "Granja" located in the Pallatanga canton, using the technical note of prevention (NTP) 293. For this a widespread fire was considered in the 5 storage tanks with combustible liquids, having three containers of premium diesel, one with "super" (high octane) gasoline and one with "extra" (regular) gasoline, each with a nominal capacity of 10000 gallons (37.85 m^3). The impact on people at different distances was then calculated using the vulnerability models of people due to major accidents according to Finney's PROBIT functions. In the end, the intervention, and alert distances due to exposure to thermal radiation from the fire were determined according to SEVESO regulations for different events. It was found that, at 23.83 °C, 76.09% relative humidity; 2911.1 Pa; The intervention distance is: 783 m, and 980 m distances for the alert zone. It was also calculated: a) 100% of burns with protective clothing at 400 m, b) 100% of burns without protective clothing at 500 m, c) 100% of 1st degree burns affected at 600 m, d) 100% of affected with burns of 2nd degree at 800 m and e) 100% of fatal burns close to 400 m.

Keywords: Thermal Irradiation · Bleve · Probit · Affected areas · Fire

1 Introduction

The control of major accidents in Fuel Service Stations (FSS) is important, due to the handling of a large volume of these dangerous substances and their potential to generate fires, explosions and toxic gas clouds [1]. A recent catastrophic example is the explosion of a gasoline fuel station in Ghana in 2015 that killed 150 people and caused thousands of people affected by the fire [2]. Likewise, a study of around 319 accidents in the last 100 years attribute 58% to unwanted events caused by the handling of combustible and explosive substances; most often due to malfunction of a piece of equipment and negligence of operating personnel [3].

Because an FSS handles, stores and dispenses liquid fuels including gasoline, aviation fuel, marine fuel, diesel and residual fuel, the danger is constant due to the ability

© The Author(s), under exclusive license to Springer Nature Switzerland AG 2024
M. Z. Vizuete et al. (Eds.): CI3 2023, LNNS 1041, pp. 98–109, 2024.
https://doi.org/10.1007/978-3-031-63437-6_8

of these substances to change phase and generate flammable vapors [4]. Although the design and construction of an FSS must be carried out in accordance with the provisions of the official codes and standards in force, many of them comply partially due to the age of their facilities [5]. On the other hand, the control of the dangers in the sale of fuels in an FSS depends not only on the acts of the worker but also on the buyer who in many cases are the ones that can generate some fire outbreak near the pumps which, although they anoint with a submersible pump and have an emergency valve, are insufficient front to the ignition sources that the customer can generate when performing some unsafe action and starting a flame and subsequently a fire [6].

In the FSS, several events can occur either a) fire, b) explosion or c) both events at the same time. But most of the reported cases define fires with explosion of scampering steam of boiling liquid (BLEVE) as the most frequent industrial accident in this type of business [7].

The time of the BLEVE and the energy it produces in the fire are important for an adequate management of the emergency since, by quantifying it, the containment and response to the incident can be established more accurately because it is possible to quantitatively know the potential of the damage it would cause to the people and materials that are in the field of incidence of the unwanted event [8]. One of the available alternatives to have a numerical percentage of the value closest to the effect of the mechanical accident that can occur in an FSS is to use Finney's PROBIT functions [9]; these PROBIT equations require a point calculation as a function of distance but for this you need to know the related magnitudes. Thus, for BLEVE, the affectation in humans and structures requires the calculation of: a) overpressure, b) mechanical impulse and c) thermal irradiation. In a BLEVE of an FSS where the handling of flammable liquids with a high thermal load content is prioritized, thermal irradiation will occur as the main magnitude [10].

BLEVE incidents, typically caused by the ignition of a flammable liquid, generally provide more reaction time in an emergency with thermal radiation compared to an explosion. This lack of awareness increases the level of risk due to the exposure of workers, occupants, neighbors, customers, and visitors of an FSS who might be within the hazardous zones. Moreover, local Fire Department regulations and legal norms do not include the quantification of harmful effects on individuals or buildings resulting from major accidents, which turns FSS into hazardous sites that need to be assessed [11].

On the other hand, Ecuadorian regulations establish that distribution centers are classified as: a) Fuel Service Station (FSS), b) tank for national shipping fuel, c) tank for international shipping fuel, d) tank for air transport fuel, e) tank for fuel for industrial use [12].

This work considers the analysis of a BLEVE of an FSS that has 5 fuel tanks with an individual capacity of 10000 gallons located in the central highlands of Ecuador, to the thermo hygrometric conditions of the Pallatanga canton in the Province of Chimborazo. The affected areas were determined considering the site of the event as the axis of the radius of affectation up to 1000 m away since there is a town within that sector. The reference limits for the intervention and alert areas were defined following the SEVESO methodology of Spain [13].

2 Methodology

2.1 Calculation of the Thermal Irradiation Received in a BLEVE

The calculation of thermal irradiation received was made or considering that the liquid of the fuel storage tanks vaporizes suddenly producing a homogeneous nucleation and produce the fire by some external ignition source [14]. To establish the conditions of the site, data of relative humidity and dry air temperature were taken at the place of location of the storage tanks following the standard UNE-EN ISO 7243:2017 [15].

The thermal irradiation received with the Eq. 1 is calculated. [16].

$$I = d * F * E \qquad (1)$$

where:
   I = the thermal irradiation received (kW/m^2),
   d = the atmospheric transmission coefficient,
   F = the geometric factor of vision,
   E = the average intensity of the radiation generated by BLEVE (kW/m^2).

2.2 Calculation of the Percentage of People Affected

Although the effects of a BLEVE also occur overpressure, mechanical impulse, and projection of fragments of the container, in this work the thermal irradiation received was considered as a magnitude of affectation. To determine the percentage of people affected by a BLEVE in a specific area, you first quantify the effects of thermal radiation based on the intensity of received irradiation and the exposure time. Most fires caused by immediate ignition of flammable liquids and gases result in BLEVE incidents [17]. The deflagration occurring in a BLEVE-induced fire is often ignited by nearby ignition sources, causing the cloud to ignite and generate high-temperature gas that can affect individuals breathing it or within its range. The distance between people and buildings to the ignited cloud has the capacity to impact in various ways, regardless of their quantity. Thus, a numerical indicator that generalizes the damage according to the event is required. The PROBIT number is used for this purpose. Figure 1 outlines the steps for calculating thermal irradiation and BLEVE impact using the PROBIT equation.

The PROBIT method utilizes an empirical linear equation, which is an improved estimation of the dose-response relationship studied by Finney (1971). This function takes the latest estimated responses as a cycle to determine a new response indefinitely. To simplify calculations, they can be directly read from the provisional regression diagram and obtain non-fiducial limits of the exact formula [18].

The use of empirical coefficients in linear equations reduces algebraic complexity and calculation time. The PROBIT transformation is merely a mathematical artifice to represent the sigmoid curve with a straight line, and there are some options for solving equations using normal distribution of tolerances, such as those proposed by Garwood (1941) and Mantel (1950); however, PROBIT has gained significant acceptance nowadays [19]. Equation 2 presents the PROBIT function.

$$Pr = -a + b * (V) \qquad (2)$$

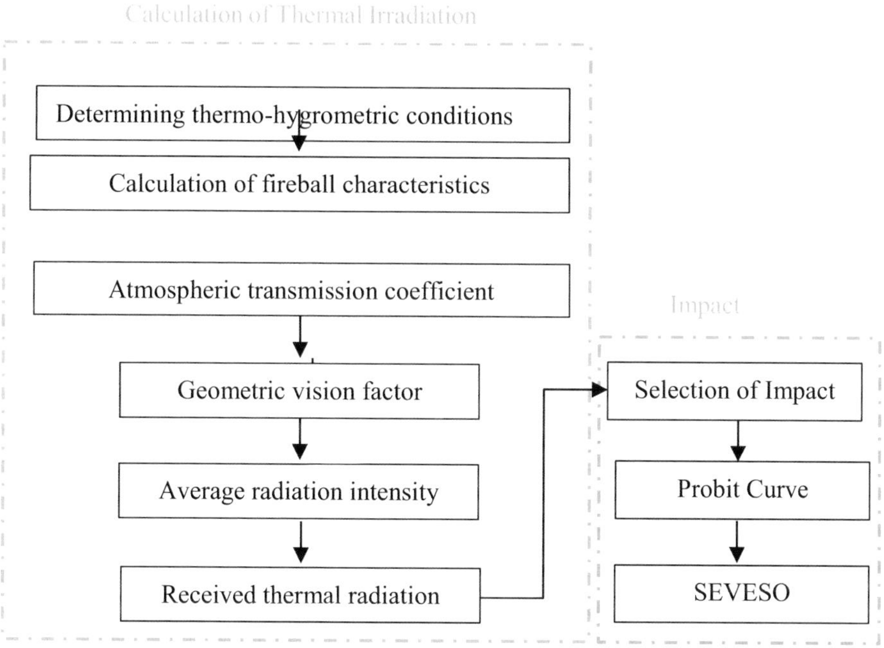

Calculation of Thermal Irradiation

```
┌─────────────────────────────────────────────┐
│  Determining thermo-hygrometric conditions  │
└─────────────────────────────────────────────┘
                      ↓
┌─────────────────────────────────────────────┐
│     Calculation of fireball characteristics │
└─────────────────────────────────────────────┘

┌─────────────────────────────────────────────┐
│    Atmospheric transmission coefficient     │
└─────────────────────────────────────────────┘
                      ↓
```

Impact

```
┌──────────────────────────┐        ┌──────────────────────┐
│  Geometric vision factor │   →    │ Selection of Impact  │
└──────────────────────────┘        └──────────────────────┘
            ↓                                  ↓
┌──────────────────────────┐        ┌──────────────────────┐
│ Average radiation        │        │    Probit Curve      │
│ intensity                │        └──────────────────────┘
└──────────────────────────┘                  ↓
            ↓                        ┌──────────────────────┐
┌──────────────────────────┐        │       SEVESO         │
│ Received thermal radiation│       └──────────────────────┘
└──────────────────────────┘
```

Fig. 1. Calculation of the impact due to thermal radiation from a BLEVE.

where:

    a = dependent constant of the type of injury and type of explosion load,

    b = dependent constant of the type of explosion load,

    V = variable representing the explosion load.

    To calculate the affectation, the effective exposure time was first determined. See Eq. 3 [20],

$$t_{ef} = t_r + \frac{3}{5} * \frac{x_o}{\mu} \left[1 - (1 + \frac{\mu}{x_o} * t_v)^{-5/3} \right] \tag{3}$$

where:

    t_{ef} = the effective exposure time (s),

    t_r = the time of reaction,

    x_o = the state at the center of the fire (m),

    μ = the escape velocity (m/s) y.

    t_v = the time to reach the distance in which the intensity of irradiation is 1 kW/m^2.

Then Finney's functions were applied in the PROBIT analysis to:

Fatal burns in people in protective clothing. See the Eq. 4:

$$Pr = -37.23 + 2.56 \, ln(tI^{4/3}) \tag{4}$$

Fatal burns without protective clothing. See the Eq. 5:

$$Pr = -36.38 + 2.56 \, ln(tI^{4/3}) \tag{5}$$

Second-degree burns. See the Eq. 6:

$$Pr = -43.14 + 3.0188 \, ln(tI^{4/3}) \tag{6}$$

First-degree burns. See the Eq. 7:

$$Pr = -39.83 + 3.0186 \, ln(tI^{4/3}) \tag{7}$$

Mortality percentage. See the Eq. 8:

$$Pr = -14, 9 + 32, 56 \, ln(\frac{t * I^{4/3}}{10^4}) \tag{8}$$

where:
t = the effective exposure time (s),
I = the intensity of thermal irradiation (W/m^2).

2.3 Delimitation of the Areas of Affectation

To identify the affected areas, the following steps were considered: a) first the calculation of the thermal irradiation generated in a BLEVE type fire in the FSS taking as the point of origin of the fire the 5 storage tanks, then b) the percentage of people affected was determined using the corresponding Finney PROBIT function, and finally c) with the results obtained and through the SEVESO Regulations of Spain, a value higher than 5 kW/m^2 was taken for an intervention area and a value lower than 3 kW/m^2 for the alert zone [21].

3 Results

3.1 Thermal Irradiation Received

Table 1 shows the result of the calculation of the thermal irradiation received in the fire of an FSS with a BLEVE type fire in conditions of the Pallatanga canton for Premium Diesel.

The calculated diameter of the fireball for the BLEVE formed in the Premium Diesel fire at the FSS is 190.10 cm, with a height of 142.57 cm, and a fireball duration time of 12.72 s. The effective exposure time is 100.34 s.

Table 1. Result of the calculation of Thermal Radiation Received for Premium Diesel

Distance (m)	x1 (m)a	db	Fc	E (kw/m^2)d	I (kw/m^2)e
0	47,52	0,47	0,44	147,70	30,98
50	56,04	0,46	0,40	147,70	27,18
100	79,10	0,45	0,30	147,70	19,83
150	111,90	0,44	0,21	147,70	13,61
200	150,57	0,43	0,15	147,70	9,41
250	192,75	0,42	0,11	147,70	6,70
300	237,11	0,41	0,08	147,70	4,94
350	282,88	0,40	0,06	147,70	3,75
400	329,60	0,40	0,05	147,70	2,93
450	377,00	0,39	0,04	147,70	2,35
500	424,88	0,39	0,03	147,70	1,91
550	473,13	0,38	0,03	147,70	1,59
600	521,66	0,38	0,02	147,70	1,33
650	570,40	0,38	0,02	147,70	1,14
700	619,32	0,37	0,02	147,70	0,98
750	668,38	0,37	0,02	147,70	0,85
800	717,56	0,37	0,01	147,70	0,75
850	766,83	0,37	0,01	147,70	0,66
900	816,17	0,37	0,01	147,70	0,59
950	865,59	0,36	0,01	147,70	0,53
1000	915,06	0,36	0,01	147,70	0,47

Note. [a] Distance, [b] Atmospheric transmission coefficient, [c] Geometric vision factor, [d] Mean Irradiation Intensity, [e] Received Thermal Irradiation. The radiation coefficient is 0.25 since the floor of the warehouse is made of pavement and does not reflect sunlight. The dry air temperature is 23.83 °C, the relative humidity is 76.09%. The absolute partial pressure of vapor in the ambient air is 2911.1 Pa. The tanks replicate the same conditions

Table 2 shows the result of the calculation of the thermal irradiation received in the fire of an FSS with a BLEVE type fire in conditions of the Pallatanga canton for "Super" Gasoline.

The calculated diameter of the fireball for the BLEVE formed in the "Super" Gasoline fire at the FSS is 178.88 cm, with a height of 134.16 cm, and a fireball duration time of 12.11 s. The effective exposure time is 120.81 s.

Table 2. Result of the calculation of Thermal Radiation Received for Super Gasoline

Distance (m)	x1 (m)[a]	d[b]	F[c]	E (kw/m^2)[d]	I (kw/m^2)[e]
0	44,72	0,47	0,44	249,97	52,71
50	53,73	0,47	0,39	249,97	45,53
100	77,89	0,45	0,29	249,97	32,24
150	111,80	0,44	0,20	249,97	21,57
200	151,39	0,43	0,14	249,97	14,66
250	194,28	0,42	0,10	249,97	10,33
300	239,19	0,41	0,07	249,97	7,55
350	285,39	0,40	0,06	249,97	5,72
400	332,46	0,40	0,04	249,97	4,45
450	380,13	0,39	0,04	249,97	3,55
500	428,25	0,39	0,03	249,97	2,89
550	476,69	0,38	0,02	249,97	2,39
600	525,38	0,38	0,02	249,97	2,01
650	574,26	0,38	0,02	249,97	1,71
700	623,30	0,37	0,02	249,97	1,47
750	672,46	0,37	0,01	249,97	1,28
800	721,73	0,37	0,01	249,97	1,12
850	771,08	0,37	0,01	249,97	0,99
900	820,50	0,37	0,01	249,97	0,88
950	869,99	0,36	0,01	249,97	0,79
1000	919,52	0,36	0,01	249,97	0,71

Note. [a] Distance, [b] Atmospheric transmission coefficient, [c] Geometric vision factor, [d] Mean Irradiation Intensity, [e] Received Thermal Irradiation. The radiation coefficient is 0.25 since the floor of the warehouse is made of pavement and does not reflect sunlight. The dry air temperature is 23.83 °C, the relative humidity is 76.09%. The absolute partial pressure of vapor in the ambient air is 2911.1 Pa. The tanks replicate the same conditions

Table 3 shows the result of the calculation of the thermal irradiation received in the fire of an FSS with a BLEVE type fire in conditions of the Pallatanga canton for "Extra" Gasoline.

The calculated diameter of the fireball for the BLEVE formed in the "Extra" Gasoline fire at the FSS is 182.60 cm, with a height of 139.95 cm, and a fireball duration time of 12.31 s. The effective exposure time is 124.22 s.

Figure 2 shows the result of the calculation of the thermal irradiation received of the five tanks in the fire of an FSS with a BLEVE type fire in conditions of the Pallatanga canton.

Table 3. Result of the calculation of Thermal Radiation Received for Extra Gasoline

Distance (m)	x1 (m)[a]	d[b]	F[c]	E (kw/m^2)[d]	I (kw/m^2)[e]
0	45,65	0,47	0,44	251,40	52,92
50	54,49	0,47	0,39	251,40	45,96
100	78,27	0,45	0,29	251,40	32,88
150	111,81	0,44	0,20	251,40	22,19
200	151,09	0,43	0,14	251,40	15,17
250	193,75	0,42	0,10	251,40	10,72
300	238,48	0,41	0,08	251,40	7,86
350	284,54	0,40	0,06	251,40	5,96
400	331,49	0,40	0,05	251,40	4,64
450	379,08	0,39	0,04	251,40	3,71
500	427,12	0,39	0,03	251,40	3,02
550	475,49	0,38	0,03	251,40	2,50
600	524,13	0,38	0,02	251,40	2,10
650	572,97	0,38	0,02	251,40	1,79
700	621,97	0,37	0,02	251,40	1,54
750	671,10	0,37	0,01	251,40	1,34
800	720,34	0,37	0,01	251,40	1,18
850	769,66	0,37	0,01	251,40	1,04
900	819,06	0,37	0,01	251,40	0,92
950	868,52	0,36	0,01	251,40	0,83
1000	918,03	0,36	0,01	251,40	0,74

Note. [a] Distance, [b] Atmospheric transmission coefficient, [c] Geometric vision factor, [d] Mean Irradiation Intensity, [e] Received Thermal Irradiation. The radiation coefficient is 0.25 since the floor of the warehouse is made of pavement and does not reflect sunlight. The dry air temperature is 23.83 °C, the relative humidity is 76.09%. The absolute partial pressure of vapor in the ambient air is 2911.1 Pa. The tanks replicate the same conditions

3.2 Results of Determination of People Affected by Thermal Irradiation

Based on the data obtained from thermal radiation received by a BLEVE type fire and applying the PROBIT method, the percentages of affected population are determined, starting with the population affected by fatal burns wearing protective clothing as shown in Table 4.

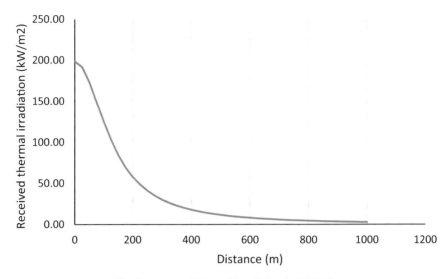

Fig. 2. Received thermal irradiation in BLEVE.

Table 4. Result of the percentage of population affected.

X (m)	Pr1	%A1	Pr2	%A2	Pr3	% A3	Pr4	% A4	Pr5	%A5
0	>8	100	>8	100	>8	100	>8	100	>8	100
400*	>8	100	>8	100	>8	100	>8	100	7,10	98
450	7,58	99,5	>8	100	>8	100	>8	100	6,33	91
500	6,88	97	7,73	100	>8	100	>8	100	5,63	74
550	6,24	89	7,09	98	>8	100	>8	100	4,99	50
600	5,65	74	6,50	93	7,42	99	>8	100	4,40	27
650	5,10	54	5,95	83	6,77	96	>8	100	3,85	12
700	4,59	34	5,44	67	6,17	88	>8	100	3,34	4
750	4,11	19	4,96	49	5,61	73	>8	100	2,86	2
800	3,66	9	4,51	31	5,08	53	>8	100	<1	0
850	3,24	4	4,09	18	4,58	34	7,89	100	<1	0
900	2,84	2	3,69	9	4,11	19	7,42	99	<1	0

Note. Pr1: PROBIT burns with protective clothing, Pr2: PROBIT burn without protective clothing, Pr3: PROBIT second-degree burn, Pr4: PROBIT first-degree burn, Pr5: PROBIT fatal burn, %A1: % Burn affected by protective clothing, %A2: % affected by burns without protective clothing, %A3: % affected by 2nd degree burn, %A4: % affected by 1st degree burn, %A5: % affected by burn death

3.3   Result of the Determination of the Affected Areas

Table 5 shows the areas affected by a BLEVE type fire in an FSS.

Table 5. Areas affected by a BLEVE according to SEVESO

Storage tank	Intervention area (m)	Alert zone (m)
Premium Diesel	298	395
Super Gasoline	376	491
Extra Gasoline	385	502
General Fire	763	980

Figure 3 shows the areas affected by a total fire in the 5 fuel storage tanks. With the use of ArcGIS Software (Arc Map 10.4.1) and based on SEVESO regulations, the radii of affectation of each zone can be seen from the top view.

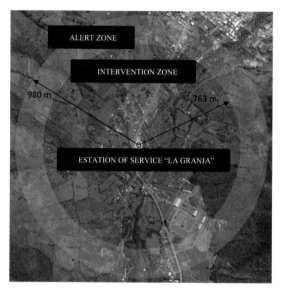

Fig. 3. Alert and intervention areas in a fire by BLEVE in a Service Station.

4   Conclusions

In a fire of the 5 fuel storage tanks: 3 of premium diesel, 1 of "super" gasoline and 1 of "extra" gasoline with a total volume of 50000 gallons the % of affected by unwanted event (fire with BLEVE) would have the following values: 100% burned with protective clothing at 400 m, 100% burns without protective clothing at 500 m, 100% affected with burns of 1st degree at 600 m, 100% affected with burns of 2nd degree at 800 m and 100% of fatal burns close to 400 m in the worst scenario.

The intervention areas were delimited based on the levels of thermal radiation recommended by the SEVESO Regulations, calculating a radius of 763 m for the intervention area and 980 m for the alert area. These distances must be considered for emergency plans because at that distance there are citadels and restaurants with a significant influx of tourists that would be directly affected by thermal radiation.

The calculated intervention zone radius value for the BLEVE fire in the 5 fuel tanks does not exhibit a proportional behavior to the amount of fuel. However, it is observed that "Extra" Gasoline represents a distance that corresponds to 50.45% of the intervention radius for all the tanks. The population is close to the intervention zone, occupying locations starting from 80 m and encompassing the entirety within a radius of 300 m. Therefore, it is necessary to consider the placement of containment measures in the event of a complete fire.

References

1. Rashid, R., Umar, R.: Control of industrial major accident hazard regulation in Malaysia: second decade in examination. Int. J. Pub. Heal. Sci. Arti. **12**(1), 138–145 (2023)
2. Dionisio, A.: Relación del sistema de riesgos con el índice de accidentabilidad de la empresa del sub sector hidrocarburos EBH Ingenieros SAC, p. 19. Universidad Nacional Mayor de San Marcos (2022)
3. Khan, F., Abbasi, S.: Major accidents in process industries and an analysis of causes and consequences. J. Loss Prev. Process Ind. **12**(5), 361–378 (1999)
4. Ureña, O., Yaryma, E., Torres, H., German, R.: Evaluación del nivel de conocimiento de la norma inen 2251: 2013, expendio de combustibles líquidos, en los centros de distribución de la provincia del Guayas, p. 11. UEES (2019)
5. Gálvez, R., Blanco, J.: Evaluación del nivel de riesgo de incendio y explosión en las estaciones de servicio de combustible del cantón Loja en el 2019, a través del método Índice Dow y la estimación de las zonas de amenaza con el software informático ALOHA, p. 10. Universidad del Azuay (2020)
6. Keyn, O., Klimova, I.: Reduction of individual risk for the employee of fuel filling stations with building the accident logical models. Bezopastnost' Truda v Promyshlennosti, Article **7**, 81–85 (2019)
7. Hemmatian, B., Planas, E., Casal, J.: Fire as a primary event of accident domino sequences: The case of BLEVE. Reliabil. Eng. Sys. Safe. Article **139**, 141–148 (2015). Art. no. 5269
8. Landucci, G., Molag, M., Reinders, J., Cozzani, V.: A modelling approach to assess the effectiveness of BLEVE prevention measures on LPG tanks. Safety, Reliability and Risk Analysis: Theory, Methods and Applications - Proceedings of the Joint ESREL and SRA-Europe Conference **4**, 3153–3161 (2009)
9. Finney, D.: Probit analysis; a statistical treatment of the sigmoid response curve, 2nd edn, p. 11. Mcmillan (1947)
10. Téllez, C., Peña, J.: Boiling-Liquid Expanding-Vapor Explosion (BLEVE): An introduction to consequence and vulnerability analysis. Chemi. Eng. Edu. Rev. **36**(3), 206–211 (2002)
11. Agromonte, J.: Diseño de un sistema de gestión de la seguridad y salud en el trabajo, bajo los requisitos de la ley 29783, en la empresa Gasolineras Piura SRL, p. 19 (2018)
12. Salazar, J.: Estudio de riesgos de incendios en el almacenamiento de combustible en el hospital, Universidad de Guayaquil. Facultad de Ingeniería Industrial, p. 18 (2022)
13. Reche, M.: Los accidentes graves en la industria química: análisis de la normativa Seveso y nuevas propuestas, p. 5. Universidad de Murcia (2018)
14. Abbasi, T., Abbasi, S.: The boiling liquid expanding vapour explosion (BLEVE): Mechanism, consequence assessment, management. J. Hazardous Mat. Artic. **141**(3), 489–519 (2007)

15. Onder, M., Kursunoglu, N., Onder, S.: Psychrometric analysis of a fully mechanized underground coal mine and establishment of acceptable climate conditions. J. Mining Sc. Artic. **57**(5), 863–872 (2021)
16. Planas, E., Pastor, E., Casal, J., Bonilla, J.: Analysis of the boiling liquid expanding vapor explosion (BLEVE) of a liquefied natural gas road tanker: The Zarzalico accident **34**, 127–138 (2015)
17. Carrera, M.: Análisis de riesgos por incendio, explosión y accidente tóxico en el área de tratamiento de la Planta de Procesamiento de Crudo de la EPEP Centro, pp. 10–12. Universidad de Matanzas (2022)
18. Martín, E., Yago, F.: Perturbaciones agrarias derivadas de la guerra ruso-ucraniana, Economistas, N°. 81, p. 38. Entorno Nacional y Geopolítica (2003)
19. Cornfield, J., Mantel, N.: Some new aspects of the application of maximum likelihood to the calculation of the dosage response curve. J. Am. Stat. Assoc. **45**(250), 181–210 (1950)
20. Turmo, E.: Modelos de vulnerabilidad de las personas por accidentes mayores: método Probit, N°. 291. Notas Técnicas de Prevención (1991)
21. Agnello, P., Ansaldi, S., Bragatto, P.: Plugging the gap between safety documents and workers perception, to prevent accidents at seveso establishments. Chem. Eng. Trans. **26**, 291–296 (2012)

Information Security Management in Higher Education Institutions in Compliance with the Organic Law for the Protection of Personal Data

Karen Estacio[✉]

Instituto Superior Tecnológico ARGOS, Guayaquil, Ecuador
k_estacio@tecnologicoargos.edu.ec

Abstract. Information security has become crucial for any organization due to a large amount of data handled daily and the need to protect it from possible external and internal threats. Organic Law for the Protection of Personal Data (LOPDP) regulates in Ecuadorian territory the protection and exercise of the rights of individuals concerning the processing of their data. The objective of this research is to analyze information security (IS) standards that have adequate controls to ensure the confidentiality, availability, and integrity of information. The ISO/IEC 27001:2022 and NIST 800–53 r5 standards were selected because they are widely recognized in the IS field and provide solid guidance for the implementation of security controls. The resulting instrument, a tool with 64 controls, will allow evaluation of IS compliance within Ecuadorian institutions of higher education to mitigate IS risks and avoid sanctions by the national data protection authority.

Keywords: information security · personal data · ISO 27000 · NIST 800-53

1 Introduction

The proper preservation of information is essential for educational institutions and it is vital to ensure its protection. When talking about security, reference is made to the need to reduce risk to acceptable levels, i.e., to a manageable level within the organization [1].

Personal data has become a valuable commodity in our days. The massive collection of personal information drives the information age economy. Through detailed, often sensitive profiles, people and their data are treated as objects, being frequently bought and sold, even by the most scrupulous actors in the data economy [2]. This is without considering those who see personal data as an industrial commodity, disconnected from the lives and concerns of the people from whom the data was obtained.

In May 2021, the Organic Law for the Protection of Personal Data (LOPDP) came into force in Ecuador. This law seeks to regulate the treatment of personal data by public and private entities, establishing the principles, duties, and rights that must be complied to the collection, storage, use, transfer, and disposal of such data, highlighting

M. Z. Vizuete et al. (Eds.): CI3 2023, LNNS 1041, pp. 110–121, 2024.
https://doi.org/10.1007/978-3-031-63437-6_9

privacy, informational self-determination and the security of personal data, promoting its responsible use and avoiding abuses or violations of people's rights [3]. The provisions related to corrective measures and the sanctioning regime came into force two years after its publication in 2021. Therefore, public and private institutions that handle the personal information of their clients may be sanctioned by the regulator since May 2023 for not complying with the provisions of the LOPDP.

The purpose of this article is to develop a guide that serves as a reference to assess the degree of compliance with information security (IS) in higher education institutions (HEI) in Ecuador, allowing to provide a set of guidelines and criteria to comprehensively assess the security measures and practices implemented in these institutions.

2 Information Security Standards

2.1 ISO 27000

The International Organization for Standardization (ISO)/International Electrotechnical Commission (IEC) 27000 series encompasses a variety of internationally recognized IS-related standards. These standards provide a solid basis for meeting the technical and operational requirements needed to reduce the risk of breaches [4].

Specifically, the 27000:2019 series of the ISO/IEC standard, focused on the Information Security Management System (ISMS), has undergone an update with the inclusion of Annex D. This annex incorporates certification clauses that move closer to proactive compliance with the provisions outlined in the General Data Protection Regulation (GDPR) in the European territory. The ISO/IEC 27000 series of standards, has represented a significant milestone in the certification process by integrating various clauses that address the most essential aspects contemplated in the personal data protection reference [5].

Table 1 shows the description of the standards that are part of the ISO/IEC 27000 family. ISO 27001 is the first standard within the ISO/IEC family of standards and provides a framework for establishing, implementing, operating, monitoring, reviewing, and improving the ISMS. IS requirements are grouped into eleven main categories in this standard [6].

Table 1. Information Security in the ISO/IEC 27000 Family

Standard	Description
ISO/IEC 27000	Standard vocabulary for ISMS
ISO/IEC 27001:2022	Certifiable Standard, specifies the requirements for the implementation of the ISMS
ISO/IEC 27002:2022	93 IS controls, provides a reference set of generic controls and best practices for the implementation of an ISMS
ISO/IEC 27003	It supports the ISO/IEC 27001 standard
ISO/IEC 27005	Provides recommendations and guidelines for information security risk assessment methods and techniques

2.2 NIST

The National Institute of Standards and Technology (NIST) has created cyber security and privacy controls, such as SP-800–53, whose purpose is to facilitate risk management in information systems. It has security controls and recommendations for IS management in federal information systems in the United States. These controls are used to ensure the confidentiality, integrity, and availability of information in federal information systems [7].

Its 1000 IS controls are grouped into 18 families of controls, designed to address a wide variety of IS threats and risks, including protection of confidential information, vulnerability management, physical security, incident management, authentication, and access management, among others [8].

3 Methods

This study was conducted using a quantitative, descriptive, non-experimental research approach. Relevant information on international information security (IS) standards was collected through desk research to select the Information Security Management System (ISMS) standard.

In the current context of the information society-based economy and increasing threats, it is imperative to increase the level of IS in the protection of personal data. This is necessary to promote the constant development of information services in society.

In the university environment, it is common for students, faculty, administrative staff, and visitors to frequently use information technology (IT) infrastructures to consume and generate data in various forms [9]. These activities include the use of personal cell phones, corporate laptops, and tablets, to the use of laboratory sensors and biometric access systems. As a result, there is a constant exchange of data between universities and institutions and the various categories of end users involved.

The LOPDP [3] in its article 4 "Personal data security breach: Security incident that affects the confidentiality, availability or integrity of personal data". Article 10 specifies "The person responsible for the processing of personal data must demonstrate having implemented mechanisms for the protection of personal data; that is, compliance with the principles, rights, and obligations established in this Law, for which purpose, in addition to the provisions of the applicable regulations, it may use standards, best practices, self- and coregulation schemes, protection codes, certification systems, personal data protection seals or any other mechanism determined to be appropriate to the purposes, the nature of the personal data or the risk of processing".

Although the LOPDP mentions the term "information security" four times within its content, it is clear that information security practices are recognized as a framework to ensure accountability in the processing of personal data, given that, Article 37 mentions, "4. 4) Those responsible for and in charge of the processing of personal data, may adhere to international standards for an adequate risk management focused on the protection of rights and freedoms, as well as for the implementation and management of information security systems or codes of conduct recognized and authorized by the Authority for the Protection of Personal Data", this measure allows strengthening security and accountability in the processing of personal data.

In addition to what is mentioned in Article 37, "the person responsible for or in charge of the processing of personal data shall demonstrate that the measures adopted and implemented adequately mitigate the risks identified", "Among other measures, the following may be included: 1) measures to anonymize, de-anonymize or encrypt personal data".

Article 70 "Serious breaches by the DPO. 4) Failure to implement mechanisms designed to maintain the confidentiality, integrity, availability, and resilience of personal data; 5) Failure to implement preventive and corrective measures in the security of personal data to avoid breaches".

Regarding penalties for non-compliance Article 72: "Penalties for serious violations. The personal data protection authority shall impose the following administrative sanctions in the event of a serious violation". "The public servant or public official shall be subject to the following sanctions". The public servant or public official who, by his act or omission, is guilty of one of the serious offences provided for by the present Law, shall be punished by a fine of between 10 and 20 times the basic unified salary of the worker in general; without prejudice to the extracontractual liability of the State, which shall be subject to the rules established by the corresponding regulations; 1) if the responsible party, the person in charge of the processing of personal data or, where applicable, a third party, is a private law entity or a public company, a fine of between 0. 7% to 1%, calculated on its turnover for the financial year immediately preceding that in which the fine is imposed".

Encryption and pseudonymization, as mentioned by the LOPDP in the context of risk management, is a preventive response to incidents, especially with the obligation to notify data subjects in the event of a security breach, so institutions or organizations must be prepared to identify when a breach has occurred, take the necessary measures to contain it, recover the affected personal data, restore operations and report the incident to the supervisory authorities [10].

To find an international regulatory framework that would allow higher education institutions to implement an ISMS assessment guide in compliance with the LOPDP, a documentary search was conducted for those international IS standards with a con-fidentiality, integrity, and availability (CID) scope. Table 2 details the content of each standard reviewed and its compliance with CID in terms of controls and approach.

Table 2. CDI in information security frameworks and standards

Standard	Content	CDI
NIST 800–53	Provides a catalog of security and privacy controls. [11]	Yes
ISO 27001	Guidelines for the selection, implementation, and management of ISMS controls. [4]	Yes
COBIT	IT management tools framework enables alignment of IT processes with business objectives. [12]	No
ITIL v4	A set of tools and best practices to help organizations manage and protect their information. [13]	No

4 Results

The implications of an IS-focused international standard on the right to data protection represent a very strong connection to the LOPDP. The ISO/IEC 27001 standard and the NIST 800–53 standard are both highly recognized internationally and comprehensively cover information security in organizations [14]. Both develop controls and security measures that can mitigate technical, legal, and legal risks.

Both develop controls and security measures that can mitigate technical, legal, and economic risks, even in areas that are not directly addressed by the aforementioned LOPDP. NIST 800–53 offers a conglomerate of IS controls that are much more specific than the ISO 27002 standard, allowing efforts to be focused on specific situations, leaving aside possible breaches that could materialize vulnerabilities in the information systems and processes that manage the personal data of internal and external users in an institution. Figure 1 details the mapping of the controls included in NIST 800–53 revision 5 and ISO/IEC 27001:2022 grouped into categories.

ITIL v4 (IT Infrastructure Library) focuses on IT service management and mentions in a generalized way that the IS is established through policies, processes, behaviors, risk management, and controls, which allow a balance between prevention, detection, and correction of risks affecting the IS that could be detected. However, it does not specifically address a set of specific controls that may be related to the management of IS processes within the institutions but rather points out a series of good practices and recommendations that allow adequate management [15].

The recommendations of the above standards can serve as a reference and guide for those responsible for making security decisions within an IES. These standards provide the ability to select appropriate security concepts and their details according to specific IS needs [16]. This means that the recommendations of the standards can be used as a

Organizational Aspects of Information Security (OA)

ISO/IEC 27001:2022	NIST 800-53 r5
5.3	AC-5
A.5.15, A.5.33*, A.8.3, A.8.4*, A.8.18, A.8.20, A.8.26	AC-3
A.5.14, A.6.7, A.8.1	AC-17
A.5.14*, A.6.7, A.7.9	PE-17
5.2, 5.3, 7.5.1, 7.5.2, 7.5.3, A.5.1, A.5.2, A.5.4, A.5.31, A.5.36, A.5.37	AT-1

Access Control (CA)

ISO/IEC 27001:2022	NIST 800-53 r5
A.5.15*, A.8.2, A.8.18	AC-6
A.5.16	IA-2
A.5.16, A.5.17	IA-5
A.5.11, A.6.5	PS-4
A.5.11, A.6.5	PS-5
A.8.3*	AC-24
A.8.5*	AC-7
A.8.5*	AC-8
A.8.2, A.8.4, A.8.9, A.8.19, A.8.31, A.8.32	CM-5

Telecommunications Security (TS)

ISO/IEC 27001:2022	NIST 800-53 r5
A.5.14, A.8.1, A.8.20	AC-18
A.5.14, A.7.9, A.8.20	AC-20
A.5.14*, A.8.16*, A.8.20*, A.8.22*, A.8.23*, A.8.26*	SC-7
A.5.10*, A.5.14, A.8.20*, A.8.26*	SC-8
A.8.20	SC-10
A.5.14, A.8.21	CA-3
A.5.2*, A.5.4*, A.5.8*, A.5.14*, A.5.22, A.5.23, A.8.21	SA-9
A.5.14, A.8.22, A.8.23	AC-4

Physical and Environmental Security

ISO/IEC 27001:2022	NIST 800-53 r5
A.7.2*	PE-2
A.7.1, A.7.2, A.7.3, A.7.4	PE-3
A.7.2, A.7.3, A.7.7	PE-5
A.5.29*, A.7.5*, A.8.14*	CP-6
A.5.29*, A.7.5*, A.8.14*	CP-7
A.5.29*, A.5.33*, A.8.13	PE-9
A.7.5, A.7.8	PE-13
A.7.5, A.7.8, A.7.11	PE-14
A.7.5, A.7.8,	PE-15
A.5.10*, A.7.5,	PE-18
A.7.5*, A.7.8*, A.8.12	PE-19
A.5.29*, A.7.11	CP-8
A.7.11*	PE-10
A.7.11*	PE-11
A.7.11*	PE-12
A.7.10*, A.7.13*, A.8.10	MA-2
A.7.13	MA-6
A.5.10, A.7.10*, A.7.14, A.8.10	MP-6
----	SA-19
11.2.8	AC-11

Operational Security (OS)

ISO/IEC 27001:2022	NIST 800-53 r5
8.1, 9.3.3*, A.8.9, A.8.32	CM-3
A.5.2*, A.8.9	CM-9
A.8.9, A.8.28*, A.8.30*, A.8.32	SA-10
A.8.6	AU-4
-----	AU-5
7.5.1, 7.5.2, 7.5.3, A.5.2, A.5.29, A.8.14	CP-2
-----	SC-5
7.3, A.6.3, A.8.7*	AT-2
A.8.7	SI-3
A.8.16*	SI-4
A.5.29*, A.5.33*, A.8.13	CP-9
A.8.15*	AU-3
A.5.25, A.6.8, A.8.15	AU-6
A.5.28, A.8.15	AU-11
A.8.15*	AU-12
A.8.15*	AU-14
6.1.2, 8.2, 9.3.2*, A.8.8*	RA-3
A.8.8*	RA-5
A.6.8*, A.8.8, A.8.32*	SI-2
A.5.6*	SI-5
A.8.19*	CM-7
A.8.19*	CM-11

Fig. 1. Controls mapping ISO/IEC 27001:2022 and NIST 800–53 revision 5 [18]. **Note:** An asterisk (*) indicates that the ISO/IEC control does not fully satisfy the intent of the NIST control.

solid basis for designing and implementing customized and effective security measures that are tailored to the specific needs and risks of the institution by the LOPDP.

The proper management of an ISMS provides a benchmark for compliance and ensures a high level of IS under Article 37 of the LOPDP. In addition to its legal perspective, [5] the relationship between these international standards and the LOPDP has also been analyzed from an efficiency approach. Although both standards share the objective of properly processing and storing confidential and sensitive data, the ISO/IEC 27001 certification process allows organizations to avoid duplication of effort, cost, and time in complying with the LOPDP provisions. This, in turn, increases business efficiency and facilitates more effective achievement of IES objectives.

As selection criteria during the literature review, both the international regulatory framework NIST 800–53 and ISO 27001 [17] present a set of guidelines for ensuring the proper management of critical information, as explained in Table 2. Unlike other standards and regulatory frameworks in the IS field, these two were selected as the basis for the development of a tool that will allow the assessment and control of IS in higher education institutions. Table 3 details the sixty-four controls selected as a result of the mapping of controls detailed in Fig. 1, which make up the evaluation tool for assessing IS management in an HEI. The adaptation and compliance approach to information security management in CDI [19] is presented as a tool to measure performance in securing information, which is an essential resource in organizations, according to their technological environment.

Table 3. Rubric for evaluating information security

Code	Control
AC-5	Separation of duties
AC-3	Account management
AC-17	Remote access
PE-17	Alternative workplace
AT-1	Awareness and training policies and procedures (PS) Minimum privileges
AC-6	Identification and authentication (organizational users)
IA-2	Authentication management
IA-5	Termination of personnel contract
PS-4	Personnel transfer
PS-5	Access control decisions
AC-24	Failed login attempts
AC-7	Notification of system usage
AC-8	Access restrictions for change
CM-5	Physical access authorization
PE-2	Physical access control
PE-3	Access control for output devices

(*continued*)

Table 3. (*continued*)

Code	Control
PE-5	Alternate storage site
CP-6	Alternate processing site
CP-7	Power equipment and cabling
PE-9	Fire protection
PE-13	Temperature and humidity controls
PE-14	Water damage protection
PE-15	Location of system components
PE-18	Information leakage
PE-19	Telecommunication services
CP-8	Emergency shutdown
PE-10	Emergency power
PE-11	Emergency lighting
PE-12	Controlled maintenance
MA-2	Timely maintenance
MA-6	Media disinfection
MP-6	Authenticity of components
SA-19	Device locking
AC-11	Configuration change control
CM-3	Configuration change control
CM-9	Developer configuration management
SA-10	Storage capacity auditing
AU-4	Audit processing failure response
AU-5	Contingency planning
CP-2	Denial of service protection
SC-5	Awareness training
AT-2	Malicious code protection
SI-3	System monitoring
SI-4	System backup
CP-9	Audit log content
AU-3	Audit review, analysis, and reporting
AU-6	Audit log retention
AU-11	Audit generation

(*continued*)

Table 3. (*continued*)

Code	Control
AU-12	Separation of duties
AU-14	Session audit
RA-3	Risk assessment
RA-5	Vulnerability scanning
SI-2	Fault remediation
SI-5	Security alerts, warnings, and policies
CM-7	Less functionality
CM-11	User-installed software
AC-18	Wireless access
AC-20	Use of external systems
SC-7	Boundary protection
SC-8	Transmission Confidentiality and Integrity
SC-10	Network Disconnection
CA-3	Interconnection systems
SA-9	External system services
AC-4	Information flow application

The process of establishing, implementing, and managing an ISMS according to ISO/IEC 27001 starts by defining clear objectives. Then, the necessary controls are implemented and operating procedures are put in place. It is essential to constantly monitor the effectiveness of the IS controls in place. This process also involves conducting periodic reviews to ensure that the system is working effectively.

This whole process is supported by a comprehensive risk analysis [4]. Figure 2 illustrates the implementation scheme of an ISMS, where it is essential to know exactly the initial situation in terms of risk management of sensitive information for the assurance of the same within IES. This is achieved through a thorough analysis of potential risks. To carry out this assessment, the aforementioned rubric will be used as a key tool.

During the selection and implementation of security measures mentioned in the rubric, HEIS must consider the necessary evidence, such as documents and tests, that will support current and future security assessments [11]. These assessments are essential to determine whether security measures were implemented correctly, whether they are working as intended, and whether they comply with security and privacy policies. These results provide vital information that allows senior leaders to make informed decisions based on the level of risk associated with the security measures.

Fig. 2. Implementation of an ISMS

5 Discussion and Conclusion

IS is crucial in HEIs, as they handle a large amount of sensitive student, faculty, research, and project information. Developing a tool to assess the management of IS in HEIs can help ensure that the necessary controls are in place to protect information and minimize the risk of potential external and internal threats.

By using the ISO 27001 and NIST 800–53 standards as a reference for the creation of this tool, you can ensure that the appropriate security controls are in place and that you are following a sound methodology that is recognized in the IS industry. In addition, by having a tool that allows evaluating IS maturity on an ongoing basis, it is possible to identify areas and processes that are not efficient and take preventive measures to protect information in terms of compliance with the LOPDP in Ecuadorian territory, to avoid the sanctions detailed in Chapter XI, corrective measures, infractions and sanctioning regime by the Authority for the Protection of Personal Data as mentioned in the law.

As a result, the rubric could play an important role in reducing the uncertainty associated with the degree of progress of the HEI in terms of security concerning critical and confidential information, providing security and peace of mind to the authorities and the users that make up the educational community.

It is important to emphasize that IS management does not fall solely on the IHE's Information Technology team. Rather, it is a task that demands the collaboration of all members of the institution. This ranges from senior leaders to faculty, students, and administrative staff.

Each individual has a vital role to play in this process, as the protection of sensitive information is a shared effort. Internalizing the importance of safeguarding information contributes to compliance with established policies and controls. This, in turn, ensures the integrity and confidentiality of the information. The collaboration of all hierarchical levels and functional areas strengthens the holistic approach to information security in the IES and reinforces a solid and proactive security culture. In addition to the above, it is proposed to carry out the application of this IS evaluation rubric in a HEI, to verify the achievement of its general objective in future research.

References

1. Samarati, P.: Protecting respondents' identities in microdata release. IEEE Trans. Knowl. Data Eng. **13**(6), 1010–1027 (2001)
2. Katulic, T., Protrka, N.: Information Security in Principles and Provisions of the EU Data Protection Law. In: 2019 42nd International Convention on Information and Communication Technology, Electronics and Microelectronics (MIPRO). IEEE, Opatija, Croatia (2019). https://ieeexplore.ieee.org/document/8757153/
3. Presidencia de República del Ecuador: Ley orgánica de protección de datos personales, https://www.finanzaspopulares.gob.ec/wp-content/uploads/2021/07/ley_organica_de_proteccion_de_datos_personales.pdf. Last accessed 1 June 2023
4. Chatzipoulidis, A., Tsiakis, T., Kargidis, T.: A readiness assessment tool for GDPR compliance certification. Comput Fraud Secur **2019**(8), 14–19 (2019)
5. Viguri Cordero, J.A.: La adopción de instrumentos de certificación como garantía eficiente en la protección de los datos personales. Rev Catalana Dret Públic. **62**, 160–176 (2021)
6. Meriah, I., Arfa Rabai, L.B.: Comparative Study of Ontologies Based ISO 27000 Series Security Standards **160**, 85–92 (2019)
7. Huang, Y., Debnath, J., Iorga, M., Kumar, A., Xie, B.: CSAT: A User-interactive Cyber Security Architecture Tool based on NIST-compliance Security Controls for Risk Management. In: 2019 IEEE 10th Annual Ubiquitous Computing, Electronics & Mobile Communication Conference (UEMCON), pp. 697–707. IEEE, New York City, NY, USA (2019). https://ieeexplore.ieee.org/document/8993090/
8. Grama, J.L.: Protecting privacy and information security in a federal postsecondary student data system. Protecting Students, Advancing Data: A Series on Data Privacy and Security in Higher Education. Eric, 3–13 (2019)
9. Díaz, Z., Ocampo, R., Díaz, K., Mora, J.: Seguridad de la información para instituciones educativas a tercer nivel basado en la ISO/ IE27001. Roca Rev Científico - Educ Prov Granma **16**(1), 546–590 (2020)
10. Zhu, W.: Personal information security environment monitoring and law protection using big data analysis. J. Environ. Public. Health. **7**, 1–12 (2022). https://doi.org/10.1155/2022/1558161
11. Joint Task Force Interagency Working Group: Security and Privacy Controls for Information Systems and Organizations [Internet]. Revision 5. National Institute of Standards and Technology. https://nvlpubs.nist.gov/nistpubs/SpecialPublications/NIST.SP.800-53r5.pdf. Last accessed 21 May 2023
12. Wolden, M., Valverde, R., Talla, M.: The effectiveness of COBIT 5 Information Security Framework for reducing Cyber Attacks on Supply Chain Management System. IFAC-Pap **48**(3), 1846–1852 (2015). https://doi.org/10.1016/j.ifacol.2015.06.355
13. ITIL Foundation: ITIL 4 Foundation Course Book. In AXELOS, pp. 120–145 (2019). https://worldaedait.com.mx/wp-content/uploads/2019/09/ITIL-4-Foundation-Material-Participante.pdf
14. Kasem-Madani, S., Malderle, T., Boes, F., Meier, M.: Privacy-preserving warning management for an identity leakage warning network. In: Proceedings of the European Interdisciplinary Cybersecurity Conference, pp. 1–6. ACM, Rennes France (2020). https://doi.org/10.1145/3424954.3424955

15. Hochstein, A., Zarnekow, R., Brenner, W.: ITIL as Common Practice Reference Model for IT Service Management: Formal Assessment and Implications for Practice. In: IEEE International Conference on e-Technology, e-Commerce and e-Service, pp. 704–710. IEEE, Hong Kong, China (2005). https://doi.org/10.1109/EEE.2005.86

16. Stankov, I., Gotseva, D.: An overview of security and risk management in business intelligence systems. In: 2020 III International Conference on High Technology for Sustainable Development (HiTech), pp. 704–710. IEEE, Sofia, Bulgaria (2020). https://doi.org/10.1109/hitech51434.2020.9363990

17. Ghaffari, F., Abouzar, A.: A new adaptive cyber-security capability maturity model. In: 9th International Symposium on Telecommunications, pp. 298–304 (2018). https://doi.org/10.1109/ISTEL.2018.8661018

18. NIST: NIST SP 800–53, Revision 5 Control Mappings to ISO/IEC 27001:2022, https://csrc.nist.gov/files/pubs/sp/800/53/r5/upd1/final/docs/sp800-53r5-to-iso-27001-mapping.docx. Last accessed 21 May 2023

19. Estacio, K.: Modelo de evaluación de seguridad de la información en centros de datos. Revista Cumbres 9(1), 39–50 (2023). https://doi.org/10.48190/cumbres.v9n1a3

Food Loss as a Global Problem Identifiable in the Production Chain: The Case of Oversupply in Agro-Products in Azuay-Ecuador

Rafael Maldonado Yépez[(✉)] [iD], Marco Antonio Gómez Parra[iD],
Richard Antonio Martínez Villegas[iD], and Diana Sánchez Cabrera[iD]

Instituto Tecnológico Superior Particular Sudamericano, Cuenca AZ 010101, Ecuador
rafael.maldonadoye@gmail.com

Abstract. The Food and Agriculture Organization of the United Nations (FAO) recognizes as "food waste" the scraps of products intended for human nutrition, the origin of which lies in the processes involved in its generation (production, post-harvest, and processing). The Ecuadorian government addresses this global problem but does not establish guidelines to find practical solutions. In the city of Cuenca, information was collected through semi-structured surveys of 78 producers from the "Chaguarchimbana" market and 62 from the "Cristo Rey" market. The information from each agroecological fair collected is sized by one-way analysis of variance using SPSS statistical software. The significance level was determined at $p < p\ 0.05$, the probability of success/failure ($p/q = 50\%$). The significant differences between samples were analyzed using Tukey's Honesty Significant Different (HSD). The ten least traded products identified were tomato, purple cabbage, and lettuce; however, the waste perceived by the producers was the same only in tomato, purple cabbage, pumpkin, and beetroot, demonstrating that even though there is a different volume of sowing and marketing and that they are geographically distant, the problem is the same in the rural parishes of the canton Cuenca. The purpose of this study is to provide a basis for decision-making by public and private entities.

Keywords: Food waste · Oversupply · Agroecological solution

1 Introduction

The Food and Agriculture Organization of the United Nations (FAO) recognizes as "food waste" the scraps of products intended for human nutrition, the origin of which lies in the processes involved in its generation (production, post-harvest, and processing) [1]. According to Chen et al. [2], globally, 33% of food is wasted or solely reported as "losses" along the phases of cleaning and handling before consumption, marketing, and other processes. In monetary terms, the lack of use generates an annual loss of food inputs amounting to four hundred billion dollars [3]. In the case of the European Union,

M. Z. Vizuete et al. (Eds.): CI3 2023, LNNS 1041, pp. 122–130, 2024.
https://doi.org/10.1007/978-3-031-63437-6_10

around 88 million tons are wasted annually [4]. Taking into account the 447.2 million inhabitants reported by the World Bank for the year 2021, a food waste of 0.539 kg of food per day is inferred, which is very similar to what is reported by Conrad et al. [5], who specifies that in America the average American throws away about 0.422 kg per day, requiring 1.21 million square meters to produce.

The consequences of this waste, together with uncontrolled wastage along the production chain and lack of marketing, not only lead to increased environmental impact but are reflected in the production of greenhouse gases [6] as a repercussion of the increment in food production. Moreover, the use of land for mostly extensive cultivation, as well as; the consumption of water sources [6, 7].

In the last two decades, efforts have been made to diversify production towards more eco-friendly ways, such as organic food, which has the advantage of increasing seed biodiversity and improving sustainability in rural communities [8]. However, the problem is still present to such an extent that one of the objectives set by the United Nations organization for the next seven years is the reduction of 50% of the waste generated per capita on a global scale [2]. It is making a vital contribution to responding to the growing global demand for food driven by state policies focused on consumer-friendly strategies like price discounts and vouchers on purchases with a unique emphasis on healthy products [9, 10]. Moreover, paradoxically, the desired increase in demand is expected to generate the opposite effect over time as it will raise the value of fruit and vegetables by segregating those buyers who were attracted in the first instance by the economic incentive [2, 9].

Latin America is not unrelated to this global problem even though in percentage terms (15%), it shows the least waste and losses; there are specific cases that stand out notably [11, 12]. Mexico reports a rate of 35.3% of food that is merely wasted annually; 96% of small and micro enterprises in this country indicate that they do not have action plans against or concreted protocols in their working days to counteract this damage, in contrast to 23% of large Mexican companies that currently have specific daily activities [12] Furthermore, in the autonomous city of Buenos Aires, 4.6 out of 10 porteños admit wasting more than 10% of the food in their homes, especially vegetables and fruit [13], only through food cooking processes.

In addition, several countries in the region have already committed themselves to conservation methods as a way to make use of products and by-products from agriculture and industry originating from the problems in the distribution processes to the use of foodstuffs [14] to the application of new techniques such as high hydrostatic pressure or freeze-drying to prolong shelf life despite the change in organoleptic properties [15, 16].

At the national level, it is estimated that an average of 72 kg of waste per person is produced annually in Ecuador [17], resulting in 1.2 million kilograms of food every year, which reflects a socio-economic paradox as evidenced by Vargas [18] who indicates that 4 out of 10 Ecuadorian households have difficulties in obtaining food, one of the main causes are economic difficulties, thus leading to one of the most urgent problems in developing countries, which is protein deficiency and associated malnutrition [19]. Although Ecuador has made significant progress in the last decade, such as increasing the web presence of small and medium-sized Ecuadorian companies to 70.50% to make their activity visible to local demand [20], the national government still has some unfinished business, mainly in terms of establishing guidelines that lead to efficient and applicable resolutions [21].

The specific case of the province of Azuay has historically been considered a center of agricultural production focused on supplying resources for regional consumption. A clear example of this is the large number of agroecological fairs that allow the commercialization of inputs directly from the producer. These open-air markets bring together farmers from the rural parishes of the canton of Cuenca in strategic places of high traffic in the urban area of the city [22, 23].

The support of the local government is evidenced by the organization of outdoor markets called "fair" by the municipal public company for economic development (EDEC EP) in order to facilitate internal trade and avoid this problem of global oversupply [24]. However, there is no record of the most susceptible products to this phenomenon, a perception of waste of each product by farmers, or even a contrast between marketing centers to discern if this is a regional trend or if they are part of isolated cases. This research seeks to answer these questions and serve as a basis for decision-making by public and private actors in the region.

2   Materials and Methods

2.1   Study Area

The area studied is the Canton of Cuenca, one of the 15 cantons of the province of Azuay, which has 15 urban parishes and 22 rural parishes focused, among other activities, on the production of food inputs for the city of Cuenca, the capital of the province of Azuay.

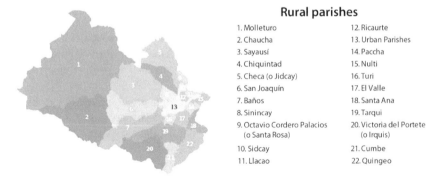

Rural parishes

1. Molleturo	12. Ricaurte
2. Chaucha	13. Urban Parishes
3. Sayausí	14. Paccha
4. Chiquintad	15. Nulti
5. Checa (o Jidcay)	16. Turi
6. San Joaquín	17. El Valle
7. Baños	18. Santa Ana
8. Sinincay	19. Tarqui
9. Octavio Cordero Palacios	20. Victoria del Portete
(o Santa Rosa)	(o Irquis)
10. Sidcay	21. Cumbe
11. Llacao	22. Quingeo

Fig. 1. The study area. Canton Cuenca is divided into urban and rural parishes as distinct political and agro-alimentary regions.

The field of study is the rural parishes of the canton of Cuenca, represented in the two largest agroecological fairs organized by EDEC EP. These outdoor markets sell their products simultaneously in the parishes of Bellavista and Huayna Capac (Fig. 1) Notwithstanding, the agro-producers come from 20 of the 22 rural parishes (Fig. 2) of the canton Cuenca to sell their products in Bellavista and Huayna Capac communities (Fig. 1).

Urban parishes

1. San Sebastián	9. Sucre
2. El Batán	10. Huayna Cápac
3. Yanuncay	11. Hermano Miguel
4. Bellavista	12. El Vecino
5. Gil Ramírez Dávalos	13. Totoracocha
6. El Sagrario	14. Monay
7. San Blas	15. Machángara
8. Cañaribamba	

Fig. 2. Distribution of the urban parishes of Cuenca

2.2 Data Collection Techniques

Information was collected to identify the food supplies with the most prominent waste and to quantify them according to the perception of the traders.

By using a mixed approach, stratified and semi-structured surveys were carried out with 78 producers from the "Chaguarchimbana" agroecological fair and 62 from the "Cristo Rey" fair. From a qualitative point of view, priority was given to identifying the ten products with the most considerable oversupply for each respondent, and with this preliminary survey, a quantitative survey was conducted using a 7-point Hedonic Scale in order to identify the degree of waste perceived by the farmer for each product previously listed. The scale was validated using Cronbach's Alpha internal consistency index by the item variance method. The inclusion criteria were: men and women between 18 and 65 years of age, mainly dedicated to agriculture, inhabitants of one of the rural parishes of the canton of Cuenca with one-year trading agro-products. Participants were asked to score from 1 (no waste) to 7 (total waste) on 42 food inputs, which included vegetables, grasses, and proteins.

2.3 Statistical Análisis

The food, which was determined at the two fairs, was contrasted connecting each other to recognize the existence of significant variance between the level of wastage assigned by the farmers; these collected values are sized by one-way analysis of variance using SPSS statistical software (Version 18.0 IBM Statistics, USA). The significance level was determined at $p < p\ 0.05$; the probability of success/failure ($p/q = 50\%$). Notable differences between samples were analyzed using Tukey's Honesty Significant Different (HSD).

3   Results and Discussions

The qualitative data of the survey applied to farmers is shown in Table 1, where the most repeated fruits are tomatoes and purple cabbage. However, it has been verified that, in both cases, seven products are constant in the top 10.

Table 1. The number of producers who corroborated product wastage in Cuenca.

Order	N° responses (78)	Chaguarchimbana Fair	N° responses (62)	Cristo Rey Fair
1	76	Tomato	57	Tomato
2	71	Purple cabbage	53	Purple cabbage
3	61	Lettuce	51	Lettuce
4	54	Broccoli	43	Pumpkin
5	49	Pumpkin	29	Lettuce
6	42	Ulluco	24	Beetroot
7	40	Chard	12	Ulluco
8	22	Beetroot	10	Pea
9	11	Spinach	7	Spinach
10	9	Chili pepper	6	Turnip
11	8	Pea	5	Chard
14	3	Turnip		–
15		–	1	Chili Pepper

The Hedonic Scale shows that, in the case of tomato, purple cabbage, pumpkin, and beetroot, there is no significant difference in the waste perceived by the same producers in the two fairs, so it can be inferred that, despite having different volumes of sowing, harvesting, and product for each farmer, it can be identified that consumers tend to avoid these products to a similar percentage extent in these geographically separated centers of sale, as shown in Fig. 1 (Table 2).

Table 2. Tomato Likert scale data grouped by Tukey's method.

Fair	N	Average	Clustering
Chaguarchimbana Fair	50	5,489	A
Cristo Rey fair	50	5,344	A

Furthermore, the seasonality of the food or even the need for re-cooling to prolong shelf life is ruled out since, as Table 3 shows, pumpkin is another product with low turnover in its commercialization, despite being a berry that can remain at room temperature for weeks.

Curious cases are presented in lettuce and spinach that, despite being equally identified as oversupplied by the farmers, in reality, they manifest that their commercialization shows variation from one fair to another. For instance, 61 people identified lettuce as an oversupplied product in Chaguarchimbana, indicating an average of 4,787. Meanwhile, in Cristo Rey, 71 farmers expressed a surplus of the same product. Nevertheless, the

Table 3. Information from the Likert scales of Calabash clustered by Tukey's method.

Fair	N	Average	Clustering
Chaguarchimbana Fair	50	4,911	A
Cristo Rey fair	50	4,356	A

number was much lower: 2,276, as shown in Table 4. This group includes all 13 products in Table 1 except for tomato, purple cabbage, pumpkin, and beetroot, as previously described.

Table 4. Information from the Likert scales of the Lettuce was grouped by Tukey's method.

Fair	N	Average	Clustering
Chaguarchimbana fair	50	4,787	A
Cristo Rey fair	50	2,276	B

Considering the specific case of EDEC, where more than 400 producers participated, the conclusion was that an average of 0.5 kg of the ten foods in Table 1 per producer is not merchandised. Taking into account that these products are exclusively sold on Saturdays and Sundays. An oversupply of 2000 kilos per month can be deduced, highlighting local vegetables and fruits from the region as the most important. The tendency can be extrapolated to the province of Azuay, in specific cases, such as markets in the region, over and above that, it could be verified through a semi-structured interview with the community manager of the Peasant Insurance of Azuay, who pointed out an overproduction of vegetables among the members of this obligatory universal scheme of the Ecuadorian Institute of Social Security (IESS).

The data collected from the EDEC agro-producers showed 0.5 kg of wastage, which is in line with the above, globally 0.539 kg, and in the case of the United States of America (USA), the figure decreases to 0.422 kg [4, 5]. Among the causes of this reduction are early strategies such as the implementation of food banks, consumption trends, or general government policies that have been replicated in countries of the region [8, 25–28]. In Ecuador, educational institutions have presented solutions focusing on waste in markets and municipal gardening wastes [29], in other words, at the last link in the production chain.

Government policies for this particular sector are practically non-existent or have arrived vastly late in the country; [20, 30] the results have not been socialized to avoid focusing public-private efforts on specific products with this problem in the province of Azuay.

Previous research recognizes the importance of the problem leading it to different perspectives, which include solid waste management, soil selection for these activities, and food management [31, 32]. However, a detailed qualitative analysis of the region will serve as a basis for future research.

4 Conclusions

This research has addressed a critical and internationally relevant problem; food waste. Among the causes in the production, distribution, and marketing chain, it is concluded that the oversupply of farmers' products, specifically in the province of Azuay, Ecuador, generates 0.5 kg of oversupplied products per day per farmer. Through meticulous analysis, with a methodology that combines semi-structured surveys, statistical analysis, and a mixed approach, significant findings have been identified, and they have direct implications for the region.

As fundamental findings, the most wasted products were identified, including tomatoes, purple cabbage, lettuce, broccoli, and pumpkin. These findings are consistent across two different agroecological fairs in the region, suggesting a pattern of consumer behaviour that can be addressed in the future. On the other hand, the perception of loss by producers is similar, as wastage was found in products such as tomato, purple cabbage, pumpkin, and beetroot, regardless of differences in geography and volume of planting and marketing.

This food loss has a direct economic impact on local producers and the regional economy, affecting the sustainability of the food supply chain.

The importance of this research lies in its contribution to the understanding of a complex and multifaceted problem that has implications for food security due to the loss of food that affects millions of people worldwide. In developing countries such as Ecuador, where malnutrition and food insecurity are persistent concerns, food loss is subjected to implications for the health and well-being of the population. In addition, environmental sustainability also has an impact, as the resources used in the production, transport, or storage of food that is then wasted contribute to natural resource depletion and climate change.

The findings of this research can serve as a basis for decision-making by public or private entities. Identifying specific products with higher wastage and understanding producers' perceptions can inform policies and strategies to reduce food loss.

This research has shed light on a global problem from a local perspective, offering a valuable and applicable perspective. However, it is important to recognize that food loss is a complex problem that requires a multidisciplinary and collaborative approach. Collaboration between governments, industry, academia, and society will be crucial to developing effective and sustainable solutions. The acquisition of agroecological practices, consumer education, and ultimately supply chain innovation are potential areas for intervention and improvement.

Ultimately, reducing food waste is not just a matter of economic efficiency or environmental sustainability; it is a matter of social justice as well as ethical responsibility. The present research contributes to this global effort, providing a solid basis for future research and action in the Azuay region and beyond.

References

1. FAO: The State of Food and Agriculture 2019. Moving forward on food loss and waste reduction. THE STATE OF THE WORLD (2019)

2. Chen, C., Chaudhary, A., Mathys, A.: Nutritional and environmental losses embedded in global food waste (2020). https://doi.org/10.1016/j.resconrec.2020.104912
3. El mundo desperdicia el 17% de los alimentos mientras 811 millones de personas sufren hambre | Noticias ONU. https://news.un.org/es/story/2021/09/1497582. Accessed 07 July 2023
4. Scherhaufer, S., Moates, G., Hartikainen, H., Waldron, K., Obersteiner, G.: Environmental impacts of food waste in Europe. Waste Manage. **77**, 98–113 (2018). https://doi.org/10.1016/J.WASMAN.2018.04.038
5. Conrad, Z., Niles, M.T., Neher, D.A., Roy, E.D., Tichenor, N.E., Jahns, L.: Relationship between food waste, diet quality, and environmental sustainability. https://doi.org/10.1371/journal.pone.0195405
6. Girotto, F., Alibardi, L., Cossu, R.: Food waste generation and industrial uses: A review. Waste Manage. **45**, 32–41 (2015). https://doi.org/10.1016/J.WASMAN.2015.06.008
7. Gustavsson, J., Cederberg, C., Sonesson, U., van Otterdijk, R., Meybeck, A.: Pérdidas y desperdicio de alimentos en el mundo. Save Food. FAO, Roma (2012)
8. Willer, H., Yussefi, M.: The World of Organic Agriculture. Statistics and Emerging Trends, 9th ed. International Federation of Organic Agriculture Movements IFOAM, Born (2007)
9. Blom-Hoffman, J., Kelleher, C., Power, T.J., Leff, S.S.: Promoting healthy food consumption among young children: Evaluation of a multi-component nutrition education program. J. Sch. Psychol. **42**(1), 45–60 (2004). https://doi.org/10.1016/J.JSP.2003.08.004
10. Steffen, W., et al.: Planetary boundaries: Guiding human development on a changing planet. https://doi.org/10.1126/science.1259855
11. Basso, N., Brkic, M., Moreno, C., Pouiller, P., Romero, A.: Valoremos los alimentos, evitemos pérdidas y desperdicios **34**, 25–32 (2016)
12. Aguilar Gutierrez, G.: Responsabilidad social corporativa en las pérdidas y desperdicios de alimentos en México. Cadernos PROLAM/USP **17**(33), 168–197 (2019). https://doi.org/10.11606/issn.1676-6288.prolam.2018.133625
13. Adriana Leal, M., Fernando Sacco, P., Dana Rondinone, F.: Desperdicio de alimentos en hogares de la Ciudad Autónoma de Buenos Aires, Argentina: comportamiento del consumidor Food waste in households of Ciudad Autónoma de Buenos Aires, Argentina: consumer behavior **12**
14. Ballesteros Gómez, C.: Estrategias para la reducción de pérdidas de productos perecederos en el proceso de distribución. caso de estudio plátano en la región de Cundinamarca, Tesis de maestría. Universidad Nacional de Colombia, Bogotá (2017)
15. LA Conservación De Los Alimentos, E.: Aplicación de la alta presión hidrostática application of high hydrostatic pressure in the food preservation aplicación da alta presión hidrostática na conservación dos alimentos. Cienc. Tecnol. Aliment **3**(2), 66–80 (2001). Accessed: 08 July 2023. [Online]. Available: www.flowcorp.com
16. Mejia Barragan, M.L., Tocagon Vásquez, W.A., Téllez Velandia, R.A.: Proceso conservación alimentos por el método de liofilización
17. United Nations News: Wasting food just feeds climate change, new UN environment report warns (2021). https://news.un.org/en/story/2021/03/1086402. Accessed 07 July 2023
18. Vargas, V.: Desnutrición crónica infantil en la prensa digital del Ecuador: Un análisis de la calidad de contenidos en Primicias, GK, El Comercio y El Universo durante el cambio de gobierno. Universidad San Francisco de Quito, Quito (2021)
19. Muller, O.: Malnutrition and health in developing countries. Can. Med. Assoc. J. **173**(3), 279–286 (2005). https://doi.org/10.1503/cmaj.050342
20. Arellano, P.R., Aguilar, F.D., Quintana Bornot, A., Carrera, L.A.: Contexts of Digital Tools in Marketing Management in MSMEs for Their Permanence in the Market. In: Zambrano Vizuete, M., Botto-Tobar, M., Diaz Cadena, A., Durakovic, B. (eds.) Innovation and Research

- A Driving Force for Socio-Econo-Technological Development, pp. 528–541. Springer International Publishing, Cham (2022)

21. Ayaviri-Nina, D., Quispe-Fernández, G., Romero-Flores, M., Fierro-López, P.: Avances y progresos de las políticas y estrategias de seguridad alimentaria en Ecuador. Revista de Investigaciones Altoandinas - Journal of High Andean Research 18(2) (2016). https://doi.org/10.18271/ria.2016.202

22. León, M.: Análisis de la producción industrial de la provincia del Azuay, Tesis de Maestría. FLACSO, Quito (1992)

23. Vista de Caracterización productiva de las ganaderías en los cantones occidentales de la provincia del Azuay. https://publicaciones.ucuenca.edu.ec/ojs/index.php/maskana/article/view/1510/1195. Accessed 08 July 2023

24. Empresa pública municipal de desarrollo económico de cuenca. Feria artesanal por fiestas de Cuenca (2022)

25. Giménez, A.M., Montoli, P., Curutchet, M.R., Ares, G.: Estrategias para reducir la pérdida y el desperdicio de frutas y hortalizas en las últimas etapas de la cadena agroalimentaria: avances y desafíos. Agrociencia Uruguay 25(NE2) (2022). https://doi.org/10.31285/AGRO.25.813

26. Cañet, F., Didonna, F.: Pérdidas y desperdicios de alimentos: puntos críticos y como evitarlos. Ambientico, 38 (2014). [Online]. Available: https://link.gale.com/apps/doc/A37021 1832/IFME?u=anon~33754acc&sid=googleScholar&xid=dc781d3c

27. Walia, B., Sanders, S.: Curbing food waste: A review of recent policy and action in the USA. Renewable Agric. Food Syst. 34(2), 169–177 (2019). https://doi.org/10.1017/S1742170517000400

28. Barnard, A.: Freegans: Diving into the Wealth of Food Waste in America (2016)

29. Jara-Samaniego, J., et al.: Urban waste management and potential agricultural use in south american developing countries: a case study of chimborazo region (Ecuador). Commun. Soil Sci. Plant Anal. 46(sup1), 157–169 (2015). https://doi.org/10.1080/00103624.2014.988587

30. Chicaiza Ortiz, C.D., Navarrete Villa, V.P., Camacho López, C.O., Chicaiza Ortiz, Á.F.: Evaluation of municipal solid waste management system of Quito - Ecuador through life cycle assessment approach. LALCA: Revista Latino-Americana em Avaliação do Ciclo de Vida 4, e45206 (2020). https://doi.org/10.18225/lalca.v4i0.5206

31. Cobos-Mora, S.L., Guamán-Aucapiña, J., Zúñiga-Ruiz, J.: Correction to: suitable site selection for transfer stations in a solid waste management system using analytical hierarchy process as a multi-criteria decision analysis: a case study in Azuay-Ecuador. Environ. Dev. Sustain. 25(6), 5753–5754 (2023). https://doi.org/10.1007/s10668-022-02220-x

32. Cordero-Ahiman, O.V., Vanegas, J.L., Beltrán-Romero, P., Quinde-Lituma, M.E.: Determinants of food insecurity in rural households: the case of the paute river basin of azuay province, Ecuador. Sustainability 12(3), 946 (2020). https://doi.org/10.3390/su12030946

Information Technologies: A Tool for Accident Prevention in Gerontological Patients

Toapanta Kevin[(✉)] [ID], Novoa Geremy [ID], Solórzano Josselyn [ID],
Quilumbaquin Mayra [ID], and Vallejo Alejandra [ID]

Tecnológico Universitario Vida Nueva, Quito, Ecuador
kevin.toapanta@istvidanueva.edu.ec

Abstract. The implementation of home automation systems in a home for an older adult can improve the quality of life and increase security. Home automation systems can include devices such as motion sensors, security cameras, smoke alarms, and activity monitors that can alert caregivers or family members in the event of an emergency. Home automation systems can also allow remote control of devices, such as lights and thermostat, which could help an older adult with reduced mobility. Installing internet-connected devices, such as a virtual assistant, can also offer assistance services for medication reminders, medical appointments, and news updates. In general, the implementation of home automation systems appropriate to the needs of an elderly person can help their independence and improve their safety and comfort at home. The implementation of technology in a home for the elderly can improve their quality of life, increase their safety and offer more comfort. The installation of these technologies must be carried out by a professional and must take into account the specific needs of the elderly, guaranteeing greater independence and care.

Keywords: Older adults · IoT · Home automation · Gerontology · Care · Technology · Prevention

1 Introduction

How can the implementation of the Internet of Things (IOT) improve health care for the elderly population in the context of global aging? At present, the aging of the population is a worldwide phenomenon that has led to a growing concern for the well-being and stability of the elderly. In this sense, the health care of this population has become a matter of great importance for health systems throughout the world.

In this context, a constant effort has been made to improve the health care of the elderly through various interventions, both preventive and treatment. The implementation of innovative technologies, the training of specialized personnel and the promotion of self-care are just some of the strategies used to improve the quality of life of this population group.

In this era of digital transformation, artificial intelligence has become an important tool for improving the health care of the elderly. With the passing of time, technology has evolved in an explosive way, the IoT will be able to make a plus in the constant improvement of housing systems for the elderly to obtain great solutions to the new challenge of gerontological care systems.

Ultimately, improving healthcare for older adults remains one of the biggest public health challenges around the world, and the IoT has a key role to play in tackling this challenge.

2 Theoretical Framework

What is directly related to the care of the elderly, it is necessary to take into account different factors that influence physical and cognitive changes, as well as the relationship of the person in their environment, to understand a little of what these factors are when implementing IoT technology must first define what well-being is and what are the factors that contribute to it, as well as the quality of life.

2.1 Health and Well-Being of the Elderly

For decades, developed countries have faced the effects of aging populations. In response, they have implemented programs and policies that have proven effective in various social, economic, and medical aspects of older people. Furthermore, by drawing on research on the condition of older people and considering the changes they may undergo, agencies responsible for the care of older people can systematically address the various aspects associated with aging [1].

It is important to comprehensively address the concepts of health, well-being, success, active aging and quality of life in the elderly. To achieve this, the implemented policies and existing regulations must be effectively implemented to promote the development of an adequate culture of aging and old age [2].

2.2 Quality of Life

The quality of life in the elderly, specifies the conditions of well-being in which they find themselves despite the risk factors that may be had, the aging process is affected by the number of people with whom the elderly live since if you do not have someone who shares your environment, the well-being index is lower in relation to those who share with a caregiver or family member [3].

2.3 Health Condition

According to the state of health of the elderly, this population in general has a deterioration that occurs due to the aging process. In the context of an elderly person, the state of health must be controlled by a doctor specializing in adult medicine. Major that will give a diagnosis for the process of a good quality of life [4].

2.4 Career

The responsibilities of a caregiver will depend on the specific needs of the person, but generally includes help with their daily activities such as: medication administration, accompaniments, mobility assistance, coordination of doctor's appointments and transportation, among others. It is important to highlight that caregivers of older adults play a fundamental role in the well-being and quality of life of the people they care for [5].

2.5 Risks of the Elderly

Currently, in the field of care for the elderly, the caregiver must take into account that the elderly is sensitive to certain medications, which can cause complications such as cognitive impairment, dementia, poisoning and allergic reactions due to the physiological changes typical of this stage of life. Loneliness and isolation are also important risk factors for the well-being of older adults. In addition, there are various diseases that more frequently affect this group, such as arteriosclerosis, cardiovascular diseases, cerebrovascular diseases, acute renal failure, severe sepsis and Alzheimer's disease.

To prevent these risks, it is essential that older adults and their caregivers are informed about the particularities of each disease and its associated risk factors [5]. One of the problematic situations faced by seniors living alone is the difficulty in carrying out daily activities due to their natural condition and the aging process of their body. These difficulties are manifested in symptoms of deterioration that affect their quality of life and their ability to reason [6].

2.6 IOT

The Internet of Things (IoT) refers to the interconnection of physical objects with sensors, software, and other technologies that allow them to connect and exchange data with other devices and systems over the Internet. In recent years, the Internet of Things (IoT) has become one of the most important technologies of the 21st century. This technology allows you to connect a variety of everyday items, appliances, vehicles, thermostats, baby monitors, etc. through embedded devices, allowing seamless communication between people, processes and things.

With the availability of affordable computing technology, cloud services, big data, data analytics and mobile technology, physical objects can automatically exchange information with minimal human intervention and can now be collected. In this hyper-connected environment, digital systems can record, monitor, and coordinate all interactions between connected objects. The physical and digital worlds are intertwined and linked.

According to [7] among one of the technologies that have been developed through the use of the IOT, are the following:

- Access low-cost, low-power sensor technology.
- Connectivity.
- Cloud computing platforms.
- Machine learning and analytics.
- Conversational artificial intelligence (AI).

2.7 Older Adult with IOT

Within the context of the elderly, depending on the severity of the disease, the treatment can be carried out on an outpatient basis or remotely at home, which benefits the health of patients, relieves congestion in hospital centers and optimizes the time of doctors and specialists. To address the problem of older adults living alone, approaches such as home monitoring and care using wireless sensor networks, remote monitoring through mobile devices, augmented reality, biomedical signal analysis, and remote medical devices have been developed.

The scenario in which the model is applied is a house with the following basic elements of the house: B. Technical components of furniture and household appliances, home automation and remote assistance tools. The basic elements of a house include physical boundary elements such as floors, stairs, walls, ceilings, windows, and doors that define the boundaries, compartments, and entrances of the house. This includes the infrastructure of services and sanitation, as well as the areas of application. [12].

It is important to mention that through a validation model it is assumed that the person has knowledge of the distribution and location of the household items, furniture, electrical appliances and other belongings that they have in their home [6].

3 Methodology

3.1 Qualitative Approach

This research is based on a qualitative approach that considers the subjective and dynamic nature of the individuals studied. It focuses on the in-depth investigation of the conflicts and needs that affect the elderly and on justifying the importance of the technological implementation of the IOT to improve the care of their quality of life.

3.2 Exploratory Investigation

Through an exploratory research approach, information will be collected on the status and quality of life of the elderly and their various risks as possible solutions adapted to the needs of this research group.

It seeks to explore and understand how this technology can help mitigate the risks that older adults face in their daily lives. Through qualitative methods, such as in-depth interviews and a literature review, the aim is to obtain an initial and general vision of how the IoT can improve the safety and well-being of older adults. Different IoT applications, such as remote monitoring devices, early warning systems, and smart assistants, will be explored with the aim of identifying the advantages and challenges of their implementation. The results of this exploratory research will provide a solid basis for further research and help define more specific research questions on how to maximize the potential of IoT in customs protection against the risks associated with their age and health condition.

3.3 Selection Criteria

Participants. Regarding the study selection criteria, research focused on the needs of the elderly was included, since they are those that require control and monitoring to improve their quality of life during the implementation of IOT.

Outcome Measures. In relation to the effects of the risk assessment of the elderly, the primary outcome measure was to consider the general health status of the participants, including medical conditions, disabilities or chronic diseases. This is important to understand how IoT implementation can impact your well-being and quality of life [10].

Among secondary outcome measures is establishing an age range that defines older adults, generally considered as people aged 65 and over. However, a broader range may also be considered depending on the objectives of the study, assessing whether participants reside at home or in an elderly care facility. This can influence how IoT devices are deployed and used in your daily life, physical and functional well-being, since they are directly related to the symptoms of fatigue, controlling the activity and usual risk situations for this type of people, ensuring that the participants are available and willing to participate in the study. This involves considering your availability of time and your willingness to share information and experiences related to IoT implementation [16].

Data Extraction and Analysis. Once the information from the most relevant and interesting studies that contribute to IOT research with older adults was compiled, data were obtained on the physical intervention, time, frequency and application of the technologies, as well as the effects it produces on the elderly. Older adults.

By considering these selection criteria, it will ensure that the study participants are representative and that the research effectively addresses the specific needs and challenges of older adults in relation to the help provided by the IoT.

3.4 Planning

During the planning phase of the project, the objective of the survey was clearly established, which is to understand the needs and level of knowledge of older adults about IoT (Internet of Things).

A brief and easy-to-understand questionnaire was created, which includes questions about the perceived needs of older adults in relation to IoT, as well as their level of knowledge and experience on the subject. Open questions can also be included to elicit additional feedback and feedback.

In the same way, it established a team to support the elderly in the adaptation process with the IOT. Different types of implementations were considered. The results of the different treatments applied were also considered and the techniques associated with the different types of applications for intervention in the elderly were identified [14].

The importance of having IoT technology to avoid catastrophes for the elderly is established, taking into account the necessary help for the caregiver. IoT equipment is necessary and paramount within the risks for the elderly and obtain an adequate benefit depending on the treatment and risk involved.

3.5 Importance

Understand the value of solving each of the needs of the elderly, implementation of measures that facilitate and guarantee the quality of life of the elderly, considering all aspects relevant to the process of implementing this technology, judging the importance of the use of technologies in the population rate of older adults in the world today in the face of present conflicts and risks.

4 Results

Thanks to the IoT and the different technologies used for the development of this system, it has been possible to thoroughly investigate implementations of stable materials, with their respective validations and friendly interfaces for the elderly. The system will allow the elderly to obtain a better quality of life and family members and doctors to have a controlled home system.

4.1 Automated Requirements

From the results obtained when carrying out the surveys to the caregivers of the elderly in the different Geriatric Centers, the following results are obtained in relation to the question: What are the types of risks in the home for the elderly? 33% of respondents state that the greatest risk is due to falls caused by slips and trips due to the design of the patient units and the lack of visibility of safety signs, 27% stated that there are security problems and thefts due to the high turnover of personnel hired in the care units, 20% due to accessibility and physical barriers due to physical or mental disability problems that make it impossible for them to access the different areas independently,

The second question arises: What is the best way to solve and guarantee the main needs of the elderly? Regarding the use of the IOT, where we can show that more than 50% of the respondents accept the use of these new technologies for monitoring health, communication and security.

The third question, how to get an older adult to interact with IOT devices? It is addressed to the group of researchers, where it was determined that they significantly simplify technology in various fields. In Geriatric Centers, for example, IOT devices simplify daily tasks by allowing the control of lighting, air conditioning and other devices with voice commands or from a mobile application.

Through the questions asked related to the use of IOT in older adults, more than 90% of the respondents stated that they agree with the use of these new technologies, since they are easy to understand, provide personalized assistance according to the requirements of the user, and in the question about the increase in vitality, it should be noted that the use of these devices improves the quality of life by simplifying daily activities, but does not increase the years of life.

The performance of the system through the devices needed by the elderly will have an optimal performance of 90%.

The system manages the security of the data and system to be implemented.

The system is stable and secures the transactions carried out by users and administrators 100% of the time throughout its useful life.

The system is capable of operating 100% of the working hours of the fairgrounds, including the whole day, covering a high level of availability.

The system will be compatible with operating systems that support web browsers.

By categorizing the elderly according to the risks determined in the matrix outlined above, it is possible to evaluate and establish optimal and effective technologies to accurately and adapted address any type of risk that may affect the well-being and safety of the elderly population. Therefore, using this matrix as a guide, one can select and apply IoT-based technologies that fit different risk profiles. To facilitate and guarantee the correct care and well-being of the elderly, a variety of equipment and technologies can be implemented. Below are some examples of equipment that can be used.

Family participation is essential to provide care and well-being to the elderly, and currently, new technologies have opened up a range of possibilities to strengthen this collaboration (Table 1).

Table 1. The following table presents IoT solutions for family participation

Family and community engagement			
Categorization of Ministry of Health systems	input technology	Characteristic of the technological system	Contribution of technology in the elderly
Prevention and detection of cognitive and physical deterioration	virtual assistants	Place a virtual assistant in an accessible place in the home, configure it according to the preferences and needs of the older adult	Offer medication and appointment reminders
Nutrition and hydration	telecare systems	Install a telecare device connected to a monitoring center and that is easily accessible	Allow quick communication with emergency services or caregivers

In the same way, the Ministry of Health indicates the Psycho-socio-Educational follow-up where the follow-ups to the patients are recorded, which are carried out physically, in Table 2 we proceed to describe instruments and contributions for the use of digital IoT tools of these record systems among others.

Table 2. The following table presents solutions for the categorization of the Psycho-socio-Educational follow-up

Psycho-socio-educational (interinstitutional articulation enters the component)			
Categorization of Ministry of Health systems	input technology	Characteristic of the technological system	Contribution of technology in the elderly
The care unit has monitoring records of the activities carried out with the elderly, the evaluation of achievements obtained and the generation of information on individual and group progress. (Evolution tab)	Virtual records through an app	Download a mobile application designed for managing medical records and configure it with the information of the elderly	It offers the option to track daily physical activity, count steps, measure distance traveled and log specific exercises to promote a healthy lifestyle
All older adults in the unit have records:	digitized files	By scanning medical records and other important documents and saving them in an electronic document management system	It allows the registration of previous medical visits, facilitates the follow-up of consultations and procedures, improves communication with health professionals and the quality of care
In the care unit, users have a record of the planning of activities of the individual care plan on a monthly basis, considering the components in file No. 12:	Digitized planning systems	Use of specialized applications or software to create care plans, medication schedules, and activity scheduling	It helps organize and prioritize daily tasks and activities, such as shopping, housework, etc

Food and nutrition are the primary basis of the health of the elderly for whom necessary solutions are implemented in the different sections of Table 3 that improve functioning and update systems for adaptations with the elderly.

Table 3. The following table presents solutions for the categorization of preventive health and nutrition.

Preventive health, food and nutrition

Categorization of Ministry of Health systems	input technology	Characteristic of the technological system	Contribution of technology in the elderly
The social promoter has a mask and alcohol	Smart device for personal hygiene monitoring	A smart device is purchased that monitors the use of masks and hand sanitizer with alcohol	It offers proximity sensors to hygiene points, records whether the social promoter is wearing a mask, and issues alerts or notifications
Application of the Get Up and Walk Test	Smart bracelet or watch	Help the elderly to select and configure a smart bracelet or watch that has the capacity to generate the multiple functionalities that the elderly require	They allow detecting movement and recording the physical activity of the elderly
List of people who are at risk of falls	Motion sensors	Install motion sensors in strategic areas of the home, such as hallways and bathrooms, which have an appropriate alert configuration	They allow alerts or notifications to be sent to caregivers when they detect unusual or sudden changes in the pattern of movement
Fall hazard identification cards	Risk Identification Smart Tag	Associate a smart tag with relevant information on the risk of falls in the elderly. Explain to care professionals and caregivers how to use and scan the tag for quick access to data	It has the ability to establish two-way communication in an emergency through alert buttons and also through the built-in geolocation system that allows tracking the location of the elderly in real time

One of the protocols is good treatment, systems for falls and loss prevention for the elderly in a care center for them (Table 4).

Table 4. The following table presents solutions for the public health categorization of human talent

Human talent			
Categorization of Ministry of Health systems	Input technology	Characteristic of the technological system	Contribution of technology in the elderly
Protocol for Good Treatment and Conflict Resolution	smart voice assistant	Place a smart voice assistant, such as an Amazon Echo or Google Home, in a central location in your home. Configure the assistant according to the preferences and needs of the older adult	Provides assistance in resolving conflicts or providing advice for good treatment
Fall Protocol	Smart fall sensor	Install a smart fall sensor in key areas of the home, such as the bedroom or bathroom	It offers emergency call and geolocation functions to facilitate immediate assistance in the event of a fall
Loss and loss protocol	GPS tracking device	Provide the older adult with a GPS tracking device, such as a watch or wearable device	It makes it easier to detect the location of the elderly in case of misplacement or loss. You can send real-time location via mobile app or online platform

The final point of this categorization system of the Ministry of Health is the infrastructure where the use of Iot is essential since in a safe, healthy and comfortable environment, elderly people will be able to have an excellent quality of life, for which the contribution of technology in Table 5.

Table 5. The following table presents solutions for the categorization of public health infrastructure.

Infrastructure, safe and accessible environments			
Categorization of Ministry of Health systems	Input technology	Characteristic of the technological system	Contribution of technology in the elderly
The care unit encourages families to allocate adequate space and furniture for the care of the elderly person, in order to preserve their intimacy and privacy	Security cameras	Place security cameras at strategic points in the home, such as the main entrance, the living room or the rest area	They provide real-time monitoring and surveillance to ensure the security of the space and the privacy of the elderly
The care unit has a physical space for technical work, meetings and workshops, equipped with computer equipment	Interactive boards	Install interactive whiteboards in work or meeting areas to facilitate communication and collaboration between the elderly, caregivers, and care professionals	They enable collaboration and information presentation during meetings and workshops, improving efficiency and user engagement
The care unit has a publicized risk management plan aimed at protecting the safety of the user in any emergency situation	Smoke sensors and carbon monoxide detectors	Place smoke sensors and carbon monoxide detectors in areas of the home prone to fires or toxic gas leaks	They alert about emergency situations allowing an immediate response and safeguarding the safety of the elderly

5   Discussion

Gathering data on the needs of older adults is essential to understanding the challenges and specific problem areas that could be addressed by implementing IoT solutions. This implies identifying the areas in which technology could provide greater support and improve the quality of life for older adults. By knowing the specific needs of this population, developers can design solutions tailored to their requirements.

There are various methodologies to obtain data on the needs of older adults. One of the most common is through surveys and questionnaires, which allow information to be collected directly from the participants. These surveys can address issues such as health, safety at home, mobility, communication and socialization, among other relevant aspects for the well-being of older adults. Additionally, in-depth interviews and focus groups can also provide more detailed and contextual information about the needs and preferences of this population.

6 Conclusion

Based on the introduction, taking the research question that I proposed can be answered by obtaining knowledge of the planning, methodology, and technologies used to develop advances that are used by older adults, which can have a very positive impact on their quality of life. These technologies allow them to have greater autonomy in their daily lives, give them access to services and resources that they did not have before, and help them stay connected with loved ones. In summary, the implementation of IoT technologies in the home of the elderly can improve their safety, comfort and quality of life, which can lead to greater independence and autonomy in their daily lives. However, it is important to take into account the need to train older adults to use these technologies and guarantee the privacy and security of their personal information.

This research project has shed light on the importance of Information Technology (IT) as a tool to prevent accidents in the elderly population, we have explored how IT can play a significant role in the early detection of risks, the promotion of safety and improvement of the quality of life of the inhabitants of the third age. We extend our gratitude to the Tecnológico Universitario Vida Nueva for the access to technological resources and the guidance of professionals to achieve the results obtained.

References

1. González, A., Piñeros G.: Sistema de IoT de detección de caídas aplicadas a los adultos mayores en Colombia. AJ (2022)
2. Vera, P., Rodríguez, R.: Tecnologías vestibles aplicadas al cuidado de la salud: construcción de un prototipo de monitoreo. C. M (2020)
3. González P., Martínez P.: Sistema Web para un dispensador de medicamentos, con integración de oxímetro para adultos mayores, con uso de IoT (2022)
4. Muñoz G.: La incorporación del adulto mayor en el uso de las TICS y su repercusión en las tecnologías de servicio de las organizaciones (2019)
5. Álamo V., Ávila Á.: Principales factores de riesgo en la tercera edad. Su prevención (2021)
6. Sánchez, A., González, G., Barreto, E.: Modelo informático integrado AmI-IoT-DA para el cuidado de personas mayores que viven solas. Revista Colombiana de Computación (2019)
7. Limachi, H.: Prototipo para la detección de caídas en adultos mayores mediante Arduino e IoT (2020)
8. Mogolon, B.: Diseño de sistema de supervisión y control de la salud en el hogar, para adultos mayores en la vereda "La venta" del municipio de Belón-Boyacá, haciendo uso de internet de las cosas (IOT) (2022)
9. Capera, P., Huertas, P.: Dispositivo IoT detector de caídas para las personas de la tercera edad (2022)
10. Trujillo, D., Abreu, B., Ferro, M.: Estrategia de intervención para mejorar la calidad de vida del adulto mayor. Revista Cubana de Medicina General Integral **36**(4), 1–10 (2020)
11. Altamirano, M., Moreta, T.: Sistema Domótico basado en IoT para el centro integral del adulto mayor de la ciudad de Ambato (2022)
12. Benalcázar, P., Jiménez, M.: Teleconsultorio para diagnóstico y tratamiento en atención primaria de adultos mayores utilizando IoT y tecnologías E-Health (2021)
13. Pinto-Fernández, S., Muñoz-Sepúlveda, M.: Uso de tecnologías de información y comunicación en adultos mayores chilenos. Revista Iberoamericana de Ciencia, Tecnología y Sociedad-CTS **13**(39), 143–160 (2018)

14. Luengas, L., Díaz, M.: Domótica para asistir adultos mayores. Ingenio Magno **10**(1), 79–88 (2019)
15. Barrios, R., Rodríguez, R.: Asistencia de adultos mayores mediante la implementación de robots en entornos domóticos (2018)
16. López, N., Gómez, A.: Tecnología y envejecimiento activo: Una revisión sistemática. Rev. Esp. Geriatr. Gerontol. **56**(2), 84–92 (2021)
17. Rodríguez-Ruiz, B., López-Noguero, F.: Uso de la tecnología en personas mayores con demencia: Revisión sistemática. Revista de Neurología **68**(8), 343–353 (2019)
18. Giraldo, C., Moreno, C., Pacheco, J., Torres, M.: Impacto de las tecnologías de la información y comunicación en la calidad de vida de las personas mayores. Ciencia, Tecnología e Innovación en Salud **2**(2), 24–31 (2018)
19. Díaz, J., Rojas, M.: Desarrollo de un dispositivo de rastreo para personas adultas mayores en una red IoT y Cloud (2021)

Digital Competency Enhancement in Personnel Training and Development: A Literature Review of Current Trends and Challenges

Luis Fernando Taruchain-Pozo[1](✉) 🆔 and Fátima Avilés-Castillo[2] 🆔

[1] Carrera de Psicología, Facultad de Ciencias de la Salud y Bienestar Humano, Universidad Indoamérica, Ambato 180103, Ecuador
fernandotaruchain@uti.edu.ec
[2] Centro de Investigaciones de Ciencias Humanas y de la Educación-CICHE, Universidad Indoamérica, Ambato 180103, Ecuador

Abstract. In a perpetually advancing technological landscape, workforce training and development centered around digital competency and computer science principles are critical for ensuring the organizational resilience and competitiveness. This study undertakes a bibliographic analysis of technology-focused personnel training and development, aspiring to identify the strategies implemented, benefits perceived, and challenges emerging in this digital era. Three foundational research questions are outlined: What strategies are employed in technology-focused continuous training? What benefits do workers acquire after technology-centric training? And what are the prevalent challenges in digital skills training? A literature search approach was deployed to answer these queries, executing a meticulous analysis of 33 significant scientific papers. The results indicate a burgeoning trend towards virtual, technology-driven training, and the usage of practical simulations in training. It was also confirmed that continuous technology-focused training creates positive impacts on both work performance and worker attitudes. However, challenges like resistance to change and technological limitations persist in certain organizations.

Keywords: Digital competency · Workforce upskilling · Adaptability · challenges · tech-learning

1 Introduction

Within the business landscape, the human component is pivotal for the success and evolution of organizations, now more than ever, situated within an environment of rapid digitalization. Employee performance is an outcome not only of their professional experience and technical prowess, but also of their societal commitment and adherence to values and principles that foster robust and constructive working relationships [1]. These relationships are essential both internally to nurture a technology-focused, positive work environment, and externally to effectively represent the organization in the digital society [2]. Beyond mere training and skills development, there's a recognized importance

M. Z. Vizuete et al. (Eds.): CI3 2023, LNNS 1041, pp. 144–154, 2024.
https://doi.org/10.1007/978-3-031-63437-6_12

of holistic training that encompasses ethical, equity, and social responsibility aspects, alongside emerging technology and computer science competencies [3]. This approach cultivates highly competitive and productive workforces, capable of adapting to the constantly evolving social and technological challenges of the work environment [4].

In the current corporate scenario, the digitization of processes plays a significant role, as does the training of personnel from a comprehensive perspective [5]. This perspective transcends the mere acquisition of technical knowledge, focusing on meeting individual and collective needs, socio-emotional dimensions of work, and the creation of a rewarding work environment that values employee wellbeing and their mastery of digital technologies [6]. In a globalized and fiercely competitive era, human talent management is a key challenge for leaders and managers, and implementing effective training processes is vital to optimize the workforce and ensure the development of skills and competencies that drive organizational success, particularly in relation to technology [7].

Digital skills also enable the automation and optimization of processes across various sectors. Professionals who understand how to use data analysis tools, programming, and automation can enhance operational efficiency and data-driven decision-making. For instance, in the manufacturing industry, the implementation of digital manufacturing systems has led to more agile production and a reduction in errors within the supply chain [8]. This digital transformation has proven to be critical in maintaining competitiveness in a globalized market [9].

Identifying training needs is the first step in devising effective strategies. This is accomplished through diagnoses that identify gaps in knowledge and skills at individual, group, and organizational levels, utilizing quantitative and qualitative technical methodologies [10]. After diagnosis, a training plan is structured, aligned with a competency model considering three key dimensions: knowledge, skills, and values, with a particular emphasis on technological competency [11, 12].

The development of professional skills and the training of personnel are vital for the growth and competitiveness of organizations in Latin America. However, there exist disparities in the investment and effectiveness of training programs across the region. A study by the Inter-American Dialogue/Laureate International Universities reveals these differences [13]. Highlighting that Brazil leads in the number of trained employees at 31.6%, whereas countries like Uruguay, Ecuador, Chile, and El Salvador present lower figures. On the other hand, countries such as Guatemala, Colombia, and Paraguay exhibit less interest in empowering their workforce through continuous training and personal development programs.

In light of the labor challenges that have arisen following the Covid-19 pandemic, the International Labor Organization emphasizes the necessity to train employees to bolster their professional skills and acquire contemporary abilities rooted in technology management [14]. Digital skills have become paramount, not only in managing basic tools but also in utilizing specialized platforms [15, 16].

Along these lines, some organizations have adopted a form of supplementary compensation known as emotional compensation. This non-monetary evaluation aims to fortify the physical, psychological, and emotional wellbeing of employees, promoting healthy interpersonal relationships and loyalty to the organization [17, 18]. Emotional remuneration includes aspects such as autonomy at work, a sense of belonging, passion for work, creativity, and professional and personal development through training programs [19].

The paper is structured in four sections, where Sect. 1 presents the introduction to the topic. Section 2 shows the methodology and Sect. 3, the results. Finally, Sect. 4 describes the conclusions of the study.

2 Methodology

This study employed a bibliographic approach to gather and scrutinize theoretical-scientific information pertinent to the research topic, which pertains to technology and digitalization's role in personnel training and development. Rigorous, standardized, and transparent methods were utilized to offer a comprehensive and unbiased synthesis of existing literature. The detailed selection process is represented in Fig. 1.

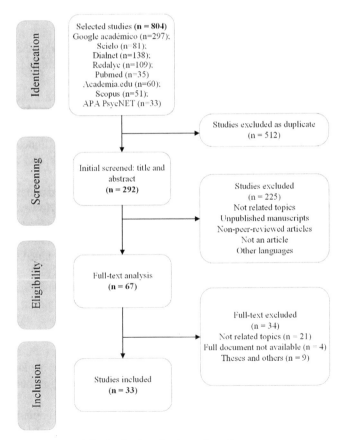

Fig. 1. Study selection prism diagram.

2.1 Review Objective

This literature review aimed to compile and analyze relevant theoretical-scientific information pertaining to technology-integrated strategies, benefits, and challenges in personnel training and development. The goal is to provide an unbiased and comprehensive synthesis of current scholarly discourse, facilitating a comparison of progress in the digitization of training practices.

2.2 Research Questions

The research sought answers to the following questions: 1) What technology-based strategies are employed for personnel training and development? 2) What benefits arise from integrating technology into staff training? 3) What contemporary challenges persist in the context of digitally-oriented staff training? These inquiries were formulated to understand better the growing relevance of digital tools and techniques in personnel training and their impact on the current work environment.

2.3 Keyword Identification

Relevant keywords related to the research topic were Keywords relevant to the research topic were identified, focusing on terms associated with both digital and technological aspects of personnel training and development. These terms align with those incorporated into the UNESCO Thesaurus portal, which includes a structured list of terms for analyzing and synthesizing a diverse range of publications.

2.4 Selection of Search Engines and Databases

Academic search engines and bibliographic databases containing quality and relevant information sources in the field of study were chosen. The search engines used include Google Scholar, APA PsycNET and Academia.edu, while the databases used were Redalyc, Scielo, Dialnet, PubMed and Scopus.

2.5 Search Terms

The following search terms were used in relation to each research question: "digital competency", "technological skills training", "workforce upskilling", "continuous tech-learning", "adaptability", "digital era", "training challenges", "digital transformation", "AI", "machine learning" and "workforce training".

2.6 Inclusion and Exclusion Criteria

Papers for review were selected based on publication date (limited to those published since 2015), thematic relevance to the integration of technology in training and development processes, and language (English and Spanish). Exclusion criteria were established to omit theses and works not directly related to the subject of study.

2.7 Compilation and Selection of Scientific Papers

The chosen search engines and databases were searched exhaustively using the identified keywords. An initial review of titles and abstracts of papers was conducted to establish relevance to the research objectives. This resulted in the selection of 33 scientific papers that met the inclusion criteria.

2.8 Analysis and Synthesis of the Literature

The selected papers underwent thorough analysis, identifying strategies, benefits, and challenges related to the process of personnel training and development in the digital era. A documentary-based synthesis of findings was performed, offering a comprehensive view of the subject.

2.9 Limitations and Ethical Considerations

Potential limitations of the study, such as reliance on bibliographic sources and the exclusion of specific document types, were addressed. The ethics applied in collecting and analyzing bibliographic data were also considered. This methodology offered a robust framework for the systematic review, permitting an analysis of the literature, synthesizing the advancements, and trends in the process of personnel training in the context of digitalization.

3   Results

Throughout the study, various research questions were addressed with the objective of analyzing the strategies and approaches used in training, as well as identifying the contemporary challenges associated with this process.

3.1 Organizational Strategies for Digital Skills Development

In relation to the first research question, the findings indicate a trend of organizations leaning towards continual innovation of strategies, targeted at fortifying digital competencies and augmenting the performance of their teams. This strategic orientation is a response to the escalating demand to deliver high-quality products and services in an increasingly competitive and dynamic marketplace. These strategies, discovered through bibliographic analysis, underscore the need for robust professional development in areas such as Big Data, artificial intelligence, virtual environments, and machine learning. Therefore, these strategies stress adaptability to changing market requirements and the utilization of innovative tools to assure the procurement of pertinent and current digital skills by staff.

The diversity of strategies uncovered in this study is outlined in Table 1. Prominent methods include employee-centric approaches that prioritize experiential learning, microlearning, and the integration of gamification. Additionally, team-focused strategies encompass social learning, adaptive learning, and effective mentoring. The significance of enhancing employee well-being, alongside the promotion of work-life balance, also surfaces prominently in the findings.

Table 1. Organizational strategies for digital skills development.

Strategies described	Approach and strategy	Relevant articles
Individual-Centered Training	Employee experience	[18, 20–23]
	Hyper customization	
	Remote learning	
	Learning by doing trends	
	Microlearning	
	Gamification (Lego Serious Play, virtual reality)	
	Employee-centric	
	Experimental learning	
Team-focused training	Social learning	[10, 16, 19, 24]
	Adapted learning (machine learning)	
	Management of mentoring and coaching	
	Storytelling	
	Blended learning	
	Lifelong learning	
	Learning management system and LCMS (learning contents management system) (artificial intelligence)	
Well-being and labor and personal equity	Wellbeing	[12, 25–27]
	Emotional management in the workplace	
	Balance of physical, psychological, and emotion-al health of the collaborator	
	Strengthening social skills and emotional intelligence	
	Motivational training	
	Meditation alternatives	

3.2 Benefits of Staff Digital Skills Training

The results demonstrate several positive aspects related to the usage of digital platforms and MOOCs in personnel training. Instead of merely noting individual improvements in workers' profiles after training, the benefits permeate the entire organization. One remarkable advantage is the instilling of a sense of commitment and labor loyalty among team members, stemming from the perception of the organization's dedication to their continual development. Furthermore, digital skills training promotes a culture of innovation and work enhancement in the company, encouraging creativity and initiative of team members to propose inventive ideas and refine their work practices.

A crucial benefit was observed in acquiring new job skills via training with MOOCs and digital platforms. Three main types of skills were identified: 'upskilling', referring to learning new knowledge within the same work field; 'reskilling', enabling workers to perform different functions than usual; and 'outskilling', equipping employees with skills in demand by the market for potential job redirection.

Moreover, digital training fosters the evolution of workers' self-concept and self-assessment, recognizing them as individuals with potential and promoting a more holistic and human perspective of their role within the organization. This challenges the classical conception of considering them as mere productive tools. Table 2 summarizes these benefits derived from the staff digital skills training process.

Table 2. Benefits of staff digital skills training.

Described benefits	Relevant articles
Commitment and labor loyalty	[28–30]
Skill Enrichment (Reskilling - Upskilling - Outskilling)	[10, 31, 32]
Organizational innovation management (MOOC)	[17, 18, 33, 34]
Competitive workforce	[15, 20, 35]
Embrace of E-Learning (Digital Platforms)	[11, 36]
Adaptability and Resilient Skill Development	[26, 37]
Strengthened Employee Self-Concept and Engagement	[12, 33, 38]

3.3 Contemporary Challenges in Staff Digital Skills Training

Notably, some work environments still perceive training, especially when technologically advanced, as an additional expense rather than valuing it as a critical investment for the development of their team's digital skills and competencies. Additionally, a certain resistance is detected among the staff to participate in digital training activities. This can be attributed to a variety of factors, including resistance to change, difficulty dedicating time outside of work hours, and generational differences in the perception of continuing education.

Another significant challenge resides in the limitations of technological infrastructure. In some organizations, these make it difficult to adapt to the demands of virtual work, and the saturation of connectivity in work environments can negatively impact job performance, particularly in terms of access and quality of online training.

Presented in Table 3 are challenges confronting organizations. These challenges encompass the necessity for resilient business investments, the resolution of technological infrastructure gaps, the integration of artificial intelligence, overcoming resistance to change, and addressing employee turnover. Identifying these challenges provides invaluable insights for shaping future strategies.

Table 3. Contemporary challenges in organizations.

Adverse factors described	Relevant articles
Business investment resilience	[39, 40]
Technological infrastructure (AI)	[41, 42]
Identification of massive open online training events	[43]
Late incorporation of artificial intelligence (machine learning and big data)	[36, 44]
Resistance to change and adaptability	[21, 45]
Employee turnover and separation	[6, 38]

The emergence of new work modalities, such as the implementation of artificial intelligence, poses additional challenges in terms of updating and adapting staff digital skills. These contemporary challenges represent significant hurdles for organizations in their efforts to provide relevant and effective digital skills training for their employees.

4 Conclusions

This study, grounded in a bibliographic review, provides valuable insights into how current human resource management practices in training are shifting towards the enhancement of digital competencies, including areas like big data, artificial intelligence, virtual environments, and digital platforms. In this era of accelerated information and digitalization, it is imperative that personnel training and development not only focus on improving technical skills and computer competencies but also on preparing workers to adapt and thrive in an increasingly technology-driven work environment.

Integrating digital competencies has become increasingly essential in personnel training and development to meet the demands of the digital era. Organizations must recognize the significance of equipping their workforce with the necessary digital, computing, and communication skills to thrive in a technology-dominated landscape. By embracing technological advancements in training programs, organizations can ensure their employees are well-prepared to navigate the complexities and opportunities presented by smart technologies.

Training has a significant impact at both the individual and organizational levels. By bolstering the technological development of their personnel, organizations cultivate an atmosphere of commitment and job loyalty. Simultaneously, training fosters a culture of innovation and constant improvement, incentivizing creativity and initiative from employees to propose innovative ideas and optimize their performance through the effective use of emerging technologies.

However, significant contemporary challenges exist that organizations must overcome. These include the perception of training as an unnecessary expenditure, resistance to change on the part of the workforce, and limitations in technological infrastructure that can impede the effective implementation of technology-based training programs. The introduction of new work modalities, such as artificial intelligence, also presents challenges in terms of updating and adapting skills. To surmount these obstacles, innovative approaches are required that ensure effective and relevant training for personnel in mastering these new technologies. In this way, organizations can maintain their competitiveness and respond adequately to the demands of an increasingly digital and interconnected market.

Considering limitations is vital. The small sample size and reliance on self-reported data could influence result generalization and objectivity. Yet, these findings provide a valuable foundation, guiding future research towards more comprehensive methodologies and broader approaches.

Considering these limitations, future investigations should expand upon this study. Examining the enduring impacts of digital competency training on job performance and satisfaction warrants exploration. Studying personalized training models, integrating real-world cases, and gauging skill transferability across diverse roles will yield a holistic understanding of sustained effects, shaping more targeted training approaches.

References

1. Aliyyah, N., et al.: What affects employee performance through work motivation? J. Manage. Info. Deci. Sci. **24**(1), 1–14 (2021)
2. Al-Awar, K.: Role of organizational change in developing team performance and support positive work environment. Int. J. Bus. Manage. **14**(6), 113–129 (2019). https://doi.org/10.5539/ijbm.v14n6p113
3. Indiyaningsih, K.M.H.: The effect of human resource competency, work culture, and utilization of information technology on employee performance. In: Proceedings of The International Seminar Series on Regional Dynamics, vol. 2, p. 13 (2020). https://doi.org/10.19184/issrd.v2i1.17468
4. Hidalgo-Parra, Y., Hernández-Hechavarría, Y., Leyva-Reyes, N.: Indicadores para evaluar el impacto de la capacitación en el trabajo. Ciencias Holguín **26**(1), 74–83 (2020)
5. Franklin, T.-O., Tubon-Nunez, E.E., Carrillo, S., Buele, J., Franklin, S.-L.: Quality management system based on the ISO 9001:2015: Study case of a coachwork company. In: Iberian Conference on Information Systems and Technologies. CISTI (2019). https://doi.org/10.23919/CISTI.2019.8760816
6. Trinidad-Hernández, M.A., Guzmán-Fernández, C., Martínez-Prats, G.: La capacitación y la rotación del personal como factor de impacto en la armonización contable. Revista de Investigaciones Universidad del Quindío **34**(2), 274–284 (2022). https://doi.org/10.33975/riuq.vol34n2.1040
7. Alvarez Morales, J.L., Ramirez Herrera, D.: Identificación de estrategias de capacitación en pymes de la Ciudad de México. Nóesis Revista de Ciencias Sociales. **31**(61), (2022). https://doi.org/10.20983/noesis.2022.1.10
8. Gerrikagoitia, J.K., Unamuno, G., Urkia, E., Serna, A.: Digital manufacturing platforms in the industry 4.0 from private and public perspectives. Applied Sciences **9**(14), 2934 (2019). https://doi.org/10.3390/APP9142934
9. Zaoui, F., Souissi, N.: Roadmap for digital transformation: a literature review. Procedia Comp. Sci. **175**, 621–628 (2020). https://doi.org/10.1016/J.PROCS.2020.07.090

10. Chapa Sosa, E., Chapa Méndez, M. del C.: Coaching gerencial: una revisión sistemática de artículos científicos disponibles en SCOPUS. Revista Internacional de Investigación en Ciencias Sociales **18**(1), 123–140 (2022). https://doi.org/10.18004/RIICS.2022.JUNIO.123
11. Frías, M.R.: Plan de capacitación para el desarrollo de competencias tecnológicas en docentes. UCE Ciencia-Revista de postgrado. **10**(3), 1–9 (2022)
12. Vásquez-Erazo, E.J., Tovar-Molina, E.A., Álvarez-Montalvo, A.-C., Tobar-Ordóñez, S.P.: Calidad de vida de los trabajadores mediante coaching y plan de carrera. Revista Cienciamatría. **8**(3), 1899–1923 (2022). https://doi.org/10.35381/cm.v8i3.922
13. Fiszbein, A., Cumsille, B., Cueva, S.: La capacitación laboral en América Latina. The Dialogue Leadership for the Americas (2016)
14. OIT: Panorama Laboral 2021. América Latina y el Caribe (2022)
15. Rowland, F.: Impacto de la capacitación en la productividad. Caso empresas chilenas del sector minería y sector alimentos. J. Manag. Bus. Stud. **3**(2), 1–20 (2021). https://doi.org/10.32457/jmabs.v3i2.1630
16. Molina-Molina, M.L., Romero-Fernández, A.J., Carrera-Narváez, P.: Modelo de coaching para el desarrollo empresarial. Revista Cienciamatría. **8**(8), 128–144 (2022). https://doi.org/10.35381/cm.v8i15.827
17. Mendoza Robles, D.L.: Capacitación laboral como derecho humano. Transdisciplinar-Revista de Ciencias Sociales del CEH **2**(4), 141–163 (2023). https://doi.org/10.29105/transdisciplinar.4-48
18. Didier, N.: Assessing job training effectiveness in Chile: a longitudinal approach. Latin American Policy **13**(2), 354–372 (2022). https://doi.org/10.1111/LAMP.12269
19. Zambrano-Pilay, E.C., Chiriboga-Mendoza, F.R., Riera-Estrada, J.: Mentoría y coaching: un enfoque empresarial: Artículo de revisión bibliográfica. Revista Científica Arbitrada Multidisciplinaria Pentaciencias. **2**(4), 14–19 (2020)
20. Batista Rodriguez, L.F., Leyva Figueredo, P.A., Mendoza Tauler, L.L.: Impact of Job Training: Its Challenges. Opuntia Brava. **12**(1), 372–384 (2020)
21. Pontes, J., et al.: Relationship between trends, job profiles, skills and training programs in the factory of the future. In: Proceedings of the IEEE International Conference on Industrial Technology (2021). ICIT 2021. IEEE Signal Processing Society, vol, 2021-March, pp. 1240–1245. https://doi.org/10.1109/ICIT46573.2021.9453584
22. Trujillo Zapata, D.A., Parra Riveros, H.: Transformación digital para PYMES de alojamiento y hospedaje de Boyacá usando metodologías ágiles. In: Encuentro Internacional de Educación en Ingeniería (2021), pp. 1–9. https://doi.org/10.26507/ponencia.1889
23. Flores-Cerna, F., Sanhueza-Salazar, V.-M., Valdés-González, H.-M., Reyes-Bozo, L.: Metodologías ágiles: un análisis de los desafíos organizacionales para su implementación. Revista Científica **43**(1), 38–49 (2021). https://doi.org/10.14483/23448350.18332
24. Restrepo Quinteto, K., Cuadra Palma, L.N.: Ecosistema Mujer: empoderamiento empresarial femenino a través del mentoring y las redes de trabajo en Chile. Revista Científica General José María Córdova **20**(39), (2022)
25. Clutterbuck, D.: Mentoring: Técnicas para motivar, desarrollar las relaciones, potenciar el talento y mejorar la productividad (2015)
26. Kim, J., Park, C.Y.: Education, skill training, and lifelong learning in the era of technological revolution: a review. Asian-Pacific Eco. Literat. **34**(2), 3–19 (2020). https://doi.org/10.1111/apel.12299
27. Syarifuddin Hasibuan, J., Lesmana, M.T., Permata Sari, A.: Employee Performance Studies: Antecedents of Work Discipline, Work Motivation, and Job Training. Int. J. Edu. Rev. Law and Soc. Sci. (IJERLAS) **1**(2), 117–128 (2021). https://doi.org/10.54443/ijerlas.v1i2.44
28. Obando Changuán, M.P.: Capacitación del talento humano y productividad: Una revisión literaria. ECA Sinergia **11**(2), 166–173 (2020). https://doi.org/10.33936/ECA_SINERGIA.V11I2.2254

29. Paredes Floril, P.R., Bustamante Villegas, J.J.: Gestión de talento humano y satisfacción laboral en bancos de Guayaquil. Journal of the Academy **5**, 44–61 (2021). https://doi.org/10.47058/joa5.4

30. Mardikaningsih, R., Putra, A.R.: Efforts to increase employee work productivity through job satisfaction and job training. Jurnal Studi Ilmu Sosial Indonesia. **1**(1), 51–64 (2021)

31. López, M.L.: Medición del Valor del Diseño a través del ROI en diseño en las pymes. ARXIU. Revista de l'Arxiu Valencià del Disseny **1**, 53–75 (2022). https://doi.org/10.7203/ARXIU.1.25336

32. Li, L.: Reskilling and Upskilling the Future-ready Workforce for Industry 4.0 and Beyond. Information Systems Frontiers (2022). https://doi.org/10.1007/s10796-022-10308-y

33. Gautam, D.K., Basnet, D.: Organizational culture for training transfer: the mediating role of motivation. Int. J. Organ. Anal. **29**(3), 769–787 (2020). https://doi.org/10.1108/IJOA-04-2020-2147

34. Alvarez Morales, J.L., Ramírez Herrera, D.: Identificación de estrategias de capacitación en pymes de la Ciudad de México. Nóesis-Revista de ciencias sociales. **31**(61), 202–225 (2022). https://doi.org/10.20983/NOESIS.2022.1.10

35. David, W., et al.: Determinación de necesidades de capacitación del personal administrativo del área de ciencias económicas y de la gestión de la UNMSM. Gestión en el Tercer Milenio **23**(45), 69–81 (2020). https://doi.org/10.15381/gtm.v23i45.18940

36. Muniasamy, A., Alasiry, A.: Deep learning: The impact on future eLearning. Int. J. Emerg. Technol. Learn. **15**(1), 188–199 (2020). https://doi.org/10.3991/IJET.V15I01.11435

37. De Los, C., Leyva, A.D., Elena, C., Fornaris, M.: La evaluación del impacto de la capacitación: retos y beneficios para las organizaciones actuales. Universidad y Sociedad **13**(6), 28–38 (2021)

38. Masilova, M.G.: Internal outsourcing as personnel process optimization in a large company. REICE: Revista Electrónica de Investigación en Ciencias Económicas **8**(16), 543–555 (2020). https://doi.org/10.5377/reice.v8i16.10722

39. Sánchez, C.M.M., Cedeño, A.B.R., Zambrano, N.M.A.M.: Diagnóstico de la gestión de talento humano como estrategia para el desarrollo organizacional. Polo del Conocimiento **6**(10), 693–704 (2021). https://doi.org/10.23857/pc.v6i10.3233

40. Alicia Hernández-Moreno, L., De Lourdes Hernández-Saldaña, M., María, Y., Tovar-Morales, T.: Capacitación laboral en herramientas digitales. Vinculatégica EFAN **7**(1), 130–143 (2021). https://doi.org/10.29105/VTGA7.2-74

41. Canossa Montes de Oca, H.A., Canossa Montes de Oca, H.A.: Retos para la inclusión laboral de personas con discapacidad en Costa Rica. Economía y Sociedad **25**(58), 50–68 (2020). https://doi.org/10.15359/EYS.25/58.4

42. Rodríguez García, O.: Home office en la nueva normalidad: retos y futuro del home office. Revista Latinoamericana de Investigación Social **3**(3), 94–99 (2020)

43. Bernal, S.P.Q., Martínez, J.G.: El diseño de ambientes Blended-Learning, retos y oportunidades. Educación y Educadores **23**(4), 659–682 (2020). https://doi.org/10.5294/edu.2020.23.4.6

44. Lyons, E.: The impact of job training on temporary worker performance: Field experimental evidence from insurance sales agents. J. Eco. Manage. Strat. **29**(1), 122–146 (2019). https://doi.org/10.1111/jems.12333

45. Aragón, C.L.M., González, A.A., Mendívil, B.C.: Las buenas prácticas de la gestión del talento humano para fortalecer el desempeño en las organizaciones. Revista de Investigación en Ciencias Contables y Administrativas. **5**(2), 45–54 (2020)

Modeling Variability in the Readings of an 8-channel Color Sensor and Its Uncertainty Estimation

Francisco Espín[1,2(✉)] , Eduardo Manzano[2] , Carlos Velásquez[1,3,4] ,
Consuelo Chasi[1] , and Paola Andrade[2]

[1] Instituto de Investigación Geológico y Energético IIGE, Quito, Ecuador
`francisco.espin@geoenergia.gob.ec`
[2] Departamento de Luminotecnia, Luz y Visión, Facultad de Ciencias Exactas y Tecnología, Universidad Nacional de Tucumán. Av, Independencia 1800 – (T4002BLR), San Miguel de Tucumán, Tucumán, Argentina
[3] Universidad Central del Ecuador, Modalidad en Línea, Quito, Ecuador
[4] Department of Applied Mathematics, University of Alicante, Alicante, Spain

Abstract. This paper presents the implementation of an illuminance meter based on the 8-channel AS7341 sensor in conjunction with the ESP32 board. The sensor mounted on a SoC board can be implemented as an illuminance meter under the conditions specified. The results obtained are compared with the Gossen illuminance meter following the distance photometry, fitting curves for 11 illuminance levels and 4 positions to calculate the coefficient of determination for all distances and supply voltages. These results determine that the cubic polynomial interpolation technique is suitable to form the electromagnetic spectrum of the radiation source. After the implementation of the experimental scheme, it was found that the correction factor is almost constant for the eleven illuminances obtained by the prototype. The uncertainty of the equipment was calculated with the values of each intensity and distance, obtaining a constant. Therefore, the experimental scheme tested has an uncertainty of 12% regardless of the distance and illumination level parameters. The repeatability of the equipment is the major source of uncertainty. A better fixation of the distances and elements of prototype can reduce this value. After the development of the experimental work, it was determined that the AS7341 sensor mounted on a SoC board is a reliable illuminance meter.

Keywords: Illuminance · LED · Uncertainty · TCC

1 Introduction

The consumption of electric power in lighting is around 2900TWh [1]. Some rules and regulations determine the minimum amount required to illuminate a space. This amount, which can be made by illuminance meters, commonly called lux meters. One of the advantages of LED technology is the ability to maintain the initial luminous flux for a longer time, so it could be used in countless ways, including as a source of radiation to know transmission or reflection properties of materials. Another application can be focused on obtaining the illuminance from the electromagnetic spectrum of the source.

© The Author(s), under exclusive license to Springer Nature Switzerland AG 2024
M. Z. Vizuete et al. (Eds.): CI3 2023, LNNS 1041, pp. 155–166, 2024.
https://doi.org/10.1007/978-3-031-63437-6_13

In general, the equation relating a given radiometric quantity Xe,λ (λ) to its corresponding photometric quantity X_v is given as [2]:

$$X_V = k_m \int_\lambda X_{e,\lambda}(\lambda)V(\lambda)d\lambda \qquad (1)$$

where λ is the wavelength, k_m is a scale factor equal to 683.002 lm/W. As can be seen, any photometric magnitude has a direct relation with the $V(\lambda)$ curve. The equipment should be as close as possible to the $V(\lambda)$, this is one of the parameters that allows us to evaluate the quality of photometric equipment. CIE S 023 calls it the f1 factor, which describes the deviation of the relative spectral responsivity of a photometer $S_{rel}(\lambda)$ concerning the photopic luminous efficiency function $V(\lambda)$. Studies [3, 4] estimate a range between 0.6% to 9% for f1 values, while in [5] the range is between 2.6% to 36.4%.

There are two ways to know the illuminance, the first is employing an illuminance meter which must have its photometric head corrected by the $V(\lambda)$ curve, the closer this correction is to the theoretical curve, the better response the sensor will have, the electrical signal obtained is scaled by a factor and its corresponding illuminance is emitted by the equipment.

The second way to obtain the illuminance is through the knowledge of the emission or reflection spectrum and weighting it by the theoretical $V(\lambda)$ curve. The advantage of this procedure is that the f1 factor is eliminated since the theoretical $V(\lambda)$ curve is used; however, the field equipment is used to obtain the spectrum and, due to its fragility, it is expensive [6].

On the other hand, the development of systems on chip (SoC) is becoming more and more frequent in research or development projects [7–9] since they allow the incorporation of sensors, facilitating measurements of different magnitudes [10]. The use of this technology reduces costs and allows obtaining valid results knowing the limitations of this technology. Using these sensors in the photometric and radiometric field presents challenges in the reconstruction of the electromagnetic spectrum of a light source, "converting" it into a lux meter and estimating its uncertainty in the illuminance measurement [11, 12].

The horizontal illuminance at a point is given by Eq. 1. The present work seeks to implement and characterize an illuminance meter and estimate its measurement uncertainty from measurements of the eight channels of the sensor (Fig. 1), perform its interpolation, and recomposition, and subsequently obtain the electromagnetic emission spectrum of an LED source. The radiant flux of the spectrum will be obtained in sensor counts in the 360 nm to 830 nm range of the source. To get the scaling factor, a Gossen class B illuminance meter previously verified employing a luminous intensity standard traceable to the International System of Units (SI) will be used as a reference.

The two devices will be mounted on an optical table and placed at the same distance. The power supply voltage of the light source will be varied to obtain different light intensities. This process will be repeated at several distances, repeating the power supply levels defined at the first distance. The measurement of the distances will be taken with a distancemeter.

The Commission Internationale de L'Eclairage (CIE)/CIE International Standard establishes the spectral sensitivity curve in the range from 360 nm to 830 nm in 1nm

Fig. 1. AS7341 sensor mounted on ESP32 board

steps, so the electromagnetic spectrum of the source is reconstructed in 1 nm steps. The piecewise interpolation technique using third-degree polynomials is employed. In previous studies, it was shown that the Piecewise Cubic Hermite Interpolating Polynomial (PCHIP) technique is the best fit for the reconstruction of an electromagnetic spectrum at correlated color temperatures in warm LED sources. For the present case, a 2300 K TCC LED is used as the illumination source [13].

Illuminance can be obtained by Eq. 2:

$$E = \frac{I}{d^2} \cos \theta \tag{2}$$

where I is the luminous intensity incident on the sensors, d is the sensor source distance and θ is the angle between the incident intensity and the sensor plane. The sensor and source are mounted and aligned on the optical table, so the factor $\cos \theta = 1$. Additionally, it is known that the luminous intensity is independent of the distance so the illuminance at any point can be calculated as:

$$I = E_{d_1}.d_1^2 = E_{d_2}.d_2^2 = \ldots = E_{d_n}.d_n^2 \tag{3}$$

For n positions. The power supply voltage of the radiation source varies from 50 V to 100 V obtaining 11 luminous intensities in total, being 5 V the voltage step. The readings

Fig. 2. AS7341 sensor mounted on ESP32 board with lamp

will be taken at four distances repeating the voltage steps previously exposed, for the two pieces of equipments. The sensors of the two devices will be placed at the same distance from the light source for comparison.

1.1 Iluminance Meter Gossen

From Eq. 2 it is possible to plot the variation of the illuminance as the luminous intensity and the distance from the sensor change.

The Fig. 3. Shows the variation of illuminance with respect to the square of the distance for eleven values of intensities, each graph showing the values taken by the illuminance meter at four distances.

1.2 AS7341 Sensor Mounted on ESP32 Board

The illuminance from the spectral irradiance $E_{(e,\lambda)}(\lambda)$ can be obtained from Eq. 1.

$$E_V = k_m \int_\lambda E_{e,\lambda}(\lambda)V(\lambda)d\lambda \tag{4}$$

Spectral irradiance may also be expressed as:

$$E_V = k_m \int_\lambda \frac{1}{A}\emptyset_{e,\lambda}(\lambda)V(\lambda)d\lambda \tag{5}$$

Since the area of the sensor A remains constant, the equation can be written

$$E_V = k_1 \int_\lambda \emptyset_{e,\lambda}(\lambda)V(\lambda)d\lambda \tag{6}$$

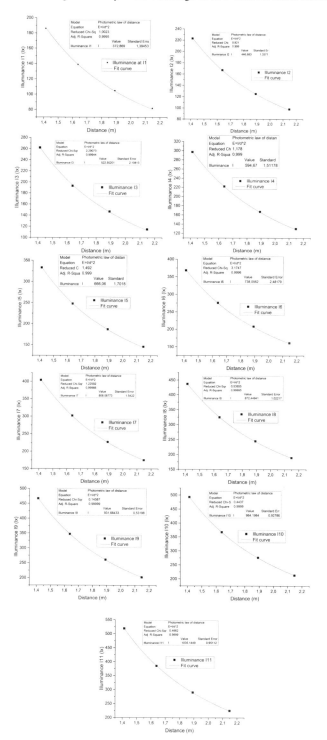

Fig. 3. Photometric law distance Gossen

where k_1 is a function of the sensor area, the calibration factor, and k_m the initial constant. Based on the results of the eight sensor channels plus two additional points of zero value, we proceed to use piecewise interpolation using third-degree polynomials.

Cubic piecewise polynomial interpolation consists of fitting the function to the form:

$$S(x) = \begin{cases} s_{1(x)} & x_1 \leq x < x_2 \\ s_{2(x)} & x_2 \leq x < x_3 \\ \quad \cdot \\ \quad \cdot \\ \quad \cdot \\ s_{n-1(x)} & x_{n-1} \leq x < x_n \end{cases} \tag{7}$$

where s_i is a third-degree polynomial defined by:

$$s_i(x) = a_i(x - x_i)^3 + b_i(x - x_i)^2 + c_i(x - x_i) + d_i \tag{8}$$

For i = 1, 2,...n − 1

$$s_i'(x) = 3a_i(x - x_i)^2 + 2b_i(x - x_i) + c_i \tag{9}$$

$$s_i''(x) = 6a_i(x - x_i) + 2b_i \tag{10}$$

The function must comply with the continuity condition in the whole interval so that in each sub-function, it is stable in the points of the readings of each channel of the sensor:

$$s_i(x_i) = s_{i-1}(x_i) \tag{11}$$

$$s_i'(x_i) = s_{i-1}'(x_i) \tag{12}$$

$$s_i''(x_i) = s_{i+1}''(x_i) \tag{13}$$

The technique used in addition to interpolate the values of $S(x)$, also interpolates the values of $S'(x)$, consequently for each xi, the slope of the polynomial of degree 3 must coincide with the slope of the function S at x_i.

Based on the above and boundary conditions, the developed program calculates the polynomial that best fits the measurements of the eight channels and interpolates to obtain the 471 values corresponding to the spectrum of the light source ranging from 360nm to 830nm with steps of 1nm. Further details on the interpolation can be found in [14, 15].

Specifically, Hermite cubic interpolation (pchip) is employed in this paper, as it is the best piecewise cubic interpolation technique that fits a warm CCL spectrum [12, 15–17].

The Fig. 4 shows the eight points measured with the AS7341 sensor, since the relative spectral sensitivity is practically zero at 360nm and 830nm, these two points are increased for the interpolation. The Fig. 4 shows in red the curve of the measurements and in black the spectrum resulting from the interpolation by parts with cubic polynomials.

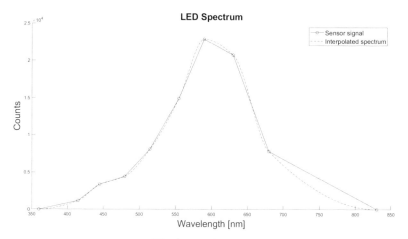

Fig. 4. Led Spectrum

The developed program takes about 200 points for each channel so that about 200 source spectra can be reconstructed. The illuminance is obtained by Eq. 2 based on the spectrum counts. The result of the illuminance variation as the luminous intensity and the distance of the eight-channel sensor changes are found in Fig. 2.

1.3 Uncertainty Measurement Estimation

The estimation of the measurement uncertainty of the equipment allows us to know the limitations of its measurements. The following is the development to estimate the uncertainty of the illuminance measurement based on the 8 channels of the sensor:

$$E_{sensor} = \frac{1}{k}(E_G + c) \tag{14}$$

$$E_{sensor} = \frac{1}{k}\left(\frac{I_{adjG}}{d^2} + c\right) \tag{15}$$

$$u^2_{(E_{sensor})} = \left[\frac{-1}{k^2}\left(\frac{I_{adjG}}{d^2} + c\right)\right]^2 u^2_{(k)} + \left[\frac{1}{k}\left(\frac{1}{r^2}\right)\right]^2 u^2_{(I_{adjG})} + \left[\frac{1}{k}\left(\frac{-2I_{adjG}}{d^3}\right)\right]^2 u^2_{(d)} + \left[\frac{1}{k}\right]^2 u^2_{(c)} \tag{16}$$

In addition to the uncertainty of the calibration model $u^2_{(E_{sensor})}$, the contributions of repeatability $u^2_{(Rep)}$ and resolution to one decimal place $u^2_{(Res)}$, are also considered, from which the total combined uncertainty is obtained.

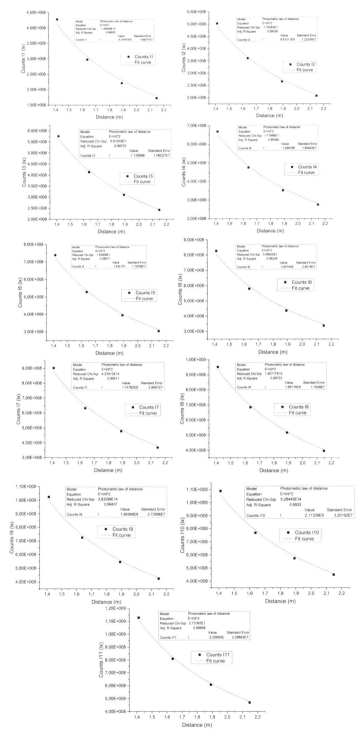

Fig. 5. Photometric law distance with AS7341 sensor

The Fig. 6. Shows the uncertainty estimate for each light intensity level and distance used in the experimental setup. It is observed that the combined uncertainty estimate is around 6%, so its expanded uncertainty under a normal distribution with a factor $k = 2$ will be 12%.

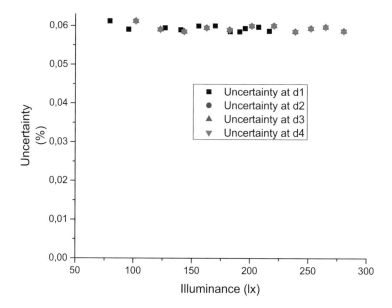

Fig. 6. Uncertainty at various distances vs Illuminance

2   Results Discussion

The data obtained with the Gossen illuminance meter follow the distance photometric, the fitting curves for 11 illuminance levels and 4 positions have a determination coefficient of 0.999 at all distances and supply voltages, which allows us to know that the experimental scheme is suitable to evaluate the illuminance readings in the different configurations.

The illuminance emitted by the eight-channel sensor can be calculated based on Eq. 2, the results show that the sensor and its board conform to the photometric law of distance, and its coefficient of determination at the most critical value only at I1 is 0.988. For the remaining luminous intensities, the coefficient of determination is higher than 0.99, which shows that the sensor mounted on the ESP32 board can be used as an illuminance meter. The results also show that the cubic polynomial interpolation technique can be used to form the electromagnetic spectrum of the source.

The correction factors of the implemented equipment are listed in Table 1 for each luminance. In the second row are the illuminance values in beads obtained from the adjustment of the sensor and shown in Fig. 5. The correction factor obtained is displayed in Fig. 3. A correction factor exists for each of the eleven illuminances. It is observed that the correction factor is practically constant, i.e. the factor applied to the equipment to transform its readings to illuminance is independent of the distance and illumination levels of the experimental scheme. The median of the eleven correction factors is 2162956.18.

Table 1. Illuminance with two equipments and standard error of sensor.

Iluminance Gossen	K*cuentas	Factor de corrección (k)	Iluminance of sensor	Standard error
372,86	8,23E + 08	2,21E + 06	380,73	1,68E + 07
446,68	9,81E + 08	2,20E + 06	453,69	1,22E + 07
522,5	1,13E + 09	2,16E + 06	521,97	1,06E + 07
594,67	1,30E + 09	2,19E + 06	600,92	1,86E + 07
666,06	1,44E + 09	2,17E + 06	666,76	1,72E + 07
738,05	1,60E + 09	2,17E + 06	740,58	2,60E + 07
808,07	1,75E + 09	2,16E + 06	808,07	2,87E + 07
872,45	1,87E + 09	2,14E + 06	863,25	1,77E + 07
931,68	1,99E + 09	2,14E + 06	921,74	2,73E + 07
984,2	2,11E + 09	2,15E + 06	977,08	3,20E + 07
1035,14	2,21E + 09	2,13E + 06	1021,31	2,30E + 07

The correction factor (k) applied to the sensor bead readings gives the illuminance at each point. The Fig. 7 Shows the residuals between the absolute illuminance values of the Gossen equipment and those obtained with the 8-channel sensor.

It is observed in Fig. 7. That the residuals increase as the intensities increase. The range of both positive and negative residual variations are similar, which could mean a distribution within these values, with the highest residuals at the lowest and highest intensities.

The uncertainty estimate of the equipment was calculated at each intensity and distance, its result is practically constant. This shows that, within the range used in the experimental scheme, an expanded uncertainty of 12% can be assigned to the measurements of the implemented equipment. This estimate is independent of distance and illumination level.

The major source of uncertainty is in the repeatability of the equipment, so optimizing the fixation of the equipment elements and improving the experimental measurement scheme will decrease its uncertainty. Likewise, since the results show that the response of the implemented equipment is similar to that of professional equipment, contrasting it with equipment of better characteristics, such as a class L illuminance meter, will improve the bias of the equipment [18].

Some factors were not considered in the implementation of the equipment, so further analysis regarding measurements of intensities not normal to the sensor plane should be studied and evaluated for uncertainty estimation.

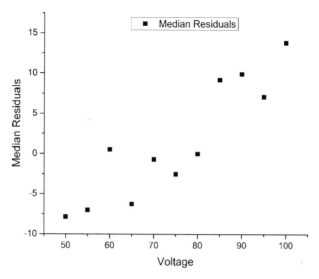

Fig. 7. Median Residuals

The study demonstrates that the AS7341 sensor mounted on a SoC board can be implemented as an illuminance meter under the conditions specified in the article. This system has the advantage that the normalized curve $V(\lambda)$ can be loaded by software, which allows irradiance measurements and from this obtains illuminance.

3   Conclusion

The spectrum obtained from the reading of an eight-channel sensor can be processed using the theoretical $V(\lambda)$ curve to obtain illuminance. The use of the theoretical curve eliminates the error due to spectral mismatch. The equipment implemented based on an 8-channel sensor is suitable for conducting measurements within the specified range.

The experimental results show that the sensor mounted on the ESP32 board can function as an illuminance meter, with an expanded uncertainty of 12%. The primary contributor to the uncertainty is repeatability; however, this could be reduced by improving element fixation and enhancing measurements on an optical bench. Another way to improve uncertainty is through calibration with a class L sensor or directly with a standard intensity lamp.

Understanding its measurement uncertainty allows for defining the applicability of the equipment.

References

1. PNUMA Aceleración de la adopción mundial de la iluminación energéticamente eficiente. Programa las Nac Unidas para el Medio Ambient (PNUMA)-Fondo para el Medio Ambient Mund Unidos por la Efic (U4E, United Effic 96 (2016)

2. Zwinkels, J., Sperling, A., Goodman, T., et al.: Mise en pratique for the definition of the candela and associated derived units for photometric and radiometric quantities in the International System of Units (SI). Metrologia 53:G1. https://doi.org/10.1088/0026-1394/53/3/G1 (2016)

3. Krüger, U., Blattner, P.: Spectral mismatch correction factor estimation for white LED spectra based on the photometer's f1′ value. In: Proc. of the CIE Centenary Conference "Towards a New Century of Light. pp. 300–307 (2013)

4. Ying, S-P., Chou, PT., Fu, H-K.: Influence of the temperature dependent spectral power distribution of light-emitting Diodes on the illuminance responsivity of a photometer. Opt Lasers Eng, **51**, 1179–1184. https://doi.org/10.1016/j.optlaseng.2013.04.002 (2013)

5. Rosas, E., Estrada-Hernández, A.: Effect of photometric detector spectral response quality on white LED spectral mismatch correction factors. Appl. Opt. **55**, 5267–5272 (2016)

6. Araguillin, R., Toapanta, A., Juiña, D., Silva, B.: Design and Characterization of a Wireless Illuminance Meter with IoT-Based Systems for Smart Lighting Applications. Latest Advances in Electrical Engineering, and Electronics. Springer International Publishing, Cham, pp. 129–140. (2022)

7. De Man, H.: System-on-chip design: Impact on education and research. IEEE Des. Test Comput. **16**, 11–19 (1999)

8. Ahammad, I., Khan, M., Rahman, M., et al.: Giga-scale integration system-on-a-chip design: challenges and noteworthy solutions. Int J Recent Technol Eng **8**, 741–746 (2020)

9. Lau, KM., Choi, HW., Lee, S-WR., et al: Cost-effective and eco-friendly LED system-on-a-chip (SoC). In: 2013 10th China International Forum on Solid State Lighting (ChinaSSL). IEEE, pp. 235–238. (2013)

10. Arm, J., Baštán, O., Mihálik, O., Bradáč, Z.: Measuring the Performance of FreeRTOS on ESP32 Multi-Core. IFAC-PapersOnLine **55**, 292–297. https://doi.org/10.1016/j.ifacol.2022.06.048. (2022)

11. Velásquez, C, Castro, M.A., Rodríguez, F., et al: Optimization of the Calibration Interval of a Luminous Flux Measurement System in HID and SSL Lamps Using a Gray Model Approximation. In: 2021 IEEE Fifth Ecuador Technical Chapters Meeting (ETCM). pp. 1–7. (2021)

12. Brusil, C., Espín, F., Velasquez, C.: Effect of Temperature in Electrical Magnitudes of LED and HPS Luminaires. (2021)

13. Illuminating Engineering Society ANSI/IES LM-79–19 Approved Method: Optical and Electrical Measurements of Solid State Lighting Products. New York. (2019)

14. McKinley, S., Levine, M.: Cubic spline interpolation. Coll Redwoods **45**, 1049–1060 (1998)

15. Boor, C.: A Practical Guide to Splines. Springer, New York, NY (2001)

16. Brusil, C., Arcos, H., Espin, F., Velasquez, C.: Analysis of harmonic distortion of led luminaries connect to utility grid. (2020)

17. Sun, M., Lan, L., Zhu, C-G., Lei, F.: Cubic spline interpolation with optimal end conditions. J Comput Appl Math 425:115039. https://doi.org/10.1016/j.cam.2022.115039. (2023)

18. Velasquez, C., Castro, M., Rodriguez, F., Espin, Falconi, N.: Optimization of the Calibration Interval of a Luminous Flux Measurement System in HID and SSL Lamps Using a Gray Model Approximation. (2021)

Home Automation System for People with Visual Impairment Controlled by Neural Frequencies

Cristian Pozo-Carrillo, Jaime Michilena-Calderón, Fabián Cuzme-Rodríguez[✉],
Carlos Vásquez-Ayala[ID], Henry Farinango-Endara[ID], and Stefany Flores-Armas

Universidad Técnica del Norte, Av. 17 de Julio, Ibarra 100105, Ecuador
jrmichilena@utn.edu.ec

Abstract. The autonomy of individuals with disabilities is increasingly common through the use of information and communication technologies, granting them some independence to carry out various activities in everyday environments such as homes and workplaces. This project focuses on the development and implementation of a home automation system for individuals with visual disabilities utilizing neural frequencies. Through this system, multiple actions within the home will be facilitated and communicated through a mobile application. The objective of this project was to enhance mobility, autonomy, and user safety within the home by interpreting concentration data, meditation levels, and electromyography signals. Specific desired actions such as controlling the lighting (on/off) and locking/unlocking the entrance door were targeted. As a result, a proper communication was achieved throughout the entire home automation system, demonstrating an efficiency of 98.75% across all the conducted tests. Overall, this work contributes to enabling individuals with visual disabilities to navigate their homes with greater ease, independence, and enhanced safety by leveraging the interpretation of neural frequencies and other relevant signals.

Keywords: Visual impairment · Neural frequencies · Mobile application ·
Meditation · Concentration · Electromyography · Home automation

1 Introduction

According to the World Health Organization, there are an estimated of 285 million visually impaired people worldwide. From a social point of view such people live strong limitations and their autonomous mobility can be extremely difficult, and in most cases may require the use of assistive devices, the most common and used device is the white cane [1, 2]. According to CONADIS, there were 471,205 individuals registered with disabilities as of January 2022. The provinces with the highest number of registered individuals with disabilities include Guayas, Pichincha, Manabí, and Azuay. Among these, 54,397 individuals have visual impairments, accounting for 11.54% of the total [3].

At home the main challenges faced by a person with visual impairment are: device management, turning on or off lights, and not being able to perform these actions easily generate forgetfulness, carelessness and insecurity inside [4].

M. Z. Vizuete et al. (Eds.): CI3 2023, LNNS 1041, pp. 167–177, 2024.
https://doi.org/10.1007/978-3-031-63437-6_14

In recent years, different systems have been developed with which the mobility of users is facilitated by providing as much information as possible, in the same way smart environments have been developed which promise to improve the lives of visually impaired people by collecting data and using them to make decisions [4–6]. For example, in China ROSmotic has been developed a system to build smart homes operated by a smartphone application controlled by touch or voice commands through the use of microcontrollers [7].

At the local level, prototype developments have also been carried out at the academy level such as [1] which establishes the design of a home automation system using mobile devices to improve the accessibility and comfort of disabled people. Another home automation system that includes hardware and free software platforms for the residence of a paraplegic person obtaining good results of independence and comfort developed by [9]. That is why the developments are not finished, we can cite works such as [10–13], which contribute significantly to the autonomous, independent, safe and comfortable mobility of people with some type of disability, especially visual.

In this context, the objective of this system is to analyze the user's neural frequencies and electromyography signals in order to control the lighting in the home, enabling the user to turn the lights on or off. Additionally, it aims to provide convenience by facilitating the locking or unlocking of the entrance door. Subsequently, the system beneficiary will engage in exercises involving concentration, relaxation, and facial gestures to effectively control the system.

The rest of the document is structured as follows: Sect. 2 shows the methodology used and the proposed electronic system for data acquisition. Section 3 shows the data obtained and the tests carried out. Section 4 the results obtained and discussion. Finally, Sect. 5 presents the conclusions of the work carried out.

2   Materials and Methods

In this section, the applied methodology is presented, based on the waterfall model in its four phases (see Fig. 1) [14]. This approach facilitates the establishment of concentration exercises, meditation practices, and facial gestures. Furthermore, the electronic system's design and its operational procedure are outlined. Lastly, the central node's implementation and the corresponding connections are demonstrated.

2.1   Methodology of Stimulation of Concentration, Meditation and Facial Gesticulations

The test aims to study the ease of a visually impaired person in concentrating, relaxing and performing facial gestures for a time of 20 s.

As a first phase, it is considered to measure the concentration of the visually impaired person, through the use of rectangular elements which will help eliminate the distractors found in the environment. The test consists about visually impaired person holding their cell phone for 20 s, during that time the concentration levels are measured, then they proceed to rest 10 s and retake the data for a time of 20 s, so you can verify the ability of concentration that the user has.

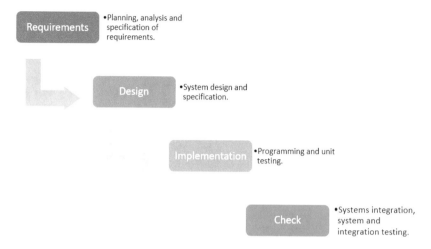

Fig. 1. Applied Waterfall Model.

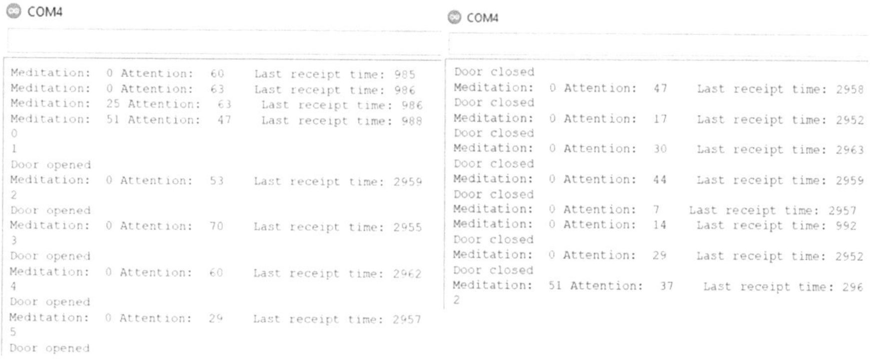

Fig. 2. Door opening and closing verification process.

Finally, the facial gesticulations tests are performed, which user is asked to raise their eyebrows for 1 s and lower them. This exercise is performed for a time of 20 s and rests 10 s. Subsequently, the concentration exercises and facial gestures are combined with the visually impaired person, to check the actions that are performed. In Fig. 2, the data obtained at the Arduino terminal during the verification process of opening and closing the main access door can be observed. Similarly, in Fig. 3, the verification process for turning on and off the kitchen light is depicted. This process is repeated for the bedroom and dining room lights, with the same validation procedure.

2.2 Electronic System Design

The biosignals needed to determine the concentration and electromyography signals are: (i) brain signals (Alpha, beta y gamma), breathing and facial gesticulations, the same ones that Mindwave Neurosky has them. The Fig. 4 shows the operation of the system.

⊚ COM4

```
4
Kitchen, light on
Meditation:  0 Attention:  63    Last receipt time: 989
Kitchen, light on
Meditation:  0 Attention:  53    Last receipt time: 985
Kitchen, light on
Meditation:  0 Attention:  69    Last receipt time: 994
Kitchen, light on
Meditation:  0 Attention:  87    Last receipt time: 981
Kitchen, light on
Meditation:  0 Attention:  84    Last receipt time: 983
Kitchen, light on
Meditation:  0 Attention:  77    Last receipt time: 990
Kitchen, light on
```

⊚ COM4

```
5
Kitchen, light off
Meditation:  0 Attention:  27    Last receipt time: 979
Kitchen, light off
Meditation:  0 Attention:  34    Last receipt time: 989
Kitchen, light off
Meditation:  0 Attention:  44    Last receipt time: 985
Kitchen, light off
Meditation:  0 Attention:  44    Last receipt time: 983
Kitchen, light off
Meditation:  0 Attention:  54    Last receipt time: 990
Kitchen, light off
Meditation:  0 Attention:  60    Last receipt time: 980
Kitchen, light off
Meditation:  0 Attention:  80    Last receipt time: 987
```

Fig. 3. Kitchen light On-Off verification process.

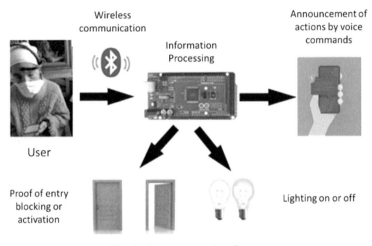

Fig. 4. System operation diagram

The system works as follows: (i) the sensor is placed to the user. (ii) the mobile application is initialized, and you wait 10 s to receive the connected system message. (iii) facial gesticulation is performed to access the control of the entrance door. (iv) the user must exceed 10% concentration to unlock the door. (v) facial gesture is performed to access the door lock and exceed 10% concentration to block it. (vi) the iv, v and vi procedure must be performed for the control of the luminaires. In Fig. 5, the visually impaired person can be observed before performing the function tests.

2.3   Deploying the Central Node

The place chosen for the central node implementation was made in the lower left part of the home as shown in Fig. 6, the place has been chosen because it does not exceed the 10 m allowed between the neural sensor and the central node, in this way the correct communication between devices is ensured.

Figure 6 show the luminaires with which the user will interact. These are the ones in the bedroom, dining room and kitchen, detailing the route of the cables from the central

Fig. 5. System placed on the visually impaired person.

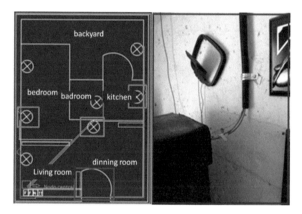

Fig. 6. Central node installation and home automation system connections.

node to these luminaires, and finally the route to the door entrance and exit of the home is indicated.

3 Data Analysis

This section focuses on the tests conducted on visually impaired individuals and establishes the thresholds for operating the proposed system.

Several tests were conducted to determine the user's concentration percentage required for operating the system. These tests took into account meditation and attention data. The results yielded minimum values of 14% and maximum values of 70%. Therefore, it is considered appropriate to establish a concentration threshold higher than 10%. Additionally, gesture tests revealed a minimum value of 25, an average of 51, and a maximum of 55. These findings indicate that the system is slightly more sensitive to the gestures performed by individuals with disabilities.

3.1 Tests Performed

The tests were performed over 3 different days, where it measures how much the visually impaired person can concentrate on a scale of 0% to 100%, in which the following results were obtained:

Test 1: concentration with cell phone in hand. A total of 5 samples were taken in which the visually impaired person had to concentrate on the cell phone that was in his hand. Table 1 shows the percentages of concentration obtained by the user with a minimum value of 60% and a maximum value of 70%.

Table 1. Concentration test, cell phone in hand

Sample	Concentration
1	60%
2	63%
3	63%
4	70%
5	60%

Test 2: concentration without cell phone in hand. In the same way as the previous year, 5 samples of the test were taken, at which observed that the concentration percentages decreased, having a minimum value of 14% and a maximum value of 47% as can be seen in Table 2.

Table 2. Concentration test, no cell phone in hand

Sample	Concentration
1	47%
2	17%
3	30%
4	44%
5	14%

Test 3: gestures with cell phone in hand. In the same way, 5 samples were taken to verify the user vales when performing facial gestures. It has a range of 4 values to work with. It has been configured to work with 3 values that are 25, 51 and 55. The tests that were carried out for three days, a total of 3 erroneous readings were obtained, which represents an 87.5% reliability at the time of changing the action to be performed. The results can be seen in Table 3.

Table 3. Facial gesticulation test with cell phone in hand

Sample	Meditation
1	21
2	25
3	25
4	25
5	51

Test 4: gestures without cell phone in hand. Finally, the test is carried out in the same way that it has been working with 5 samples of the user, in Table 4 you can see the values that have been obtained with the test.

Table 4. Facial Gesticulation Test Without Cell Phone in Hand

Sample	Meditation
1	25
2	25
3	55
4	51
5	21

3.2 Data Classification

About 4 tests carried out in 3 different days was defined the values to work with the system. In the first test was possible to establish that the visually impaired person should concentrate with a value greater than 10% as it does not have lower values than them. The consideration of 10% has been taken as the person with visual impairment could have his cell phone in his hand. While the part of facial gesticulations has been able to differentiate 4 values (21, 25, 51 and 55) of which it has been possible to discriminate 3 valid values that are 25, 51 and 55, as the value of 21 is a not admitted value to perform facial gesticulation (this value is done while making a blink).

4 Results

To obtain the final results of the proposed system, continuous testing was conducted over several days, with 8 repetitions per day, taking into consideration the condition of the person with a disability. This approach demonstrated the functionality of the system.

The results are presented in Fig. 7 where the visually impaired person can be seen using the home automation system, in which it has been divided into 2 parts, the first one is to turn on the kitchen luminaire reaching the necessary concentration, then the second part is to turn off the luminaire of said room, where facial gesticulation is performed to change mode and reaching the necessary concentration to turn off the light.

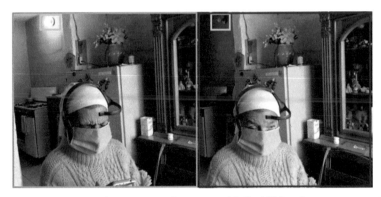

Fig. 7. Use of home automation system by the person with disabilities. Orange sector: on kitchen luminaire. Purple sector: off kitchen luminaire.

Similarly, Fig. 8 shows the operation of the mobile application, which informs the visually impaired person the actions that are being carried out inside the home. It is done thanks to the voice engine that it has. It can visualize several actions that occur within the home such as the central node connection, the locking and unlocking of the entrance door and finally the on/off home luminaires.

As a result of the interaction of the visually impaired person with the home automation system, an efficiency of 98.75% was obtained, due to false positives found when performing facial gestures. The values utilized for calculating the efficiency are presented in Eq. 1.

$$Efficiency = (Pcnc_{sn} + Pcnc_{lum} + Pae_{offlum} + Pcnc_{pi} + Pab_{dpi})/5 \tag{1}$$

where:

$Pcnc_{sn}$: Tests communication between the central node - neural sensor.

$Pcnc_{lum}$: Tests communication between center node – luminaires.

Pae_{offlum}: Tests for announcing on – off states luminaires.

$Pcnc_{pi}$: Tests for central node - entrance door communications.

Pab_{dpi}: Tests for announcing locking - unlocking of entrance door.

$$Efficiency = (93, 75\% + 100\% + 100\% + 100\% + 100\%)/5$$

$$Efficiency = 98, 75\%$$

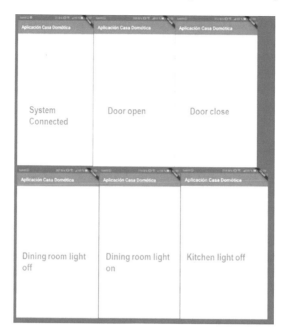

Fig. 8. Notification of actions within the mobile app

4.1 Discussion

The contribution of this work lies in providing comfort and convenience for individuals with visual impairments, enabling them to achieve independence in autonomously managing their household. This approach contributes to the ongoing improvement and integration of novel solutions, such as the one presented by [1], facilitating mobility for persons with disabilities in obstacle detection during their movements. Following the same author's line of research [2], this work employs electromagnetic devices that emit short pulses for echo analysis to detect obstacles.

A closely related study is that of [13], which examines electroencephalographic (EEG) signals for controlling the on and off functions of a lamp for individuals with physical disabilities in their upper extremities. This study achieved a success rate of 79%. This paves the way to elevate this investigation to another level and test it with individuals with visual impairments. Comparative results reveal that the proposed system is notably more effective for those with visual impairments, as it allows for enhanced concentration by eliminating visual distractions, resulting in an efficiency rate of 98.75%. Furthermore, the system contributes by offering voice command confirmation options, as explored in the work of [10]. This feature permits individuals with visual impairments to receive voice confirmation for executed actions via EEG signals.

The challenges are substantial, given the high costs associated with EEG signal analysis devices. Moreover, the complexity of data analysis and classification for processing poses a hurdle. Accessibility may be limited, even for individuals with limited financial means, making it challenging to integrate these devices into their daily lives. This aligns with the findings of [15], which delves into advanced research related to a hybrid brain-computer interface using a bipolar channel for home automation control.

It is crucial to note that academia actively engages in seeking solutions to social issues, necessitating the development of solutions that serve vulnerable groups, such as individuals with disabilities. Such individuals can greatly enhance their quality of life and autonomy through the strategic use of technology.

5   Conclusions

A home automation system was implemented for individuals with visual disabilities using neural frequencies, providing enhanced mobility within the home through a novel method of controlling lighting and the locking/unlocking of the entrance door. This ensures home security and reliability, as endorsed by Ms. Nohemí Trejo, the beneficiary of the project.

The system determined that the user's minimum concentration threshold is 10%, and it successfully discriminated reliable values of facial gestures. With these established values, the desired actions can be executed without any difficulty for individuals with visual impairments.

The implementation of the prototype promotes ease of mobility for people with visual disabilities by minimizing unnecessary travel within the home to control the lighting. It also enhances security by utilizing a mobile application that provides real-time communication on the state of the luminaires and mitigates oversights.

The overall efficiency of the home automation system is 98.75%, instilling confidence in its usage. As a result, the challenges faced by visually impaired individuals within their homes have been significantly reduced by harnessing the power of new technologies.

References

1. Di Mattia, V., et al.: An electromagnetic device for autonomous mobility of visually impaired people. European Microwave Week 2014: Connecting the Future, EuMW 2014 - Conference Proceedings; EuMC 2014: 44th European Microwave Conference 472–475 (2014) https://doi.org/10.1109/EUMC.2014.6986473
2. Di Mattia, V., et al.: A K-band miniaturized antenna for safe mobility of visually impaired people. Mediterranean Microwave Symposium (2015)
3. Vargas, T.: Personas con discapacidad en el Ecuador. Universidad Tecnica de Ambato 4 (2021)
4. Chinchole, S., Patel, S.: Artificial intelligence and sensors based assistive system for the visually impaired people. Proceedings of the International Conference on Intelligent Sustainable Systems, ICISS 2017, 16–19 (2018). https://doi.org/10.1109/ISS1.2017.8389401
5. Shahu, D., Shinko, I., Kodra, R., Baxhaku, I.: A low-cost mobility monitoring system for visually impaired users. Proceedings of International Conference on Smart Systems and Technologies 2017, SST 2017 235–238 (2017)
6. Salat, S., Habib, M.A.: Smart electronic cane for the assistance of visually impaired people. 2019 5th IEEE International WIE Conference on Electrical and Computer Engineering, WIECON-ECE 2019 - Proceedings (2019) https://doi.org/10.1109/WIECON-ECE48653.2019.9019932

7. Daniel Zuriel Villegas, G., Erick Berssain Garcia, V., Norma Elva Chavez, R.: ROSmotic: a Scalable Smart Home for Blind People Controlled with an App. Proceedings - 2017 International Conference on Computational Science and Computational Intelligence, CSCI 2017 1365–1370 (2018) https://doi.org/10.1109/CSCI.2017.238
8. Hernán, A., et al.: Estudio y diseño de un sistema domótico utilizando dispositivos móviles para mejorar la accesibilidad de las personas discapacitadas. Memorias de Congresos UTP 45–50 (2016)
9. Acosta Herrería, L.D.: Sistema domótico incluyendo plataformas de hardware y software libre para la residencia de una persona con paraplejia. Universidad Técnica del Norte (2016)
10. Jurado Lozada, M.A., Alban Mollocana, G. del R.: Sistema domótico de apoyo para personas con discapacidad motriz mediante tecnología móvil y reconocimiento de voz. Universidad Técnica de Ambato. Facultad de Ingeniería en Sistemas, Electrónica e Industrial. Carrera de Ingeniería en Electrónica y Comunicaciones (2018)
11. Calderón Grijalva, J.J.: Diseño de una herramienta para la toma de apuntes y lector de textos para estudiantes con discapacidad visual. Universidad Técnica del Norte (2018)
12. Fuertes Sotelo, E.L.: Diseño de un sistema de trazado de rutas para la movilidad de personas con discapacidad visual, en la Universidad Técnica del Norte. Universidad Técnica del Norte (2019)
13. Jiménez Guevara, M.P.: Sistema electrónico de iluminación (ON-OFF) mediante el control de señales cerebrales basado en tecnología EEG. Universidad Técnica del Norte (2017)
14. Digital Talent Agency.: Metodologías de gestión de proyectos. 1–16 Preprint at (2018)
15. Yang, D., Nguyen, T.H., Chung, W.Y.: A bipolar-channel hybrid brain-computer interface system for home automation control utilizing steady-state visually evoked potential and eye-blink signals. Sensors (Switzerland) 20, 1–15 (2020)

Application of CNC Systems
for the Manufacture of Electronic Boards

Edwin Machay$^{(\boxtimes)}$ ⓘ, Patricio Cruz ⓘ, and Darwin Tituaña ⓘ

Instituto Universitario Vida Nueva, Quito, USA
edwin.machay@istvidanueva.edu.ec

Abstract. In the present project, a CNC system was designed for a horizontal and circular positioner in the manufacture of printed circuits, which will serve to make electronic boards with greater precision where a referential, exploratory methodology was used based on computerized systems that help to a more efficient control of the production system making and manufacturing. In addition, it will allow the attainment of greater control and precision of the tracks routing through the machines by numerical control this decreases the manufacturing time, another of the problems for which the research was developed is the use of ferric acids for the burning of the plate this is harmful since the students of the technical careers of the Vida Nueva University Institute nowadays continue to use the traditional method, this can cause direct effects on the skin and eyesight, if the substances are handled incorrectly. Giving a solution to the aforementioned problems, it was built a system capable of making electronic cards in less time and without the use of chemical agents using mechanical prototypes for circular and horizontal movement provided with a mobile metal table that rotates with the product and a mobile trolley composed of pinion-rack for the free movement of the tool for the machining and routing of the tracks which is mounted on two screws that move manually on the Z and Y axes.

Keywords: Systems · CNC · Electronics · Arduino · circuits

1 Introduction

In the current project a study of CNC machines for the manufacture of PCB boards was made, which still implements the traditional method of using ferric chloride acid that is used in factories for the manufacture of boards in printed circuits which is a harmful polluting for the environment. Thus, the following research question was defined: How does the implementation of CNC systems help the PCB board manufacturing process? the purpose of this research is to reduce the use of said acid (FeCl3) since it is not being handled properly and can "irritate the eyes, skin and respiratory tract and can be corrosive by its ingestion" and to avoid this type of risk to those who manipulate it without a respective equipment of personal protection. Therefore, an innovative CNC system has been built which allows the manufacture of printed circuits in Bakelite's replacing conventional methods, in an autonomous way and with greater precision based on applied engineering calculations for the movement of the system based on the constant speed

© The Author(s), under exclusive license to Springer Nature Switzerland AG 2024
M. Z. Vizuete et al. (Eds.): CI3 2023, LNNS 1041, pp. 178–189, 2024.
https://doi.org/10.1007/978-3-031-63437-6_15

stepper motors through G and M codes which designate the movements and angular positioning that help to maximize their effectiveness and reduce the period of time in manufacture without the use of chemical agents.

2 Methodology

2.1 CNC System Design

For the design and elaboration of the mechanical plan of a computerized numerical control system, CAD software was used where the minimum and maximum distances that the CNC system must have, were determined [1].

As for the mobility of Z axis it was made with an intended structure design to be on stepwise balance with the motor and the Dremel so it is resistant in order not to generate effort overload at that specific point due to the vibrations generated in the moment of milling which can cause damage (See Fig. 1).

Fig. 1. The image shows the construction of the work table of the CNC machine.

Electronic Design. The electronic design has the purpose of making a functional electronic circuit in the simplest and most efficient way possible and in a shorter time. That is why it is responsible for carrying out its design and ensuring that it does not have any error, otherwise it must be detected immediately and give a solution, so before getting to build it, tests in simulators must be done, [11] indicates The simulators have several elements or components which will be of great help to verify their operation and observe the behavior of the circuit that has been designed (See Fig. 2).

Computer Control. The computerized design CAM (Computer Aided Manufacturing) consists of the use of software in which a numerical control (NC) in order to carry out a manufacturing process through machining [3]. AutoDesk mentions that to perform an

Fig. 2. The image shows the designed elements of the CNC system for engraving and cutting PCB boards.

NC machining it is necessary to use instructions with G and M codes that are elaborated through a computer this can generate the toolpath for the machining to obtain the product this process comes hand in hand with the CAM design, since it is necessary to carry out the design of the product in a software for 3D modelling and then, by means of a CAM design, perform its machining (See Fig. 3).

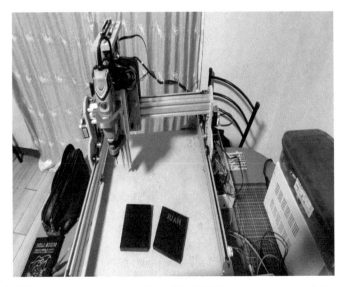

Fig. 3. The image shows the operation of the CNC or computer numerical control.

Programming.   Programming is a way of instructing computers to perform specific tasks based on logic since they can perform tasks more quickly and efficiently. Computers must continuously execute commands in a way they understand, these standards are grouped into procedures called procedures [9]. An inspection-relevant program is divided into two parts: the internal representation of the program, represented in machine language or executable code, and the external representation or documentation. Or write the documentation in a format that customers can see, read and understand [17]. Programming is based on logic and is an intellectual faculty in both human and artificial intelligence. Programming is based on a step-by-step system that controls system functions. Therefore, the concept of programming was born from the need to write programs or instructions for computers [5]. A program can be stored in memory and run when needed (See Fig. 4).

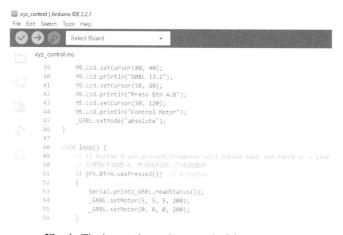

Fig. 4. The image shows the control of the x,y,z axes.

Reference Voltage and Calibration.   An idea of the calibration result is obtained using the following formula, here the value is multiplied by 70% since it is to be operated in full steps, it gives the reference voltage for the normal operation of the motor by steps [7].

$$Vref = \text{Imax}(8 \cdot Rs)$$
$$Vref = 0.4 \cdot (8 \cdot Rs)$$
$$Vref = 0.32(0.7) \tag{1}$$
$$Vref = 0.22V$$

Arduino Program.   It is an Arduino firmware version 1.1 called GRBL, the code that communicates with the CNC Shield and its control software [20].

Circuit Vectorizer.   There are countless programs of vectorization of graphics and sources of which it was chosen to use a free software called FlatCam which allows the configuration and orientation of GRBL from an electronic design software, thus allowing the vectorization of the electronic way and an effective communication with the CNC machine [23].

Control Software CNC. There is a free CNC control program called Open CNC Shipper that uses CNC numeric codes to provide communication between the control card and the PC [8]. It is often used in electronic fingerprint research due to features such as height mapping, which allows for an accurate etching process as the Bakelite mill passes with the probe (See Table 1).

Table 1. Analysis of research variables.

Variable	Conceptual definition	Dimensions	Indicators
System CNC for plate cutting and engraving PCB	Greater possibility of making electronic boards with a short time and less contact to acid exposure	CNC	Computer numerical control programming for indication of controls or movements
		G and M codes	Commands of movements or operations performed by related software programming
		Servo motor	Rotary actuator that authorises rotary control with precision both its angular position and its acceleration and speed
		Arduino	Programming software using the C++ language applied as a microcontroller
		Proteus	Schema capture software such as simulation and routing automatic electronic circuits
		Microprocessor	Union for the main development that is the brain in charge of running all programs
		Bakelite	Projection of an aspect in the function and production by means of graphical object two-dimensional symbols

3   Results

3.1   Optimization of PCB Circuit Manufacturing Processes

The project follows the guidelines of a quantitative approach since quantifiable data was collected to analyzed the results, considering what was proposed by Kenneth & Walter.

Based on a series of hypotheses therefore to carry out quantitative studies it is essential to have a theory already built, since the scientific method used in it is deductive.

Regarding the methodological design, the cross-sectional exploratory study is developed starting from a problem. In this regard, Study Techniques [26] Transactional or cross-sectional research designs collect data at a single moment, at a single time. Its purpose is to describe variables, and analyses their incidence and interrelationship.

We began with the bibliographic collection of the most relevant data of the application of computerized numerical systems in the industry in this way we identified the advances of this technology. Then we proceeded to investigate the implementations in the country where we could know that the CNC are already applied in most of the manufacturing and production industries having a great acceptance for its valuable contribution in the optimization of manufacturing time of any type of product.

For the collection of the data the starting point was the problem identified in the project. For this purpose, the time used in the manufacture of electronic or polish PCBs was quantified where it was observed that the time used for the total elaboration of the circuits takes an estimated time of 3 h following the conventional process (See Table 1).

Table 2. Analysis of time spent manufacturing PCB circuits.

Dimension PCB	Estimated time
5 x 5 cm	1 h
10 x 10 cm	2 h
20 x 20 cm	2,5 h
40 x 40 cm or more	3 h

Table 3. Analysis of manufacturing time with the implementation of the CNC

Dimension PCB	Estimated time	CNC System
5 x 5 cm	1 h	50 min
10 x 10 cm	2 h	
20 x 20 cm	2,5 h	1,5 to 2 h
40 40 cm or more	3 h	

The design of the machine is made up of two parts: design of the mechanical and electronic model. Which was tested both in the programming, as well as the electronic components that it contains for a correct operation where a programming software such as Arduino was applied because it works with a C++ language in programming and Proteus Design Suite for the simulation of electronic components, managing to properly verify the operation of the programming system. Regarding the design of the model programs for the mechanical design modelled in 3D were used, either as CAD programs, both in the base, supports and mechanical movements that the CNC performs in a certain way the construction of its parts with greater precision.

Today there is a large number of electrical devices that make up our daily life and that have become an indispensable part in which most of the electronic cards are built by using acids, where they are submerged in large amounts of chloride. Ferric (FeCl3) because it has a diluent. Application to copper with a constant use of it, leading to a longer production time as can be seen in practices carried out by higher education students and teachers of the institution in the area of electro mechanics or related careers, therefore this process of construction of plates PCB the use of acids that, also due to a bad handling of it can cause damage to both the skin and eye-sight.

In this way, the construction of this project is autonomous and without the application of acids in its manufacture, which also shortened processes in the elaboration both for printing the tracks of the electronic circuit and for burning the Bakelite that gave rise to it (See Fig. 5).

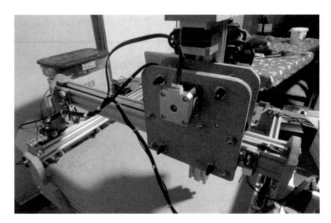

Fig. 5. The image shows the CNC machine for PCB boards.

With the arguments reviewed and studied above it was possible to apply a possible solution with the implementation of a CNC system with the same principles of a 3D printer which allowed to unify and automate the process for cutting and engraving of electronic boards in a better time and without the use of chemical agents such as ferric acid.

The following will show the results obtained with the implementation of the CNC system, where the reduction of the manufacturing time can be observed (See Table 2).

4 Discussion

4.1 Bibliographic Background of the Research

At present, the manufacturing processes of electronic boards by means of PCB burning is called screen printing method, which uses a chemical component that is an acid base to burn the residual copper in the PCBs, such a process for the implementation of the circuit to the board as, tells us in this process, we leave the PCB in a ferric acid bathing chamber to remove all the excess copper from the Bakelite, that is, remove what is not printed with the ink [7].

At the same time, according to the study based on other research of manufacturing by traditional methods in which the institution for the development of manufacturing PBC plates, which involves a process rooted in the use of said acid (FeCl3) that when handled improperly, would end up being a risk to those who handle it without a respective personal protective equipment, ignoring its elaboration process and for a better explanation of the damage it produces [20] tells us.

The traditional use of acid for chemical dissolution in PCB design in the technology industry has been severely compromised due to the negative impact on the environment and damage to ecosystems caused by the use of this type of acidic solution. One type of fluid is drained, an obvious example is ferric acid, which due to its HFeO2 nomenclature, indicates that it is a highly corrosive acid that can cause skin damage if not handled properly. Injuries can pose a risk to personnel handling acid without personal protective equipment, especially if they do not understand the issue regarding the design and implementation of electronic circuits.

The innovative processes already treated in industrialized countries such as the technology of CNC automatic machine systems to perform the cutting and engraving tasks of the aforementioned electronic rails without the use of acid to eliminate the copper residues present in Bakelite. The most favored of this project will be those with knowledge in electro-mechanics and related professions, since the theory of algorithms, parameters and mathematical formulas is implemented in the construction of CNC systems, such as: calculation of forces and resistances selection of failure criteria, wear loads, load balancing, vibration, etc., in order to know the new technologies for the manufacture of Bakelite circuits, and the various applications of computer numerical control systems "CNC".

4.2 Results Obtained in the Research

The use of printed circuit boards (PCBs) has become ubiquitous in modern electronics due to their cost-effectiveness, reliability and compactness, as such, there is a need for efficient and accurate methods of cutting and engraving PCB boards to create custom designs [26]. This is where a computer numerical control (CNC) machine comes into play, which is capable of accurately cutting and engraving PCB boards. In this functional proposal, we will describe the key features that a CNC machine must possess for efficient and accurate cutting and engraving of PCB boards.

Functionality. A CNC CPB plate cutting and engraving machine will depend on the characteristics and technical specifications of the particular machine, however, generally speaking a CNC machine is capable of performing operations such as CPB plate cutting and engraving, in which the machine can cut plates and sheets of various materials including metals wood, plastics and other materials. Engraving, the machine can engrave designs and texts on the plates and sheets using tools such as milling cutters, punches and lasers so that milling is performed such as edge beveling, drilling holes and creating complex patterns, which in turn, performs roughing of various sizes and shapes on the plates and sheets, so a CNC CPB plate cutting and engraving machine is capable of performing a wide variety of operations to create customized and detailed parts from plates and sheets of different materials.

Learning.   Only two levels of headings should be numbered. Lower level headings remain unnumbered; they are formatted as run-in headings. The operation of a CNC (Computer Numerical Control) machine can provide knowledge and skills in several aspects, such as: Computer aided design (CAD), to be able to produce parts with a CNC machine, requires a precise and detailed design in a CAD software, therefore, learning to use CAD design software is essential to size the CNC machine. CNC programming with which it will proceed to operate a computerized numerical control machine, it must be understood how the machines are programmed to perform the operations of cutting and engraving materials.

CNC programming involves writing code in G language to control the machine. Knowledge in mechanics and electronics to operate a CNC machine, for knowledge about its functionality and mechanism of its moving parts, since the machine has components such as motors, controllers and cutting tools and parts that move according to their axes. Problem solving and maintenance in handling, in a computerized numerical control machine also implies the ability to give solutions to technical problems and perform regular maintenance of the machine in addition to making improvements if necessary. Manufacturing skills, operating a CNC machine also provides skills in producing custom parts and components, as well as the ability to work with a variety of materials and tools.

Therefore, operating a CNC machine can provide skills in computer-aided design, programming, mechanics and electronics, problem solving and maintenance manufacturing skills. These skills can be applicable in a variety of fields, including manufacturing, engineering, architecture, and prototype production. There is always room to improve a CNC PCB board cutting and engraving machine some improvements that could be made, is to improve in greater accuracy and speed of modern CNC machines, in addition, there is room to improve its ability to cut and engrave high precision. Speed is also an important factor to consider as a higher cutting speed can increase productivity. Increased automation capability, as it is critical to the ability to automate CNC machine scheduling and tuning tasks that can help reduce downtime and increase productivity, this could include incorporating sensors and feedback systems to improve CNC machine accuracy and adaptability such as programming software improvements: CNC machine programming software can be improved to make it easier to use and to enable faster and more accurate programming.

Algorithms and optimization software could also be developed to maximize cutting efficiency and reduce material waste so that greater versatility in cutting materials can be achieved for modern CNC machines, which can cut and engrave to a wide variety of materials, but there is still room to improve the machine's ability to work with harder or more brittle materials, but above all safety is important in any CNC machine, which, new safety features could be incorporated to reduce the risk of accidents and injuries to operators. Whether in terms of accuracy, speed, automation, programming software, material versatility or safety, improvements in these areas could increase the productivity, efficiency and quality of the products that are produced with the CNC machine.

Optimization.   Such as the operatic part in building a CNC PCB board cutting and engraving machine can be improved in several ways, some of which are.

Materials.   The use of high-quality materials such as stainless steel, aluminum, copper and other corrosion-resistant materials can increase the durability of the CNC machine and reduce the risk of structural failures or deformations.

Increased Rigidity.   To improve the accuracy of the CNC machine, it is important that the structure is as rigid as possible. One way to achieve this is by using structural profiles of high strength and rigidity, and making sure the components are properly aligned. Drive system improvements, CNC machine drive system can be improved using high-quality motors and more efficient power transmission systems. This can help improve the cutting and engraving speed, as well as the accuracy and reliability of the machine.

Optimized Design.   The design of the CNC machine can be optimized to reduce its size and weight, making it easier to transport and handle. Parts can also be de-signed for easy assembly and maintenance to reduce the risk of errors at the time of assembly. Incorporation of new technologies new technologies such as artifi-cial intelligence, machine learning and the Internet of Things can be used to im-prove the functionality and efficiency of the CNC machine. For example, the incorporation of sensors and feedback systems can help improve cutting and engraving accuracy. In short, improving the construction of a CNC PCB board cutting and engraving machine can be achieved using high-quality materials im-proving the rigidity of the structure, improving the drive system, optimizing the design and incorporating new technologies. Each improvement can contribute to improving the accuracy, speed, reliability and functionality of the CNC machine. And in turn you can also help nature in various ways, such as reducing waste with the cutting and engraving precision of the CNC machine, you can produce more accurate and tailored parts, which reduces material waste.

This can help reduce the amount of waste that is generated during the manufacturing process, which in turn reduces the amount of natural resources employed and the use of sustainable materials for the CNC machine, as it can cut and engrave sustainable materials such as wood, cardboard and other biodegradable materials. These materials are renewable and biodegradable, which means that they do not have a negative impact on the environment as the production of custom parts, which can be produced in small quantities, which helps reduce the amount of waste generated by mass production. Custom production can also help reduce the environmental impact of producing unused parts by applying energy-efficient, in terms of energy, unlike older cutting and engraving machines. This means that the CNC machine can reduce energy consumption and greenhouse gas emissions as well as reducing production times, which can help reduce energy consumption and greenhouse gas emissions. In addition, reduced production time means more parts can be produced in less time, which can help reduce the environmental impact of mass production.

5   Discussion

As shown by the investigation of the process of a CNC in the cutting and engraving of PCB boards, these entail a shorter time spent in their manufacture with a higher pressure since it is carried out by a machine tool whose movements and roughing speed are controlled at the same time avoiding accidents to which they are exposed by traditional methods. After the completion of the project, the following can be concluded:

This study proposes the implementation of CNC systems for the manufacture of PCB boards, replacing the traditional method with ferric chloride acid. Solving the problem, a CNC system has been developed that reduces the time in the manufacturing process of printed PCB circuits, offering a safer alternative reducing health risks and environmental impact.

The theoretical framework explains the necessary theoretical foundations for the implementation of CNC systems in the manufacture of electronic boards. In addition, key aspects such as mechanical design with CAD software, electronic design, computer control using G and M code instructions, and programming to instruct and control CNC machine tasks are addressed. This fundamental knowledge enables optimization of the PCB manufacturing process.

The methodology used in this study follows a quantitative and cross-sectional exploratory approach starting from the design of an autonomous CNC machine to manufacture PCB boards without the use of corrosive acids. The functional and programming tests were carried out using software such as Arduino and Proteus Design Suite, in the same way the mechanical design was carried out with CAD programs.

CNC machines are efficient and precise tools to carry out cutting and engraving processes of electronic plates through automated systems, this was achieved using the management of numerical codes by computer, that is why this type of system and innovation benefit both the industry as well as the environment, reducing polluting waste.

Considering what is supported by the theoretical foundations, the methodology, as well as the discussion of the data obtained from this investigation, the implementation of CNC systems for the improvement and optimization of the PCB manufacturing process is considered favorable.

References

1. Pratomo, A., Perdana R.: Arduviz, a visual programming IDE for Arduino. International Conference on Data and Software Engineering (ICoDSE), 1–6 (2017)
2. Equipo Eeditorial E, https://concepto.de/lenguaje-de-programacion/, last accessed 2023/11/06
3. Martinov , G., Zakharov, A.: The Specifics of Building a Cross-Platform OPC UA Server for a CNC System. International Conference on Industrial Engineering, Applications and Manufacturing (ICIEAM), 702–707, (2023)
4. Etecé, E, https://concepto.de/diseno/, last accessed 2023/01/03
5. Fadesa, https://fadesaing.com/prototipos-y-preseries/, last accessed 2023/09/04
6. Ferros Planes, https://ferrosplanes.com/mecanizado-cam/, last accessed 2023/01/05
7. Flores, L.: Radial Basis Function Methods for Solution. First edition. EDP, México (2018)
8. Francisco, http://solidworksavanzado.blogspot.com, last accessed 2023/06/06

9. Heskelinen, H.: Improving the productivity of complex electronic systems design by utilizing applied design methodologies. IEEE Aerospace and Electronic Systems Magazine, 6–28 (2018)

10. HArduino, https://www.arduino.cc/, last accessed 23/10/06

11. Hernández, R.: Design and construction of a self-balancing wheeled robot for teaching automatic control. 2nd Edition. ICBI Basic Sciences and Engineering, San Luis Potosí (2022)

12. Hwlibre, https://www.hwlibre.com/nema-17/?msclkid=6e28dfa8d08b11ec910cbd70997 28ae8, last accessed 2023/05/07

13. Inter, https://www.inter2000mecanizados.com/post/que-es-el-cnc-y-como-esta-compuesta-una-maquina-cnc, last accessed 2023/08/07

14. Kangju, L., Weitang, L., Yefeng, S., Yuan, L.: Research on self-maintenance strategy of CNC machine tools based on case-based reasoning. 3rd International Conference on Industrial Artificial Intelligence, 1–5 (2021)

15. Kannadaguli, M.: IoT Based CNC Machine Condition Monitoring System Using Machine Learning Techniques. 9th International Conference on Communication Systems and Network Technologies (CSNT), 61–65 (2020)

16. Kuo, S., Raihany, C., Peng, Y.: Sound Detection of CNC Milling Machine by Embedded System. International Symposium on Computer, Consumer and Control (IS3C), 130–133 (2020)

17. Lawson, V., Phister, M., Rogers, C.: Automated Rotor Assembly CNC Machine. Systems and Information Engineering Design Symposium (SIEDS), 1–5 (2020)

18. Leskow, https://concepto.de/electronica/, last accessed 2023/04/05

19. Lukas, K.: A hands-on guide to bringing your robotics ideas to life using Arduino. Packt Publishing, 23–45 (2023)

20. Marlon, V.: Design and construction of a cnc for cutting and engraving electronic tracks on bakelites. 1st Edition. Instituto Vida Nueva, Quito (2020)

21. Ma, T., Cui, Z,. Yao, H.: Virtual optimization design for mechanical structures and its application, International Conference On Computer Design and Applications, 80–155 (2019)

22. Rastogi, N,. Rastogi, H., Shrivastava, N.: Design and Implement an Economical Automatic CNC Wood Lathe Machine. 4th International Conference on Recent Developments in Control, Automation & Power Engineering (RDCAPE), 6–10 (2021)

23. Rivadeneira, J.: Implementación de un módulo CNC de fresado, Quito EPN (2022)

24. Sapiee, N.: Design and Development of Printed Circuit Board (PCB) for Smart Calorie Counter System. 9th International Conference on Electrical and Electronics Engineering (ICEEE), 90–94 (2022)

25. Sun, S., Zhao, P., Zhang, T., Li, S.: A G3 Continuous Tool Path Correction and Smoothing Method for CNC Machining. 18th Conference on Industrial Electronics and Applications (ICIEA), 1798–1802 (2023)

26. Tecnicas de estudios ORG, https://www.tecnicas-de-estudio.org/investigacion/investigacio n38.htm, last accessed 2023/07/08

27. Wijaya, W., Syahroni, F., Mulyadi, C., Sani, W.: Two Axis Simple CNC Machines Based on Microcontroller and Motor Driver 27 Shield IC L293D. 14th International Conference on Telecommunication Systems, Services, and Applications (TSSA, Bandung, Indonesia, 1–5 (2023)

28. Zhang, Y.: Research on the Design and Development of Remote CNC Tool System Based on Computer. International Conference on Computer Applications Technology (CCAT), 45–48 (2023)

29. Zheng, T., Wang, Z.: Research on fault prediction and diagnosis method of PCB circuit. International Conference on Artificial Intelligence and Computer Engineering (ICAICE), 387–390 (2020)

Cuencabike Mobile Application for the Promoting of Bicycle Tourism in the City of Cuenca, Ecuador

Pablo Crespo López[ID] and Marco Guamán Buestán[✉][ID]

Instituto Tecnológico Superior Sudamericano, Cuenca, Ecuador
magbmail@gmail.com

Abstract. When analyzing the current situation of bicycle lanes in the city of Cuenca, it can be observed that within the competent institutions, there are plans and projects managed to improve mobility through non-motorized vehicles. In addition, there is a vision to continue creating bicycle lanes, although the information has not been widely disseminated to the public. One of the reasons behind this situation is the lack of updated information and its inefficient structure for end users. The objective of this research was to develop the mobile application Cuencabike for the promotion of bicycle tourism in the city of Cuenca, Ecuador. In the methodology used, the visual cartographic analysis stands out as the most applicable method of the research employing maps. As a result, a technological proposal is obtained that allows tourists and inhabitants of the city to obtain detailed information about the bicycle paths and points related to the tourist sites of the city. Through a development with the IONIC framework and the Google Maps MyMaps tool, the CuencaBike application was generated, which integrates information on all current routes and those being planned. Each of the routes generated has points of tourist interest, as well as sections where you can view the information of the different mechanics, parking, and public bike dispensers that facilitates users to have all this information organized and updated.

Keywords: Bicycle paths · Bicycle tourism · Mobile application

1 Introduction

The research focuses on tourism development in the city of Cuenca, located in the province of Azuay-Ecuador. A technological proposal is proposed that allows tourists to obtain detailed information on the use of bicycle paths and points related to bicycle tourism. The CuencaBike application contributes to the productive sector by promoting a different kind of tourism for the city of Cuenca, since it has maps that promote bicycle tourism, it can be used by national and foreign tourists, but also by people who live in Cuenca. Economically, it contributes to the promotion of businesses such as restaurants, theatres, and museums, among others, because it allows the reactivation of their activities through bicycle tourism. The application makes it possible to locate these points in the city and to easily trace an optimal route through the bicycle lanes. It highlights

M. Z. Vizuete et al. (Eds.): CI3 2023, LNNS 1041, pp. 190–202, 2024.
https://doi.org/10.1007/978-3-031-63437-6_16

the contribution to the wellbeing of the population, the use of bicycles as a means of transportation helps to exercise and contributes to the health of citizens.

1.1 Theoretical References

Gonçalves [1] argues that bicycle tourism is performed by people who visit the city on vacation or holiday trips. It can be performed by short routes in specific locations or in an environment that allows us to know the touristic points of a region. Ritchie (2010) mentions that bicycle tourism is growing in the most important cities of the countries, which provides an opportunity to take advantage of this tourist product and turn it into an economic benefit for merchants.

Cycle tourism works as organized rides among friends or under the tutelage of a tour operator company, it is important to mention that this should be guided by a map or with signs and signs. In the case of the city of Cuenca, information on bicycle paths and bicycle tourism points is digitized by the municipal mobility company (EMOV), but there is no public information, which means that both tourists and residents are unaware of the routes, uses, benefits and processes for the enjoyment of bicycle paths.

Within the research conducted at the national level, we have the thesis work conducted in the city of Guaranda, within this, it was diagnosed that in recent years there has been a significant growth in the use and circulation of motorized vehicles, this has caused mobility difficulties within the urban area. Therefore, the construction of bicycle lanes for the use of bicycles as a means of transportation is a solution [2]. In this same line, in the city of Santo Domingo, a preliminary proposal was developed on bicycle lane routes with the justification of satisfying a need and raising awareness about the future consequences it would entail, where it is mentioned: "…the problem of motorized mobility is solved since many users see feasible the use of a much more ecological means of transport but the only lack is an exclusive space for it" [3].

On the continent, there are several studies focused on mobile applications and urban mobility. One of them is the study conducted in cities such as Rio de Janeiro, Sao Paulo and Medellin, which investigated the management of information provided by private company applications such as Uber, Moovit, Waze and Stava as effective tools to support mobility planning in the cities mentioned. In addition, the authors propose: "a model that allows outlining parameters that help cities to develop a vision regarding the potential of data to generate actions and public policies for urban mobility" [4].

Studies developed by the University of Oviedo in Spain conclude that mobility is an indispensable aspect of the deployment of vehicular networks. The most outstanding contribution of this study is the creation of an intelligent mobility plan based on fuzzy logic in vehicular communications. The most encouraging results of the study were that a 12.27% decrease in pollutant emissions was achieved. More efficient mobility and environmental sustainability are obtained [5].

Along the same line, the bicycle is seen as an ecological and economical means of transportation, since it does not pollute, and its maintenance is not costly; it is used to move within and outside the city. Some advantages of bicycle use described by an organization in the city of Quito called Biciaccion:

- It is convenient and fast.

- Low cost of use and maintenance.
- Requires small parking areas.
- Occupies less space on public roads.
- Significant physical and mental health benefits.
- No noise.
- Does not pollute.
- No need for fuels.
- Allows greater contact with people and the environment.
- Generates a culture of citizenship.
- The use of bicycles allows us to generate more humane cities.

Bicycle lanes are lanes created for the exclusive use of bicycles, strategically implemented on the sides of the lanes, this generates a solution to environmental pollution and a solution to vehicular traffic. Therefore, those in charge of road planning must consider bicycle lanes as an alternative to urban mobility [6]. It was shown that these strategies designed within cities help to reduce mobility problems, to promote recreation, health and sports, to provide new spaces for socialization and encourage the proper use of public space [7]. Another benefit of the bicycle route is the opportunity to raise awareness among citizens about the pollution we generate, the sedentary lifestyle and the inadequate diet we lead, thus promoting a healthy lifestyle through physical activity and raising awareness about the use of sustainable and environmentally friendly transportation [8].

The Ministry of Tourism, in 2012, issued the technical standard for adventure tourism of bicycle tourism. This document lists the minimum equipment requirements to develop a bicycle tourism activity. In response to this, the municipality of Cuenca implemented several projects for the creation of more urban bike paths; the goal is to have more than 100 km of roads for the use of bicycles.

1.2 State of the Art

Research on bicycle tourism in recent years has pursued several goals, the most important of which are sustainability and social, economic and environmental impact. For Xue [9] cities with potential for cycle tourism should consider in their planning an overall sustainable environment. Likewise, for Bieliński [10] it is important to know the factors that maximize the performance of bicycle sharing systems (BSS) created by some cities, starting from the fact that, bicycles solve the problem of air pollution and congestion in transportation. According to Gazzola [11], bicycle tourism is the clearest expression of sustainable tourism, therefore, the purpose is to know the relationship between the tourists' fondness for bicycles and the desire to get to know new territories.

Every human activity has an impact on the environment, and the case of bicycle tourism is no exception. Therefore, Zhao [12] mentions that the creation of bicycle paths in tourist cities, requires special treatment in land planning, thus, favoring the development of urban transport. To convert the bicycle as a means of transportation, Fortunato [13] refers to the fact that disused railway sections can be used to visit the natural resources of a city. According to Nilsson [14], cities with bicycle infrastructure have the opportunity to create an impact on the economy of businesses such as bicycle sales and repair. To create an ideal environment for cyclists who wish to go sightseeing.

Bakogiannis, et. al [15] refer to the minimum requirements that businesses must meet to cater to cycle tourists. In this sense, Ciascai [16] adds that it is essential to know the concerns and motivations of the cyclist when choosing a tourism product.

Other purposes of bicycle tourism are addressed by researchers Procopiuck [17] and Carra [18]. The former refers to understanding the association between cyclists' activities and tourism in the city. The second, proposes to meet the need of cyclists to know the itineraries and cycling routes, considering that these should benefit the local economy. Additionally, Setyowati [19] mentions that despite the infrastructure of the cities it is necessary to know the factors that influence people's little attention to cycle tourism.

To achieve the aforementioned purposes, the researchers mention the following adversities. For both Setyowati [19] and Zhao [12], the ecological transportation provided by bicycles has not received much attention from users, consequently, the impact on urban traffic is unknown. On the other hand, Gazzola [11] stresses that there are still no business models related to cycle tourism. This according to Procopiuck [17] derives from the lack of association between cycling and cycle tourism activities in cities. In addition to this, Carra [18] emphasizes the deficit of a cycle tourism itinerary that allows the enjoyment of historical and scenic peculiarities. Previously, Lee [20] also noted the low importance of tourist attractions in cycle tourism destinations.

That is why Fortunato [13] mentions that the strategy for the economic reactivation of some places is to implement bicycle tourism projects, for which it is necessary to know the best methodologies for such projects. Similarly, Ciascai [16] refers to a lack of knowledge of the concerns generated by a bicycle tourism product and its social, economic and environmental impact. Although some cities have environments for cycle tourism, for Xue [9] and Bielinski [10] the effectiveness of these is unknown. In search of quality in bicycle tourism services through technology, Chen [20] highlights that, despite all the information that a smartphone can generate, there is still a lack of understanding of how to design a system of location-based recommendations for users. Figure 1 summarizes through a timeline the adversities found in recent studies.

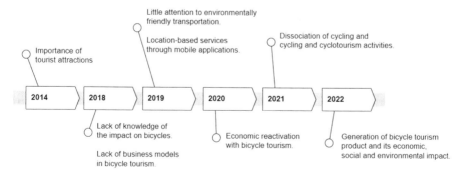

Fig. 1. Timeline of state-of-the-art adversities.

A review of recent research solutions shows that for sustainability a convergence of public transport with cycling infrastructure is necessary but clarifying that cycle tourism is a long-term solution [11]. Tourists who choose to use bicycles on their vacations always seek to soak up the local culture, but, likewise, require having places to repair their bicycles and cycling routes [15]. The proper development of cycle tourism in a city, requires a feasible environment, in this case, BSSs play an important role [10]. Finally, for a cycling route to be exploited by tourism, it should be considered to include the natural scenic places of the location [9].

1.3   Research Problem and Objective

The city of Cuenca has a road project that describes the need to "pacify" the roads. This situation of vulnerability is pronounced by pro-bicycle organizations, which denounce the lack of infrastructure and facilities for the use of bicycles, in addition to claiming the rights and obligations of all road users. This project is focused on the regulations of the Mobility Plan and Public Spaces of Cuenca 2015–2025 and the Organic Law of Land Transportation, Transit and Road Safety (July 24, 2008), Art. 30.4.- This article states that municipalities:

> ...within the scope of their competencies in matters of land transportation, transit and road safety, in their respective territorial circumscriptions, shall have the powers following the Law and the ordinances they issue to plan, regulate and control transit and transportation, within their jurisdiction, observing the national provisions issued by the National Agency for Regulation and Control of Land Transportation, Transit and Road Safety; and, they shall report on the local regulations that in matters of traffic control and road safety are to be applied (p.19).

The objective of this research is the development of a mobile application for the management of bicycle routes for the promotion of tourist sites in the city of Cuenca. The application visualizes the information on the bicycle paths allowing the dissemination of the routes. As for the information presented in the application, this was obtained from the municipal mobility company and includes the description of the bicycle routes, parking points, and mechanical and public bicycle parking.

2   Materials and Methods

Mixed research is a practically new approach that combines quantitative and qualitative methods in the same study. The purpose of mixed research is not to replace quantitative research or qualitative research; this research tries to minimize the potential weaknesses and unite all the strengths of the two types of research [22].

Mixed research clarifies that it is a method in which researchers adopt quantitative and qualitative techniques, methods, approaches, concepts or language within the same research [23]. Exploratory research is carried out when the research topic is little studied, especially when there are many doubts, or it has not been taken as a research topic subsequently [24]. But, on the other hand, Cueva [25] says that exploratory research is applied when investigating trending topics, that the variables are investigated, and the tone of subsequent research is established.

2.1 Research Techniques

The following techniques were used to carry out the research:

- An interview to obtain information from an official of the municipal mobility company on the current status of bicycle lanes and routes related to the exclusive use of bicycles.
- The information was obtained through an interview with the person in charge of the mobility area, who provided the data digitized in MyMaps in KML format.
- A survey was applied because they are non-experimental cross-sectional investigations [26].
- Another technique used for the research was the cartographic method of research which consists of "…the application of maps for the description, analysis and study of phenomena to obtain new knowledge, characteristics and investigation of their spatial interrelationships and their prediction" [27]. Within this technique, visual analysis is specified as the most applicable method of research employing maps. It is used at the first stage for the general knowledge of the studied phenomenon to choose later the methodology of the work. So also, the method called cytometric analysis consists of measurements and calculations of quantitative characteristics of phenomena, with an assessment of the accuracy of the final results [27].

2.2 Research Instruments

The following instruments were created for the application of the aforementioned techniques:

- An interview script consisting of a written record of the questions. According to Folgueiras Bertomeu [28]: In the structured interview it is decided beforehand what type of information is wanted and based on this a fixed and sequential interview script is established. The interviewer follows the marked order, and the questions are intended to be answered briefly. The interviewee must follow this pre-established script.
- Abarca, Alpízar, Sibaja, and Rojas explain that the survey instrument used is a questionnaire: "…refers to the instrument of another research technique called the survey. However, some authors refer to the qualitative research questionnaire, to differentiate it from the survey instrument, for those cases of highly structured interviews" [29].
- The instrument to perform the cartographic analysis was the checklist which is defined as: "a structured instrument, which contains a list of established evaluation criteria or performances, in which only the presence or absence of these is qualified employing a dichotomous scale, that is, it accepts only two alternatives" [30].

To know the vision of the city's inhabitants about bicycle tourism and bicycle routes, a survey was conducted. To conduct the survey, a sample of fifty-four people, with ages ranging from fifteen to fifty years old, was sought from the population living in the city of Cuenca. This technique was carried out through an online form that was sent to the participants through instant messaging applications.

3   Results and Discussion

The following are the results of the survey, which reveal a notable lack of knowledge on the part of citizens regarding bicycle routes. In fact, a significant 39% of those surveyed indicated that they had no information on the subject (see Table 1). The lack of information among citizens becomes more prominent when addressing the issue of bicycle lanes for tourism. In this context, there is an increase in lack of knowledge, reaching 44% (see Table 2). Finally, most of the people surveyed agree that a mobile application would be useful for obtaining information both about the bicycle paths and about the tourist sites near these routes (see Table 3).

Table 1. People' knowledge level of the bicycle lanes and routes within the Cuenca city.

Knowledge level	N	%
No information	21	39
Some information	8	15
Know	8	15
Know well	12	22
Know very well	5	9

Table 2. People' knowledge level of the bicycle touring routes in the Cuenca city.

Knowledge level	N	%
No information	24	44
Some information	9	17
Know	9	17
Know well	9	17
Know very well	3	5

Table 3. People's opinions about the development of a mobile application for cycle tourism in Cuenca city.

Agree/disagree	N	%
Strongly Disagree	6	11
Disagree	1	2
Neutral	0	0
Agree	14	26
Strongly Agree	33	61

Figure 2 illustrates that 35 bicycle lanes run through the city of Cuenca. The longest bike path runs from north to south, from the Panamericana Norte to Avenida 10 de Agosto. For the colonial part of the city, some routes go from Calle Larga, Calle Luis Cordero and the San Sebastián sector. The shortest route is located in the sector of España Avenue.

Fig. 2. Routes in the city of Cuenca. Source: EMOV

Figure 3 presents the 40 free points that can be used for bicycle parking in the city of Cuenca. It can be seen that in the historic center, there are fourteen parking spots.

Through Google Maps' My Maps tool, the municipal mobility company maintains information on the routes that cross the city's streets. The Table 4 shows a comparison of the characteristics of the mapping tools: Google Maps and MapBox. It is defined that the platform that adapts to the requirements of the project is Google Maps.

MapBox allows to modify the background colour, the type of icons for each entry and the colour of the routes. Google allows to create routes on the MyMaps platform, but MapBox does not.The accuracy of MapBox when exporting is low while Google Maps is accurate, so it is not recommended to export routes from MapBox. MapBox has a usage of 50,000 free visualizations, however, as MyMaps will be used to obtain routes it is free of charge.

Fig. 3. Parking lots in the city of Cuenca. Source: EMOV

Table 4. Comparison between MapBox and Google Maps' My Maps.

Feature	MyMaps (Google)	MapBox
Customization	No	Yes
Route Creation	Yes	No
Export Maps	Yes	Yes
Import Maps	Yes	Yes
Costs	Free	By use

Through an application developed in IONIC with the help of information provided by the EMOV in the Google Maps MyMaps tool, CuencaBike was generated. This integrates information on all current routes, each of these generated routes has identified points where the tourist sites of the city are highlighted. It also has sections where you can view the information of the different mechanics, parking, and BiciPública dispensers that facilitate users to have all this information organized and updated.

Figure 4 describe through the C4 Model the design process of the cyclo tourism application through cross-platform technology development technologies.

The prototype of the CuencaBike application is detailed in Fig. 5.

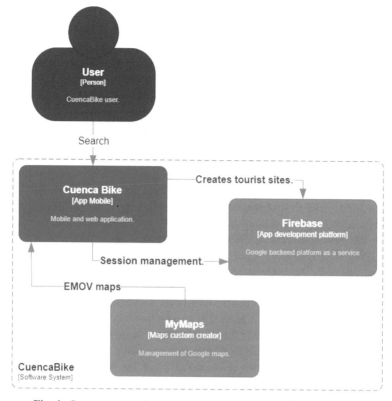

Fig. 4. System context in the C4 model of the CuencaBike application

Fig. 5. Screenshots of the mobile application CuencaBike

4   Conclusions

The information provided by the EMOV showed that the information displayed on official sites about the routes is outdated, so an alternative was presented consisting of a mobile application with all the information which is publicly accessible. The publication of the routes on Google Maps' My Maps makes it possible to maintain the graphic information of the bicycle routes.

The results of the application of the survey to know the vision of the inhabitants of the city about bicycle tourism and bicycle routes, it was found that within the population of Cuenca, there is a need to know about the routes and bicycle paths enabled within the city, also allows to know the need to establish communication and information centers about bicycle tourism in the city. Along the same lines, it was found that the population of Cuenca agrees with the creation of a mobile application to promote bicycle tourism and provide information about bicycle paths.

When diagnosing the current situation of bicycle lanes in the city of Cuenca, it can be observed that within the competent institutions (EMOV), there are plans and projects managed to improve mobility through non-motorized vehicles. In addition, there is a vision to continue creating bicycle lanes. However, the information has not been disseminated to the general public. One of the reasons for this situation is that the information is outdated and is not available on a platform accessible to ordinary users.

In the process of discriminating the characteristics of mapping technologies through a literature review, it was possible to recognize several mapping development tools, of which Google's My Maps was chosen because it has a free-to-use layer and also allows the generation of routes.

The development of the cyclo tourism application through multiplatform technologies was analyzed using the Unity platform, but after analyzing the documentation of each of the tools, it was concluded to use IONIC due to its compatibility in the development of multiplatform applications, one of its strong points is the direct connection in real-time with the Firebase database.

References

1. Gonçalves-Junior, L., Corrêa, A., Da Silva Carmo, D., Clayton, S., Toro Arévalo, S.: Bicycle diaries: educational processes in the Emotions Route. Estudios Pedagógicos **42**(1), 323–337 (2016)
2. Vistín Vistín, N.: Diseño de una ciclovía en la ciudad de Guaranda, provincia de Bolívar. PUCE, Quito (2018)
3. Silva Veloz, E., Zambrano Alcívar, J.: Estudio preliminar para la implementación de una ciclovía en la ciudad de Santo Domingo. PUCE, Quito (2018)
4. Sabino, A.B., Reis-Martins, P., Carranza-Infante, M.: Experiencias y retos del uso de datos de aplicaciones móviles para la movilidad urbana. Revista de Arquitectura **22**(1), 82–93 (2020)
5. Sánchez, J.: Redes vehiculares aplicadas a la movilidad inteligente y sostenibilidad ambiental en entornos de ciudades inteligentes. Universidad de Oviedo, Oviedo (2017)
6. Díaz García, J.: Sistema de información del mapa de conocimientos de los asesores especializados del Centro de Desarrollo Empresarial del Centro Universitario UAEM Texcoco. Universidad Autónoma del Estado de México, Ciudad de México (2014)

7. Suero, D.: Factibilidad del uso de la bicicleta como medio de transporte en la ciudad de Bogotá. Avances Investigación en Ingeniería **1**(12), 54–62 (2010)
8. Melo, E.: Diseño de una cicloruta ecoturística en la isla San Cristóbal Galápagos. Universidad Tecnológica Equinoccial, Quito (2015)
9. Xue, K., Deng, Y., Zhang, H., Pandiyan, S., Manickam, A.: Cycling environment investigation and optimization of urban central road in Qingdao. Comput. Intell. **37**(3), 1217–1235 (2021)
10. Bieliński, T., Kwapisz, A., Ważna, A.: Bike-sharing systems in Poland. Sustainability **11**(9), 2458 (2019)
11. Gazzola, P., Pavione, E., Grechi, D., Ossola, P.: Cycle tourism as a driver for the sustainable development of little-known or remote territories: The experience of the Apennine regions of Northern Italy. Sustainability **10**(6), 1863 (2018)
12. Zhao, J., Ye, Y., Li, Y., Yao, J.: Research on Integrated Bicycle Network Planning Method in Tourism City. In: Zhang L., Ma J., Liu P., Zhang G. (eds.) CICTP 2019, pp. 1639–1650. Nanjing (2019)
13. Fortunato, G., Bonifazi, A., Scorza, F., Murgante, B.: Cycling Infrastructures and Community Based Management Model for the Lagonegro-Rotonda Cycling Route: ECO-CICLE Perspectives. In: Bevilacqua C., Calabrò F., Della Spina L. (eds.) SMART INNOVATION, SYSTEMS AND TECHNOLOGIES, vol. 178, pp. 1697–1705. Reggio Calabria (2021)
14. Nilsson, J.: Urban Bicycle Tourism: Path Dependencies and Innovation in Greater Copenhagen. J. Sustain. Tour. **27**(11), 1648–1662 (2019)
15. Bakogiannis, E., et al.: Development of a Cycle-Tourism Strategy in Greece Based on the Preferences of Potential Cycle-Tourists. Sustainability **12**(6), 2415 (2020)
16. Ciascai, O.R., Dezsi, Ş, Rus, K.A.: Cycling Tourism: A Literature Review to Assess Implications, Multiple Impacts, Vulnerabilities, and Future Perspectives. Sustainability **14**(15), 8983 (2022)
17. Procopiuck, M., Segovia, Y.N.S., Procopiuck, A.P.V.: Urban Cycling Mobility: Management and Urban Institutional Arrangements to Support Bicycle Tourism Activities—Case Study from Curitiba. Brazil. Transportation **48**(4), 2055–2080 (2021)
18. Carra, M., Botticini, F., Pavesi, F. C., Maternini, G., Pezzagno, M., Barabino, B.: A Comparative Cycling Path Selection for Sustainable Tourism in Franciacorta. An Integrated AHP-ELECTRE Method. In: Cantisani, G., Le Pira, M., Zampino, S. (eds.) TRANSPORTATION RESEARCH PROCEDIA, vol. 69, pp. 448–455. Rome (2023)
19. Setyowati, E., Handayani, D.: Analysis of Influencing Factors on Using Rental Bikes at Shopping Tourism Sites in Surakarta. In: Chan W.T., Ismail M.B., Sriravindrarajah R., Hajek P., Gan B.S., Han A.L., Hidaya B.A., Kristiawan S. (eds.) MATEC WEB OF CONFERENCES, vol. 195, p. 04014. Solo Baru (2018)
20. Lee, C.F., Huang, H.I.: The Attractiveness of Taiwan as a Bicycle Tourism Destination: A Supply-side Approach. Asia Pacific Journal of Tourism Research **19**(3), 273–299 (2014)
21. Chen, C.C., Tsai, J.L.: Determinants of Behavioral Intention to Use the Personalized Location-based Mobile Tourism Application: An Empirical Study by Integrating TAM with ISSM. Futur. Gener. Comput. Syst. **96**(1), 628–638 (2019)
22. Jiménez, L.: Impacto de la Investigación Cuantitativa en la Actualidad. Convergence Tech **4**(1), 59–68 (2020)
23. Johnson, R.B., Onwuegbuzie, A.J.: Mixed Methods Research: A Research Paradigm Whose Time Has Come. Educ. Res. **33**(7), 14–26 (2004)
24. Fonseca, A.R., Gonzalez, F.: Cross-country Determinants of Bank Income Smoothing by Managing Loan-loss Provisions. J. Bank. Finance **32**(2), 217–228 (2008)
25. Cueva Cruzado, J.C.: Gestión de crédito y cobranza en una empresa de bebidas malteadas del distrito de Lima en el año 2014. Universidad César Vallejo, Piura (2015)
26. Hernández Sampieri, R., Fernández Collado, C., Baptista Lucio, P.: Metodología de la investigación. 7th edn. McGraw-Hill Interamericana, México (2018)

27. Lizmova, N.: Análisis de mapas como un método de investigación de fenómenos naturales y socioeconómicos. Revista Luna Azul **24**(1), 74–80 (2007)
28. Folgueiras Bertomeu, P.: La entrevista–Técnica de recogida de información: La Entrevista. Universidad de Barcelona, Barcelona (2016)
29. Abarca, A., Alpízar, F., Sibaja, G., Rojas, C.: Técnicas cualitativas de investigación. Universidad Costa Rica, San José (2013)
30. Lezcano, L., Vilanova, G.: Instrumentos de evaluación de aprendizaje en entornos virtuales. Perspectiva de estudiantes y aportes de docentes. Informes Científicos Técnicos-UNPA 9(1), 1–36 (2017)

Comparative Analysis of MH Versus LED Prototype Luminaires: Case Study for Lighting Public Areas

Gustavo Moreno[1], Diana Peralta[1], Jaime Molina[1] (ID), Carolina Chasi[2](✉) (ID), and Javier Martínez-Gómez[1,3] (ID)

[1] Universidad Internacional SEK (UISEK) Ecuador, Quito 170134, Ecuador
[2] Instituto de Investigación Geológico y Energético (IIGE), Quito 170518, Ecuador
cchasi@eeq.com.ec
[3] Departamento de Teoría de la Señal y Comunicación, (Área de Ingeniería Mecánica) Escuela Politécnica, Universidad de Alcalá, 28805 Alcalá de Henares, Madrid, Spain

Abstract. The primary aim of this study is to validate prototype Light Emitting Diodes (LED) luminaires designed for outdoor applications. It involves replacing 7 metal halide luminaires (MH) with 7 LED luminaires, covering an area of 1129 m. The results reveal significant improvements in illumination, as demonstrated by a comparative distribution and illuminance graph. The LED luminaires achieve an average illuminance of 27.9 lx with a relatively low standard deviation of 15.8, whereas the MH luminaires only attain an average of 25.7 lx, accompanied by a high deviation of 24.7. Moreover, the study delves into the environmental and economic implications of this transition. By applying Ecuadorian electrical system factors, it was determined that the LED luminaires result in a substantial 64.8% reduction in energy consumption, equating to 3,740 kWh saved annually. This leads to considerable monetary savings of 505 dollars, and a noteworthy reduction of 1,279 kg of CO_2 emissions every year, contributing to a more sustainable and cost-effective lighting solution for the 7-luminaire system.

Keywords: LED Lamp · Comparative Illuminance · Metal Halides Lamp · Energy Saving

1 Introduction

In Ecuador, 4.92% of the country's total electrical energy consumption, according to the National Balance of Electrical Energy, is consumed in public lighting [1]. In this sense, lighting technology by Solid State Lighting (SSL) that used a Light Emitting Diodes (LED) is called to replace incandescent-, fluorescent-, low- and high-pressure sodium (HPS), and metal halide (MH) technologies, which are used massively for the illumination of outdoor spaces. These lights have a wide range of wavelengths, consume high amounts of electrical energy, and produce heat which make these systems inefficient [2].

M. Z. Vizuete et al. (Eds.): CI3 2023, LNNS 1041, pp. 203–215, 2024.
https://doi.org/10.1007/978-3-031-63437-6_17

On the other hand, the advantages of LED luminaires compared to conventional lamps are very low energy consumption, there is low heat emission to the environment, maintenance is almost null, and it has a long life (more than 50,000 h according to the manufacturer) [3]. Hence, this source of light seems to be more suitable due to the high energy-conversion efficiency and environmental sustainability. Particularly, the positive effect that LED lamps have on the environment is that it is not made up of toxic materials such as high-pressure sodium lamps or fluorescent lamps, so they can be completely recycled. Nonetheless, it is crucial to mention that in Ecuador, there is not public lamps recycling.

Technological change is setting an international trend, especially in countries with greater economic resources. For instance, studies conducted in the US and Europe found that savings in energy consumption of about 60% can be achieved by an exchange of past used luminaires by LED lighting sources [4–6]. However, in less developed countries, the exchange rate is much lower, as suggested by the study conducted by Mónica Sabogal (2015) carried out in Colombia, where the urgent need to change the country's lighting models is evidenced, implementing LED technology with a direct benefit in energy consumption, achieving electricity savings of between 50% and 90%, projecting a return on investment in a very short term [7]. In addition, according to Hermoso-Orzáez, et al., 2016, electrical installations need to be modified in order to take advantage of the use of LED luminaires. Particularly, they generate large power-on currents. In this way, they recommend energizing the luminaires using magneto-thermal protection circuits with slow trip curves that will tolerate the large short-term inrush currents.

Some of the factors that influence the rates mentioned above are lack of semiconductor industry; difficulties in manufacturing processes; relatively small country sizes, which makes it difficult to achieve economies of scale; and other factors. In this way, although LED technology is probably the most appropriate to reduce lighting costs, and in turn reduce the environmental damage related to this activity, this change could not occur with the desired speed. With this background, the development of prototypes of LED luminaires has been proposed that can, from some fundamental components (LED, Driver, Heatpipes), be manufactured using simple and low-scale techniques.

Although several studies have been developed to compare the efficiency and effectiveness of the exchange of MH by LED technology, there is a lack of information of this change in Ecuador.

In Ecuador, the change of lighting in public lighting is currently being promoted by requesting distribution companies to purchase only LED technology luminaires for public lighting, This motivates the academy and the national industry to develop equipment with solid state technology with better features and that meet international quality standards in the photometric and electrical parameters of the prototypes.

Thus, the present work focuses on the results of field tests for prototype LED luminaires and a comparison with previous lighting based on metal halide technology. Specifically for this case study, the LED system was evaluated by the heat pipe method for heat dissipation from the LED junction and that have been developed using the additive manufacturing technique by 3D printing for the main anchors as a means of lowering production costs. The tests have been carried out at the SEK Ecuador International University campus located in the city of Quito. After the prior art study, three study scenarios are proposed, in order to compare the lighting technology with MH and the prototype LED luminaire developed, in the three scenarios an analysis of the energy consumption of the lighting system in operation as well as its photometric parameters is carried out.Methodology.

To determine the advantages of proposing the change of MH luminaires for the LED luminaire prototype, we proceeded to analyze different studies in which comparisons of the two luminaire technologies are made, as well as the review of simulations carried out in lighting programs for public lighting, which detail the advantages and disadvantages of each of the luminaire technologies.

1.1 Comparative Analysis of MH Luminaires and LED Luminaires

It is necessary to perform a comparative analysis between both technologies and LED luminaires in order to determine how viable it is to replace MH luminaires with LED luminaires in public lighting systems in Ecuador. First, we will analyze the luminous flux curves over the operation time of the two technologies to be compared.

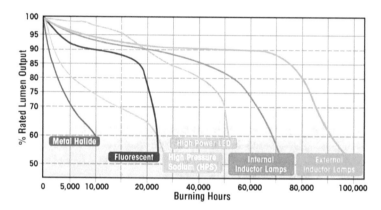

Fig. 1. Flujo Luminoso a lo largo del tiempo de los diferentes tipos de luminarias

In image 1, on the vertical axis we have the percentage of lumens, while on the horizontal axis we have the operating hours of the different luminaire technologies. Also in image 1 we can see how the luminous flux curve of the Metal Halide luminaires, blue curve, decreases the luminous flux after hours of operation, losing 70% of its luminous flux when reaching 5000 h of use, however we can see how the luminous flux curve of the light blue LED luminaires (High Power LED) after 50000 h of operation loses the same percentage of luminous flux of 70% as the MH luminaires, It is also observed how the luminous flux curve of LED technology luminaires does not decline abruptly after losing 20% of its luminous flux, but rather it gradually decreases, at 50,000 h of operation it has lost 30% of its luminous flux [8].

Another parameter to take into account is the lighting time of the luminaires, for example, mercury luminaires take 4 to 5 min to turn on, because it requires this time to acquire its operating temperature, on the other hand, LED luminaires turn on immediately optimizing energy consumption, if we talk about electricity consumption LED luminaires are more efficient than mercury luminaires, the following table shows the comparison of mercury luminaires and LED luminaires.

	Mercurio AP	Sodio AP	LED
Filamento	Si	Si	No
Factor de potencia	0,8	0,92	0,98
Temperatura de funcionamiento (°C)	300	350	40
Vida útil (Horas)	10000-12000	10000-28000	>50000
Tiempo de encendido (min)	4-5	5-10	0
Temperatura del color (°K)	3500 y 4500	2000-3500	3000-6000
CRI	40 - 45	<50	>65
Eficacia (lm/W)	31-52	80-150	>60
Parpadeo	mucho	Poco	Ninguno
Contenido de mercurio (mg)	10-100	10-50	Ninguno
Billo	Mucho	Mucho	Ninguno
Distorsión armónica	<35%	<35%	< 10%

Fig. 2. Comparative MH vs LED

For the performance verification process in the LED prototypes, three experiments were established: the first was conducted on an integrating sphere where power, luminous flux, stability, spectral distribution, and chromaticity diagram were obtained. The second consists of a field lighting test where the amount of lux delivered by a system composed of several luminaires is measured. For the third experiment, electrical consumption tests were performed in the laboratory. The tests were repeated on 7 LED luminaires and 7 MH luminaires as a comparative control group.

1.2 Experiment 1

In order to obtain actual values from the control group of the 7 MH luminaires and the 7 LED prototype luminaires, a luminous flux test was requested on the LED prototype luminaire. An integrating sphere from the National Institute for Renewable Energy (INER) as shown on Fig. 1. A total of 78 measurements during a period of 78 min, and with minimum temperature of 25.6 °C a maximum of 26.1 °C and an average of 25.8 °C. The trials are conducted following the references established in IESNA LM 78–07 Approved Method for Total Luminous Flux Measurement of Lamps Using an Integrating Sphere, IES LM 54–12 IES Guide to Lamp Seasoning, and IESLM 79–08 approved method: Electrical and photometric measurements of solid-state lighting products.

Other Equipment

– Stabilized source 04- E013_FEST_04
– Direct current source 01-FDC_01
– Digital power meter 01-MDP_01

– Flow reference pattern 01-PRF_01
– Temperature datalogger 01-RT_01

Fig. 3. Integrating Sphere, National Institute for Renewable Energy (INER)

1.3 Experiment 2

For the second experiment, the exterior parking lot of the Carcelén campus at the SEK International University of Quito was established as a test field with an area of 1129 square meters. Its geographical location is established at latitude -0.09 and longitude -78.484, it was analyzed: morphology of the place, visual comfort. The initial testing period for LED luminaires is established for one month, during which various types of weather conditions typical of the area have been presented, i.e. heavy rains and sunny days.

As can be seen in Figs. 2 and 3, the number of MH metal halide luminaires present in the designated area is equal to 7 and they are unevenly distributed. To maintain the parameters of the comparative study, it was decided to replace only those luminaires in operation, maintaining the same position in each one of them (Fig. 4).

For the lighting measurements, a grid was defined with measurement points every 3 m as can be seen in Fig. 3, establishing 106 points. The base lighting measurement (MH) was carried out at 7:30 p.m., choosing a night that had the necessary characteristics, that is, without environmental light disturbances (cloudy night and without vehicles or other nearby light sources). Once the measurement of all the points was finished, a second measurement was made to achieve greater reliability in the illuminance values of the area chosen for the study.

Subsequently, the MH lamps were replaced by prototype LEDs, as shown in Fig. 3, and data was collected again, that is, two measurements in each of the 106 points of the

Fig. 4. Place of the field tests

previously established grid. Similarly, measurements were made at 7:30 p.m. and a night that did not present environmental lighting disturbances was chosen. Figure 5. Shown the illuminance measuring instrument.

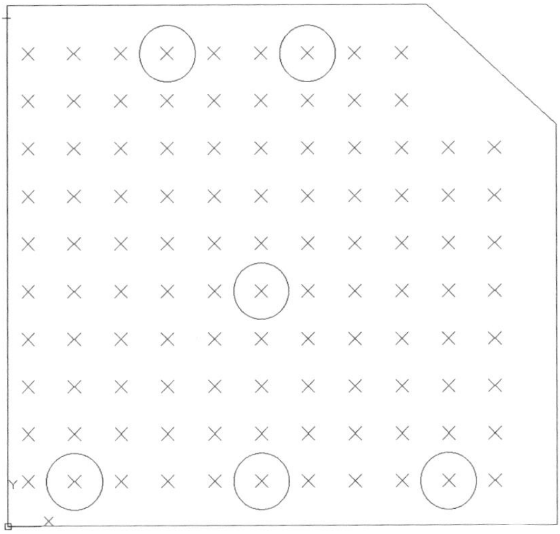

Fig. 5. Luminaire installation plan (Legend: X's mark the points chosen for taking illuminance measurements, meanwhile the areas framed by a circle indicate the place where the luminaires are installed).

The instrument used in the illuminance tests is a Super Scientific Datalogger model 850007 lx meter.

Fig. 6. Installation of prototype LED luminaires

Fig. 7. Illuminance measuring instrument.

1.4 Experiment 3

For the third experiment, being the energy consumption of the implemented lighting system an important parameter to evaluate in the comparison of the two technologies of the case study, it is important to consider the variables of: voltage, alternating current and active power consumed by each LED and MH luminaire were measured. Measurements were made using a Fluke model 115 brand multimeter as shown in Fig. 5, measurements were made in the laboratory (Fig. 8).

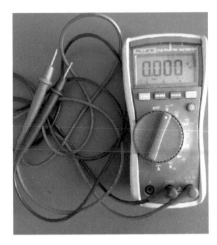

Fig. 8. Voltage and current measuring instrument

2   Results

According to the proposed methodology, it was studied the behavior of the traditional MH public luminaries in Ecuador vs LED luminaries with the goal of analyzing which technology requires less energy and produce less effects in the environment. For this purpose, three experiments were developed. In the experiment 1, the luminous efficacy was stablished by using an integrating sphere. Moreover, in the experiment 2, the surface illuminance of the abovementioned technologies was measured and compared. Finally, the energy consumption of the two systems was evaluated by measuring the power consumption.

2.1   Experiment 1

By using the integrating sphere to measure the optical radiation, the total luminous flux was determined. The used method allows the reduction of uncertainties due to a smaller calibration chain. In this case, the detector is a spectroradiometer which is in charge of getting spectral information from LED luminaires. In Table 1, it is presented the obtained results. It was found that the luminous efficacy of 113,77 lumens of output per watt of energy.

Additionally, Fig. 6 shows the spectral distribution of the analyzed luminaires. These results were obtained by using a colorimeter as a detector. It is able to quantify the color emitted by the LED lamp and calculate the chromaticity coordinates that are presented (Fig. 9).

Table 1. Luminous Flux at integrating sphere

Parameter	Value
Luminous flux (lm)	7201,7
Voltage (V)	219,6
Current (A)	0,2971
Power (W)	63,3
Luminous efficacy (lm/W)	113,77

Spectrum

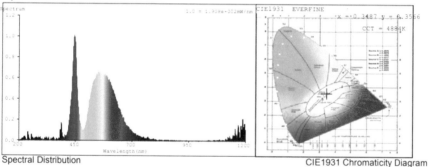

Spectral Distribution   CIE1931 Chromaticity Diagram

Colorimetric Parameters

Chromaticity Coordinate: x = 0.3487 y = 0.3566 / u' = 0.2119 v' = 0.4876 (duv=1.07e-03) Dx,Dy:0.0002,0.0023
CCT= 4884K   Prcp WL: Ld=573.1nm   Purity=11.6%
Peak WL: Lp=446nm FWHM: =21.7nm Ratio:R=14.6% G=82.7% B=2.7%

Render Index: Ra = 70.9

R1 =69	R2 =75	R3 =77	R4 =72	R5 =69	R6 =64	R7 =80
R8 =60	R9 =0	R10=38	R11=68	R12=38	R13=69	R14=87

R15=65

LEVEL:OUT   WHITE:ANSI_5000K

Fig. 9. Spectral distribution

2.2 Experiment 2

In the experiment 2, it was possible to compare the results of surface illuminance emitted by luminaires MH and luminaires LED by using an electronic lux meter. The measures were obtained in lux units, and the values are presented in Table 2.

Furthermore, Fig. 7 presents the average and normal distribution of the illuminance values reported for the Light Emitting Diodes (LED) and metal halide (MH) luminaires. Both, Table 2 and Fig. 7, demonstrates that LED luminaires emit more light than MH with a difference of approximately 2 lx. Also, Fig. 7 show that the measured values in the case of Prototype luminaires LED are closer to the mean value (Fig. 10).

Table 2. Surface illuminance

	Luminaires MH (lux)		Prototype luminaires LED (lux)	
	Measurement 1	Measurement 2	Measurement 1	Measurement 2
Mean	25,8	25,7	27,9	27,9
Maximum	151	144	76	74
Minimum	2	2	6	6

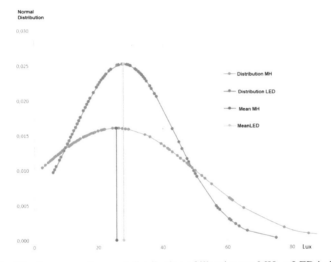

Fig. 10. Average and normal distribution of illuminance MH vs LED in lux

2.3 Experiment 3

Finally, in the experiment 3, it was determined the voltage, current, and power consumed by each of the studied luminaires. Particularly, the measures of the consumed power allow obtaining the total consumption. Hence, it was possible to identify that MH luminaires require approximately 184% more power than LED luminaires which results in higher energy consumption and higher harmful emissions to the environment (Table 3).

3 Discussion

From the first experiment conducted in the integrating sphere a luminous efficiency of 113 lm/W with a luminous flux of 7201 lumens and a consumption of 63.3 W. The values are in accordance with commercial LED luminaries varying from 90 to 130 lm/W [9].

As can be seen in Table 2 and Fig. 7, a higher average can be seen in the LED illuminance levels with 27.9 lx versus 25.8 lx in the MH luminaires, which denotes an illuminance improvement of 8.1%. On the other hand, the minimums recorded in illuminance using LEDs are 6 lx, these being higher than those presented by MH luminaires

Table 3. Electric measurements

Item	Prototype LED luminaire			Luminaire MH		
	Voltage (V)	Current (A)	Power (W)	Voltage (V)	Current (A)	Power (W)
1	115,9	0,58	68,03	221,4	0,71	157,19
2	124,8	0,51	64,64	221,8	0,62	137,51
3	123,0	0,53	66,05	222,1	0,65	144,36
4	123,9	0,53	66,78	219,3	0,85	186,40
5	123,7	0,53	66,67	217,0	1,73	375,41
6	124,1	0,53	66,64	217,6	0,84	182,78
7	123,9	0,52	65,29	217,9	0,62	135,09
		TOTAL	464,10		TOTAL	1318,74

with 2 lx. Similarly, there is a significant difference in the maximum illuminance, being higher those of the MH luminaires with 151 lx versus 76 lx of the LEDs. In this way, when analyzing the previous data and the distribution graph using the standard deviation, 15.8 for LED versus 24.8 in the MH luminaires, great improvements can be seen for the LED system with a higher average and better uniformity in the illuminance of the test area avoiding the zebra effect that is one of the problems of light pollution as pointed [10].

From experiment 3, it can also be observed a lower use of electrical power by the LED system with 464 watts compared to 1318 watts of the MH system. This means an energy saving of 64.8%, which, in terms of Energy, with a usage rate of 12 h per day, translates to 3,740 kilowatt hours per year for the 7-luminaire system. It was reported that energy savings of 19–44% can be achieved by using LED technology. Hence, this study surpasses this finding, and it can be used as a base for decision making in order to change the MH systems that still prevalence in Ecuador. Moreover, it is possible to increase the energy savings by using smart control over the luminaries.

The energy savings achieved in this research can be translated into an economic saving of $ 505 per year, considering a rate of $ 0.135 per kilowatt hour of energy for consumption between 700 and 1000 kWh per month [11]. In addition, it is crucial to mention that further cost savings are achieved using additive manufacturing technique by 3D printing for the main anchors.

In turn, energy savings translate into a lower CO_2 emission into the atmosphere with 1 279 kg using the factor of 490 gCO_2 per kWh established for the electricity generation system of Ecuador [12], found in their life cycle assessment that around 26% or 17% lower average of environmental impacts can be reached by LED luminaire [13].

It is crucial to mention that modern LED systems permit higher levels of illuminance with low electricity consumption. Hence, this LED characteristic make people feel safer and that the level of visual comfort is adequate according to the activity to be developed where it is implemented [14].

In general, this study was set out with the aim of assessing the importance of the exchange of MH luminaires by LED luminaires. In this way, demonstrated that the main advantage of LED systems are energy efficiency and associated cost savings, which was corroborated by the development of this study [15].

In addition, it is important to take into account that there is plenty of improvements to be done on LED lighting which will led in further increase of efficiency and decrease of costs. Nonetheless, this work has the limitation that do not analyze the light pollution on human health. In this sense, it was found that not only MH lamps but also LEDs would produce negative effects on human health such as scotopic and melatonin suppression. This can affect the circadian system, of the human being as well as produce visual fatigue. Hence, it is important to design lighting systems based not only on the spaces but also on the activities to be carried out in them.

4 Conclusions

This work contributes to the validation of prototypes of LED luminaires through field tests that prove again that this technology has better lighting and energy characteristics than traditional luminaires.

Evidence is presented of the economic savings presented by prototype LED luminaires with an electricity use of only 35.2% compared to the initial system; and the improvement in the environmental impact avoiding that approximately 100 kg of $CO2$ are released monthly into the atmosphere.

Although feasibility is demonstrated, it is recommended to perform deterioration tests on luminaire materials based on a commercial use of the prototypes.

There are mechanical factors of the luminaires that must also be evaluated in order to establish the life of the LED luminaire. With these factors of dust and water resistance of solid state luminaires will complement the study of the useful life of these.

In Ecuador, in tourist areas, it has been decided to change the traditional technology luminaires for LED, without a thorough study in which it is analyzed if the power of the LED luminaire is enough to not produce the zebra effect or, on the contrary, to produce light pollution to the night sky. It is important to emphasize that the lighting system should be redesigned rather than simply replaced.

References

1. Vanesa Carolina, B.L., Trujillo Guerrero, M.F., Pozo Palma, P.M., Chasi Gallo, C.C., Chavez Tapia, G.S.: International Conference on Information Systems and Computer Science (INCISCOS) (2018), pp. 107–114 (2018)
2. Nguyen, K.Q., Tran, P.D., Nguyen, L.T., To, P.V., Morris, C.J.: Aquac Fish **6**, 432 (2021)
3. Ferreira, V.J., Knoche, S., Verma, J., Corchero, C.: J. Clean. Prod. **317**, 128575 (2021)
4. Orzáez, H., Jesús, M.: Technological led´s change in public projects lighting. "swot" analysis: strengths and weaknesses (2016)
5. Valentová, M., Quicheron, M., Bertoldi, P.: Int. J. Green Ene. **12**, 843 (2015)
6. Rode, I., Moriarty, M., Beattie, C., Mcintosh, J., Hargroves, K.: Technologies and Processes to Reduce Carbon Intensity of Main Road Projects (2014)

7. García-Tenorio, F.A., Simisterra-Quintero, J.J., Barre-Cedeño, K.N., Bautista-Sánchez, J.V., Chere-Quiñónez, B.F.: Sapienza: International Journal of Interdisciplinary Studies **3**, 245 (2022)
8. Profesional, C., Hermes, J., Hijar, L., Alcántara, F., Lima -Perú, M.: Facultad de Ingeniería (2018)
9. Bosco, L.D., Catuogno, G.R., Flores, H.D.: IEEE Congreso Bienal de Argentina (ARGEN-CON), pp. 1–4 (2020)
10. Rueda-Punina, V.J.: FIGEMPA: Investigación y Desarrollo **14**, 111 (2022)
11. Luna, V., Arias-Cazco, D., Aguila Téllez, A.: in (2015)
12. Novasinergia revista digital de ciencia, ingeniería y tecnología **5**, 58 (2022)
13. Davidovic, M., Kostic, M.: Energy **254**, 124299 (2022)
14. Fryc, I., Czyżewski, D., Fan, J., Gălățanu, C.D.: Energies (Basel) **14** (2021)
15. Morgan Pattison, P., Hansen, M., Tsao, J.Y.: C R Phys **19**, 134 (2018)

Control System for Electrical Conductivity in Hydroponic Cultivation of Lettuce

Paul S. Espinoza[✉] [ID], Marco V. Pilco [ID], Vicente Reinaldo Cango Aguirre, and E. Fabian Rivera [ID]

Instituto Tecnológico Superior Universitario Oriente, Joya de los Sachas 220101, Ecuador
pespinoza@itsoriente.edu.ec

Abstract. This study investigates the proportional integrator (PI) controller for the automation of a hydroponic cultivation system of crespa lettuce, with the nutrient film technique (NFT). With the research conducted, in recent times in the sectors where the different plants have been cultivated, the soil has been degraded, for that reason it must be fertilized with organic and chemical products, during the harvest it must be moistened with fungicides to dispense with diseases and pests within the crop, there is a water irrigation system to keep the soil moist during growth and development. The hydroponic cultivation system consists of transmitting a nutrient fluid through pipes where the plants are housed for their development, the amount of nutrients required by the plants for their growth is injected into the crop allowing the variable electrical conductivity (EC) is 1.2 and 1. 8 uS, for this a PI control system must be made, in closed loop using the programmable logic controller logo V8, for the design of the controller samples are taken of the behavior of the variable of the EC at the time of adding the nutrients and the integral process is applied to determine the mathematical model of the system with this the tuning of the constants of the PI controller is made, applying methods such as: Ford, Astrom, Hay and Zou Brigham to determine the constants of the controller perturbations are applied to the process to verify the control action of the hydroponic system.

Keywords: NFT system · Electrical conductivity · Lettuce cultivation

1 Introduction

Water scarcity causes uncertainty for the future regarding the economic development of agriculture and livestock. Planting areas are reducing significantly and food demand is increasing. In this sense, aquaponics is an alternative, due it combines aquaculture and hydroponics for food production [1]. The consumption of fruits and vegetables of nutritional quality increases due to their high amounts of bioactive compounds with antioxidant activity, to prevent diseases and improve health [2], evaluate the performance of crispy lettuce cultivars grown under the hidroponic system [3]. Jambu (Spilanthes oleracea L.) is a leafy vegetable with medicinal properties used in food, cosmetics, and medicine [4]. Threats to global food security have generated the need for novel food

M. Z. Vizuete et al. (Eds.): CI3 2023, LNNS 1041, pp. 216–228, 2024.
https://doi.org/10.1007/978-3-031-63437-6_18

production techniques to feed an ever-expanding population with ever-declining land resources [5]. The new technological advances present in agriculture allow obtaining improvements in production either by monitoring through low-cost environments such as LoRa; through the use of hydroponic systems or innovative systems for local production where vegetables such as lettuce are produced [6]. Lettuce is basically considered a leaf vegetable because it has the high culinary quality of a fresh salad. It is grown worldwide in various agricultural systems, in open field and greenhouses, in soil and in aquaculture [7].

Vegetables are nutrients that humans need. Hydroponics based on nutrient bed technology (NFT) is a system commonly used on small farms to produce vegetables that can accelerate harvest time with high quality products. One of the most important parameters of hydroponics is the supply of sufficient nutrients, depending on the age of the plant. In this study, a hydroponic system based on NFT with a fuzzy logic control system is designed. The set point value of electrical conductivity (EC) of the nutrient solution is determined according to the needs of the plant [8].

Hydroponics agriculture comes up as a solution to limited agricultural land that can lead to a decline in agricultural production capacity. In Hydroponics agriculture, there is a challenge of precision agriculture, especially for some sensitive plants, e.g., bok choy and lettuce [9]. Soilless crops are presented as a cultivation alternative in the presence of marginal soils with scarcity characteristics of the province of Santa Elena (PSE). Hydroponics has a high worldwide productivity per unit area, water savings and crops throughout the year [10]. Nutrition and water play an important role for hydroponic plants. This hydroponic system has the advantage that it can be used to overcome the problem of an increasingly narrow land shortage the most important thing in a hydroponic planting system is the process of monitoring the amount of plant nutrients [11]. The levels of electrical conductivity (EC) can alter water and nutrient uptake by plants, influencing their metabolism and yield. This experiment was carried out to verify the effects of EC on the yield and the development of the crisp head lettuce [12].

A nutrient film technology (NFT) hydroponic system was evaluated for the development of French lettuce and romaine lettuce, under artificial climatic conditions provided in a greenhouse. Planting took place in autumn and production in winter. The yield of French lettuce obtained was 0.55 kg/piece (23.6 t/ha/cycle) and 0.40 kg/piece (23.6 t/ha/cycle) at planting densities of 4.3 and 5.9 seedlings/m2. The romaine lettuce yield obtained was 0.55 kg/piece (32.4 t/ha/cycle) at a planting density of 5.9 seedlings/m2. Water consumption was 13.8 L per plant, 93.0 of this consumption being for evaporation and the rest for dry matter. As Jacques Hernandez says, the crop cycle was 27 days at planting and 59 days from planting to harvest [13]. Using a biol with values of 10 2000 Nitrogen, 219.10 Phosphorus and 1103. 80 Potassium with a time of 45 days of fermentation. The biol obtained was proceeded to make dilutions of 5, 10, 15 and 20%, which served as a replacement for the standard nutrient solutions for the hydroponic cultivation of Lactuca sativa var. Longifolia, "lettuce" the results indicate that the 20% concentration of the biol is the one that generates the best yields in total weight, foliar and root length in lettuce plants [14].

An analysis carried out in the parish of El Quinche found lead levels that do not exceed the recommended amount for conventionally produced lacquer crops, when comparing the results obtained with various international concentrations. According to food standards (CODEX), the average lead content of common lettuce is 0.28 mg/kg and the confirmed concentration is 0.3 mg/kg, and after analyzing the presence of cadmium in lettuce, the concentration in common lettuce was found to be 0.05 mg/kg. Salad According to studies by Pilar Fuentes, the samples are well below the limit set by CODEX (0.20 mg/kg) [15].

2 Methodology

2.1 Electrical Conductivity Controller (EC)

The electrical conductivity meter can be adapted to different conductive electrodes, configured according to the type of cell electrode, from 1 to 2000 uS with a factor of 10, for this series, from 0.01 to 20 mS, in this case, the salinity of the water can be monitored. Aquaculture plants have 1.5–2.2 mS, the 1.8 mS recommended by aquaculture professionals corresponds to 18 mS for the conductivity transmitter.

The following table describes the operation of the conductivity converter, considering that the regulator is designed to measure up to 4 mS, the test probe gives values up to 2 mS/cm, the variation of the output current is linear as shown in the following figure (Table 1).

Table 1. Transmitter measurement

Analog value	mS/cm	mA	Slope of the line
0	0	4	0,4
100	4	5,6	0,4
200	8	7,2	0,4
300	12	8,8	0,4
400	16	10,4	0,4
450	18	11,2	0,4
475	19	11,6	0,4
500	20	12	0,4
600	24	13,6	0,4
700	28	15,2	0,4
800	32	16,8	0,4
900	36	18,4	0,4
1000	40	20	0,4

For cultivation, the optimum current is 11.2 mA, which corresponds to a measurement of 18 ms/cm, which indicates a certain amount of nutrients dissolved in water so that

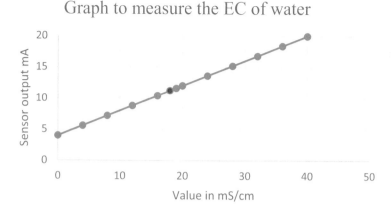

Fig. 1. Graph transmitter measurement

the plant can absorb them and not suffer damage during the development phase, if the conductivity is too high, the plant tends to burn the leaves and rot over time, if the conductivity is below the permitted level, the lettuce will wilt and if the level does not rise, the plant will die.

2.2 Obtaining the Mathematical Model

First-order systems plus dead time are often used to represent industrial processes, as shown in the following equation:

$$G_{(S)} = \frac{k * e^{-Ls}}{s} \tag{1}$$

Equation 1 is used to express mathematically in a very simple way how the actual behavior of the process or system over time is when a stimulus is applied to its input. In this way, a hydroponic system controller can be implemented.

For the sample values of the EC variable are taken, taking into account that the nutrient reservoir is 80 L of water without nutrients ready to be distributed to the hydroponic cultivation of lettuce, in 3.55 s the nutrients are injected through a dispenser corresponding to 60 ml of the formula, which for the system is the fertilizer 20 20 20, the data collected is done in an experimental way until the 18 uS is exceeded, which are used to determine the mathematical model of the hydroponic cultivation of lettuce (Table 2).

Integrating Process. Camacho describes process integration as a system characterized by the continuous growth of a process variable, i.e., its output is a slope when the control value is changed. This usually occurs when there is an imbalance between the input and output energy flow of the process [6] (Fig. 2).

Table 2. EC variable data

Measurement of electrical conductivity				
Num	Time(seconds)	Cant	Measurment (mS/cm)	Motor
0	0	0	2.44	0
1	3	60 ml	2.44	1
2	6	60 ml	2.44	1
3	9,55	60 ml	3.85	1
4	13.1	60 ml	5.15	1
5	16.65	60 ml	6.33	1
6	20.2	60 ml	7.51	1
7	23.75	60 ml	8.79	1
8	27.3	60 ml	10.05	1
9	30.85	60 ml	11.29	1
10	34.4	60 ml	12.44	1
11	37.95	60 ml	13.65	1
12	41.5	60 ml	14.82	1
13	45.05	60 ml	16.02	1
14	48.6	60 ml	17.20	1
15	50.96	60 ml	18.00	0
16	52.15	60 ml	18.40	0
17	55.7	60 ml	19.60	0

Considering the characteristics of the integration process shown in Fig. 1, the mathematical model can be represented by the following equation:

$$G_{(S)} = \frac{k * e^{-Ls}}{s} \tag{2}$$

The following equation is used to obtain m:

$$m = \frac{O2 - O1}{t3 - t2} \tag{3}$$

$$m = \frac{18 - 2,44}{50,96 - 6}$$

$$m = \frac{15,56}{44.96}$$

$$m = 0,3461$$

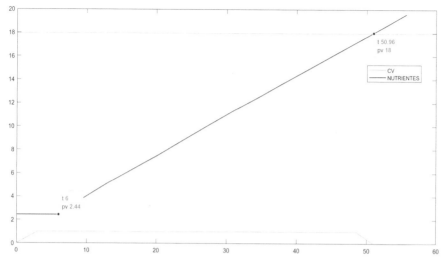

Fig. 2. Graph C.E variable

The time-out is represented by:

$$L = t2 - t1$$
$$L = 6 \tag{4}$$

The profit from the K integration process:

$$K = \frac{m}{t3 - t2} \tag{5}$$

$$K = \frac{0,3461}{18}$$

$$K = 0,01923$$

Substituting the values obtained in the transfer function equation yields the following formula:

$$G_{(S)} = \frac{0,01923}{s} e^{-6s} \tag{6}$$

The given equation is a transfer function used to model the behavior of a plant in a physical environment in hydroponics, it can be used to find the constants of a PID controller (Fig. 3).

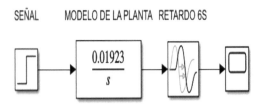

SEÑAL   MODELO DE LA PLANTA RETARDO 6S

Fig. 3. Plant representation

2.3 Pid Controller Tuning

The parameters defined in Eq. (6) are used to calculate the tuning constants of the aquaculture plant PID controller used to supply nutrients to the water recycling system and to investigate the four controller tuning methods. Perform a comparative analysis and choose the best configuration that used the specified controller. The calculated constants are shown in Table 3 (refer to the "Process Integration Methods" section for the equations used to find constant values).

Table 3. Tuning constants for hydroponic crop controller.

Método	Kc	Ti	Td
Ford	10, 4	12	2,2
Astrom - Hagglund	8, 14	12	3
Hay	3, 47	19,2	4,8
Zou –Brigham ($\lambda = 3L$)	4, 95	142	2, 78

2.4 Pid Controller

In order to simulate the mathematical model of the hydroponic lettuce cultivation, it is necessary to use computer tools that provide a graphical and user-friendly environment that facilitates the analysis of the proposed control systems for the system. 4 configuration methods are considered (Fig. 4 and Fig. 5).

The optimum values for hydroponic cultivation are determined by the Zou-Brigham method, where it stabilizes in less than 250 s, and others take a little longer to stabilize, the adjustable variable being the electrical conductivity of the system.

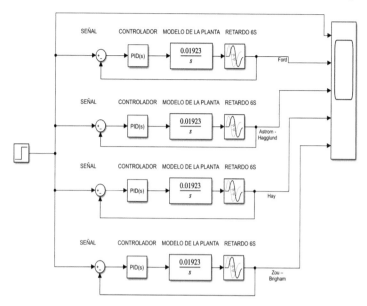

Fig. 4. PID controller with four tuning methods

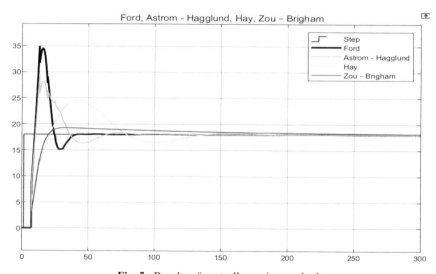

Fig. 5. Results of controller tuning methods

3   Results

3.1   Pi Control System

When creating the control models, the values of the configuration constants corresponding to the Zou - Brigham configuration method are given, the Logo V8 automaton has the constants of the PI controller, which is used for hydroponic growing. The information found about the controller is used to determine the EC of the control system as described below (Fig. 6):

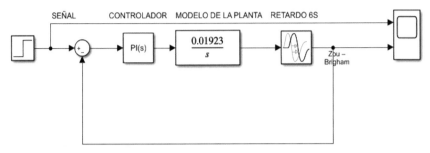

Fig. 6. PI control system

When using the Zou - Brigham tuning method for the PI controller, it has a settling of the variable of 250 s (Fig. 7).

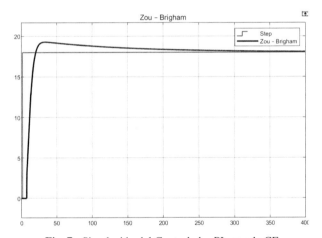

Fig. 7. Simulación del Controlador PI, para la CE

Considering the responses of the simulated controllers, any controller can be chosen to implement the hydroponic cultivation system, because the process is slow acting and does not require greater precision, the PI controller is chosen for the injection of nutrients to the crop (Fig. 8).

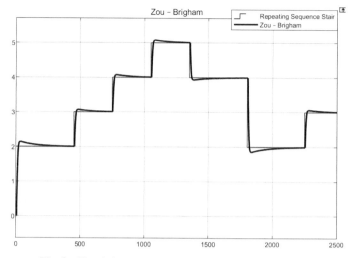

Fig. 8. Simulation of the PI controller with signal variation

3.2 Pi Control System with Disturbance

In a hydroponic crop control system, disturbances can occur when nutrient-free water is introduced into the system, the controller must operate at the same time (Fig. 9).

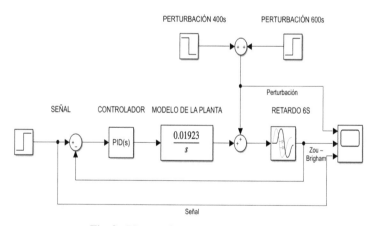

Fig. 9. PI control system with disturbance

The following graph represents the control system with disturbance that allows the PI controller to interact when there is a change in the variable to be controlled (Fig. 10).

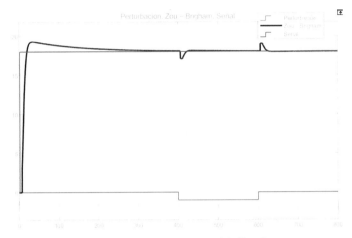

Fig. 10. Graph control system with disturbance

3.3 Testing

The implementation of the Kc and Ki constants in the Logo V8 automaton, the set point value is set to 450 in analog values, which corresponds to the value of 18 uS in electrical conductivity (Fig. 11 and Fig. 12).

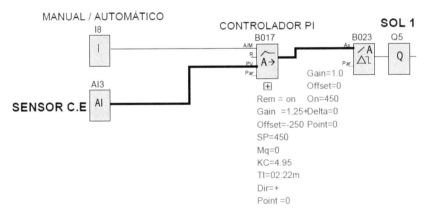

Fig. 11. PI Controller in the V8 Logo

The electrical conductivity control system allows to maintain in the range required by the plant for its development, throughout the production time, which during an average of 9 months for harvesting. The implementation of the hydroponic cultivation of lettuce with the PI controller in the PLC logo V8, as a result the growth of plants in optimal conditions is evident as shown in the following figure (Fig. 13).

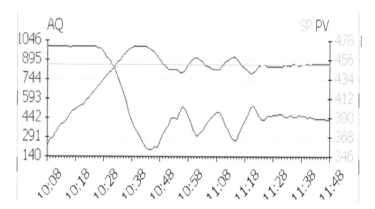

Fig. 12. PI controller operation

Fig. 13. Lettuce planting

4   Conclusions

The use of the electrical conductivity transmitter allows visualizing the changes produced by the EC variable, when introducing nutrients, which will be used to obtain the necessary data for the mathematical model that governs the hydroponic crop control system. By tuning the constants of the integral proportional controller, it allows the system to act every time there is a change of values in the variable, the Set Point that is located in the controller is 450 which corresponds to the value of 1.8 uS which is a range allowed so that the plant can absorb all the nutrients without problems.

References

1. Calderón-García, D.M., Olivas-García, J.M., Luján-Álvarez, C., Ríos-Villagómez, S.H., Hernández-Salas, J.: Investig. Cienc. Univ. Autónoma Aguascalientes **27**, 5 (2019)
2. Lara-Izaguirre, A.Y., Rojas-Velázquez, Á.N., Alia-Tejacal, I., Alcalá-Jáuregui, J.A.: Agrociencia **56**, 207 (2022)
3. Gualberto, R., Alcalde, G.L.L., Silva, C.L.: Colloq. Agrar. **14**, 147 (2018)
4. da Silva Cunha, H.P., et al.: Comun. Sci. **13**, 1 (2022)
5. Glob. Food Secur. **18**, 35 (2018)
6. Rivera Guzmán, E.F., et al.: Sensors **22**, 6743 (2022)
7. Carrasco, G., Izquierdo, J.: L A empresa hidroponica de mediana escala: la tecnica de la solucion nutritiva recirculante ("NFT") (2018)
8. Fuangthong, M., Pramokchon, P.: Int. Conf. Digit. Arts Media Technol. ICDAMT, IEEE, Phayao, pp. 65–70 (2018)
9. Herman, Surantha, N.: 7th Int. Conf. Inf. Commun. Technol. ICoICT, pp. 1–6. IEEE, Kuala Lumpur, Malaysia (2019)
10. Pertierra Lazo, R., Quispe Gonzabay, J.: La Granja **31**, 118 (2020)
11. Firdaus, S.M., Ivan, Fikri, M.R.: Monit. Control. Smart Hidroponics Using Android Web Appl. (2021)
12. Costa, P.C., Didone, E.B., Sesso, T.M.: Condutividade elétrica Soluç. Nutr. E produção alface em hidroponia (2022)
13. Hernández, C., Hernández, J.L. (2018)
14. Medina Vásquez, K.M.E., et al.: Arnaldoa **29**, 137 (2022)
15. Pila Fueres, C.Y.: (2018)

Building a WhatsApp Chatbot with Node.js: File Processing and AI Response Generation

Juan Minango[1]($\boxtimes$)(ID), Marcelo Zambrano[1](ID), Wladimir Paredes[2](ID), and Karla Ayala[1]

[1] Instituto Tecnológico Superior Universitario Rumiñahui,
Av. Atahualpa 1701 and 8 de Febrero, 171103 Sangolquí, Ecuador
juancarlos.minango@ister.edu.ec
[2] Consejo de Aseguramiento de la Calidad de la Educación Superior del Ecuador, Quito, Ecuador

Abstract. This paper presents the development and implementation of an AI-powered chatbot system tailored for educational institutions. The system utilizes a diverse dataset of conversations specific to the educational domain, enabling accurate understanding and response generation for frequently asked questions, inquiries, and specific queries related to the institute's programs, admissions, courses, and facilities. The dataset is pre-processed to ensure clean and formatted text, incorporating techniques such as tokenization, stemming, and stop-word removal. The underlying model of the chatbot is based on the GPT-2 transformer model, known for its exceptional language generation capabilities and contextual understanding. Additionally, the Logic Learning Machine (LLM) algorithm is employed to generate intelligible rules, enhancing the chatbot's decision-making process. The implementation involves the integration of various libraries and technologies, including 'whatsapp-web.js' for establishing a WhatsApp session, 'openai' for AI-driven completions, and 'twilio' for response delivery via messaging service. The system demonstrates its effectiveness in accurately understanding and responding to user queries, contributing to improved communication and information dissemination within the educational domain. Overall, the chatbot system presents a valuable tool for educational institutions, enhancing user experience, and streamlining communication processes.

Keywords: Chatbot · Educational institutions · AI-powered · Logic Learning Machine (LLM)

1 Introduction

In recent years, chatbots have emerged as a popular tool for providing efficient and personalized assistance in various domains. With the advancement of artificial intelligence (AI) technologies, chatbots have become increasingly sophisticated, enabling organizations to automate customer interactions and streamline

M. Z. Vizuete et al. (Eds.): CI3 2023, LNNS 1041, pp. 229–240, 2024.
https://doi.org/10.1007/978-3-031-63437-6_19

their operations. In the field of education, chatbots have proven to be valuable assets, providing information and support to students, parents, and staff [1, 2].

The concept of chatbots can be traced back to the 1960s when ELIZA, a natural language processing program, was developed. Since then, chatbot technology has evolved significantly, leveraging AI techniques to understand and respond to user queries in a conversational manner [3]. AI plays a crucial role in enabling chatbots to interpret user inputs, extract relevant information, and generate appropriate responses.

One of the key components of AI that powers chatbots is neural networks. In particular, recurrent neural networks (RNNs) and transformer-based models like GPT (Generative Pre-trained Transformer) have gained prominence for their ability to capture context and generate human-like responses [4]. These models leverage deep learning techniques to process and understand natural language, allowing chatbots to handle complex conversations.

Neural networks are a fundamental component of AI that enables chatbots to understand and respond to user queries. In the context of chatbots, recurrent neural networks (RNNs) and transformer-based models like GPT have emerged as powerful tools for natural language processing [5].

RNNs are particularly effective in handling sequential data, making them well-suited for processing conversational text. They have a feedback mechanism that allows them to maintain an internal state, enabling them to capture context and dependencies across different parts of a conversation. This makes RNNs capable of generating responses that consider the entire conversation history, leading to more coherent and contextually appropriate answers [6, 7].

On the other hand, transformer-based models, such as GPT, have revolutionized the field of natural language processing. GPT is a state-of-the-art language model that uses a transformer architecture to process and generate text [8]. Transformers rely on self-attention mechanisms, which allow them to capture long-range dependencies and understand the context of a sentence or a document more effectively. This enables chatbots powered by transformer-based models to generate highly accurate and contextually relevant responses.

Both RNNs and transformer-based models leverage deep learning techniques, such as training on large datasets and fine-tuning with specific objectives, to improve their language understanding capabilities [9]. These models can be trained on vast amounts of text data to learn patterns, semantics, and linguistic nuances, enabling them to handle complex conversations and generate human-like responses [10].

By utilizing neural networks like RNNs and transformer-based models, chatbots can provide more intelligent and context-aware interactions, enhancing the overall user experience and making them valuable tools for various applications, including customer support, information retrieval, and educational assistance [11].

In this paper, we present the development of a custom chatbot designed to assist in providing information about Instituto Tecnologico Universitario Rumiñahui (ISTER). Our chatbot incorporates AI techniques, including natural language processing and neural networks, to offer a seamless and interactive experience for users. We explore the capabilities of AI-powered chatbots in enhancing the information dissemination process and improving user engagement. By leveraging the advancements in AI technology, our chatbot aims to provide accurate and relevant information, streamline communication, and enhance the overall experience for users.

The remainder of this paper is organized as follows. Section 2 provides an overview of the methodology employed for developing the custom chatbot. Section 3 discusses the implementation details, highlighting the AI techniques and neural networks utilized. Section 4 presents the evaluation results and performance metrics of the chatbot. Section 5 discusses the implications and potential future enhancements. Finally, Sect. 6 concludes the paper and summarizes the key findings and contributions.

2   Methodoly

In this section, we provide an overview of the methodology employed for developing our custom chatbot. The development process involved several key steps, including data collection, pre-processing, model selection, training, and evaluation [12]. The methodology aimed to create a robust and efficient chatbot that effectively assists users in obtaining information about our educational institute.

2.1   Data Collection

To develop a chatbot that accurately understands and responds to user queries in the educational domain, we collected a dataset from ISTER's web page as it is shown in Fig. 1. The dataset includes frequently asked questions, common inquiries, and specific queries about ISTER, its programs, admissions, courses, and facilities. This data collection process ensures that the chatbot will have a broad understanding of user intents and variations in queries related to the ISTER.

The collected dataset will be used to train a language model, specifically a chatbot, using techniques such as natural language processing and machine learning. By training the chatbot on this dataset, it will be able to learn patterns, understand user intents, and provide accurate responses to queries about ISTER and related topics.

Training a language model involves processing the dataset, extracting relevant features, and training the model to understand the context, intent, and nuances of user queries [13]. This process helps the chatbot generate appropriate and informative responses based on the input it receives.

2.2   Data Pre-processing

In the process of developing the chatbot, the collected dataset underwent thorough pre-processing to clean and format the text. This pre-processing stage is

Fig. 1. Ister's web page.

crucial for ensuring the quality and effectiveness of the trained language model. Here are the steps involved:

1. Removal of Irrelevant Information: Irrelevant information such as HTML tags, special characters, or noisy data was removed from the text. This step helps in eliminating unnecessary noise and improving the quality of the dataset.
2. Text Normalization: Text normalization techniques were applied to achieve consistent representation of words. This involved converting all text to lowercase, handling contractions (e.g., "can't" to "cannot"), and removing punctuation marks.
3. Tokenization: Sentences were split into individual words or tokens. This step helps the model understand the structure of the text and process it at a granular level.
4. Stemming or Lemmatization: Stemming or lemmatization techniques were used to reduce words to their base form. This helps in reducing the dimensionality of the vocabulary and capturing the essence of the words.
5. Stop-word Removal: Stop words, such as "the", "and", or "is", which do not carry significant meaning, were removed from the text. This improves the efficiency of language processing by focusing on more informative words.

In addition to these pre-processing steps, VectorStore was utilized in the development process. VectorStore played a role in managing and working with the vector embeddings derived from the pre-processed text [14]. It allowed for efficient storage, retrieval, and manipulation of the vector embeddings, enabling tasks such as semantic search and similarity matching.

By incorporating VectorStore into the development pipeline, it enhanced the capabilities of the chatbot by providing efficient indexing and retrieval of relevant information based on semantic similarities captured in the vector embeddings.

2.3   Model Selection

For the chatbot's underlying model, we explored different options based on the requirements of our educational institute. After careful consideration, we selected a transformer-based model, specifically the GPT-2 model, known for its exceptional language generation capabilities and context understanding.

2.4   Training

The selected model was trained on the pre-processed dataset using supervised learning techniques. We employed transfer learning to leverage the pre-trained GPT-2 model's language understanding abilities and fine-tuned it on our specific educational dataset [15]. The training process involved optimizing model parameters, such as learning rates and batch sizes, to improve performance and generate high-quality responses.

2.5   Evaluation

To assess the effectiveness of our custom chatbot, we conducted rigorous evaluations using various metrics, including response accuracy, contextual coherence, and user satisfaction. We performed both manual evaluation, where human evaluators assessed the quality of responses, and automated evaluation, where we compared the generated responses against ground truth answers from domain experts.

The methodology outlined in this section ensured the development of a custom chatbot that leverages neural network models and AI techniques to provide accurate and contextually appropriate responses to user queries. The subsequent sections of this paper will delve into the implementation details, challenges faced, and the achieved performance of our custom chatbot.

3   Implementation Details and AI Techniques

In this section, we provide a comprehensive overview of the implementation details of our custom chatbot, focusing on the AI techniques and neural networks utilized. We discuss the key components, architecture, and algorithms employed to create an intelligent and interactive chatbot for our educational institute.

3.1   Chatbot Architecture

The chatbot architecture comprises several modules that work together to facilitate smooth communication between the user and the chatbot. These modules include natural language understanding (NLU), dialogue management, and natural language generation (NLG). The NLU module processes user input, extracting intents and entities, while the dialogue management module maintains the conversation flow. The NLG module generates coherent and contextually appropriate responses based on the input and system state.

3.2 AI Techniques for Natural Language Processing

To enhance the chatbot's language processing capabilities, we employed various AI techniques, including named entity recognition (NER), part-of-speech (POS) tagging, and sentiment analysis. NER identifies and categorizes named entities such as names, locations, and organizations. POS tagging assigns grammatical tags to words in a sentence, enabling the chatbot to understand the syntactic structure. Sentiment analysis determines the sentiment expressed in user queries, allowing the chatbot to respond accordingly.

3.3 Neural Networks for Chatbot Development

To enable the chatbot to understand and generate human-like responses, we utilized advanced neural network models. Recurrent Neural Networks (RNNs) see Fig. 2 and Long Short-Term Memory (LSTM) see Fig. 3 networks were employed for sequence modeling and capturing context dependencies. These models excel in understanding and generating text sequences, ensuring the chatbot's ability to provide coherent and meaningful responses.

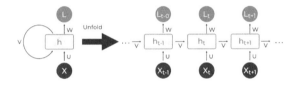

Fig. 2. RNN Network Implemented.

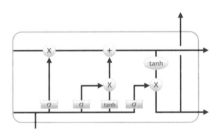

Fig. 3. LSTM Network Implemented.

3.4   GPT-2 for Language Generation

For the language generation component of the chatbot, we leveraged the power of the GPT-2 (Generative Pre-trained Transformer 2) model. GPT-2 is a transformer-based language model that utilizes the attention mechanism to capture dependencies and generate high-quality text. By fine-tuning GPT-2 on our educational dataset, we achieved more accurate and contextually appropriate responses.

4   Implications and Results

In this section, we present the implementation details of a chatbot designed to provide accurate and efficient responses to user queries in the educational domain. The chatbot leverages the capabilities of the WhatsApp Business API, Twilio messaging service, and the OpenAI API for natural language processing. Additionally, the implementation incorporates the use of the VectorStore library for data preprocessing and the Logic Learning Machine (LLM) model for rule-based machine learning. The primary goal of the chatbot is to enhance the user experience by delivering comprehensive and context-aware information about our educational institute's programs, admissions, courses, and facilities.

4.1   Model Selection

After careful consideration of various options, a transformer-based model called GPT-2 (Generative Pre-trained Transformer 2) was chosen as the underlying model for the chatbot. GPT-2 is renowned for its exceptional language generation capabilities and its ability to comprehend and generate contextually relevant responses. Additionally, the Logic Learning Machine (LLM) model, which is based on the Switching Neural Network (SNN) paradigm, was incorporated into the implementation. The LLM model, developed by Marco Muselli, a Senior Researcher at the Italian National Research Council CNR-IEIIT in Genoa, offers an efficient approach to generating intelligible rules, making it a valuable addition to our chatbot's functionality.

4.2   Integration with WhatsApp and Twilio

The implementation integrates the WhatsApp Business API and Twilio messaging service to facilitate seamless communication with users. The 'whatsapp-web.js' library is utilized to interact with the WhatsApp API, while the 'twilio' library enables message sending through the Twilio platform. Upon receiving a user query, the chatbot processes the request using the OpenAI API and generates a response based on the GPT-2 model and LLM rules. The response is then sent back to the user via both WhatsApp and Twilio SMS services. In Fig. 4 is shown its logo.

Fig. 4. Twilio.

4.3 File Processing and Response Generation

To handle user requests involving file uploads, the implementation includes functionality to read and process text files and PDF files. The 'pdfjs-dist' library is employed to extract text from PDF documents, while the 'fs' module enables reading of text files. The extracted text is preprocessed using the VectorStore library, and subsequently, the OpenAI API is utilized to generate a response based on the processed data. The response is sent back to the user via both WhatsApp and Twilio, ensuring a seamless and consistent user experience.

In the development of the chatbot the following steps where followed:

Importing Required Modules. The necessary modules and libraries are imported, including 'whatsapp-web.js' for interacting with the WhatsApp API, 'qrcode-terminal' for generating QR codes, 'fs' for file system operations, 'openai' for accessing the OpenAI API, 'twilio' for sending messages via Twilio, and 'pdfjs-dist' for PDF processing. The 'dotenv' library is used to load environment variables from a '.env' file.

Configuration and Initialization. The WhatsApp client and Twilio client are created using the provided credentials from the environment variables. A WhatsApp QR code is generated and displayed for authentication.

File Reading and Processing. Utility functions are defined to read text files and PDF files asynchronously. The 'readTextFile' function reads the content of a text file and returns it as a string. The 'readPDF' function reads the content of a PDF file and extracts the text from each page, concatenating them into a single string. The 'runCompletion' function determines the file type based on the file extension. If it's a PDF file, it uses 'readPDF' to extract the text; otherwise, it uses 'readTextFile'. The extracted text is then passed to the OpenAI API for language processing and completion. The response is limited to a maximum of 200 tokens.

Sending Response via Twilio. The 'sendResponseViaTwilio' function is responsible for sending the response to the user's phone number via Twilio. It uses the Twilio client to create a new message with the response text and sends it to the specified phone number. This code provides the foundation for a chatbot that can process user requests, generate responses based on the provided data files, and send the responses back to the user via both WhatsApp and Twilio SMS.

In Fig. 5 shows the established session with WhatsApp by generating a QR code. This QR code acts as a gateway for users to authenticate the chatbot with their WhatsApp accounts. Users simply need to scan the QR code using the WhatsApp mobile app, enabling the chatbot to send and receive messages on their behalf.

Fig. 5. QR code.

Once the session is established, the Node.js server initializes the WhatsApp client using the 'whatsapp-web.js' library. The client is responsible for managing incoming and outgoing messages. Upon successful initialization, the server logs a message indicating that the client is ready to send and receive messages.

The chatbot continuously listens for incoming messages from WhatsApp users. Using event listeners, it captures incoming messages and performs appropriate actions based on their content. One such action is triggered when a user sends a message starting with '#' as is shown in Fig. 6.

When a '#' request is detected, the chatbot extracts the file path from the received message and prepares to process the dataset. The code supports both PDF and text file formats. For PDF files, it utilizes the 'pdfjs-dist' library to extract the text content from each page and concatenate them into a single string. For text files, it reads the content directly.

With the file content extracted, the chatbot proceeds to generate an AI-driven completion response using the OpenAI API. The 'text-davinci-003' model, a variant of the GPT-3 model, is employed for this purpose. The extracted text

Fig. 6. Message starting with '#'.

is used as the prompt for generating the completion response. The generated response is limited to a maximum of 200 tokens to ensure concise and relevant replies as is shown in Fig. 7.

```
async function runCompletion () {
  const fileContent = fs.readFileSync('./file.txt', 'utf8');
  const completion = await openai.createCompletion({
      model: "text-davinci-003",
      prompt: fileContent,
      max_tokens: 200,
  });
  return completion.data.choices[0].text;
```

Fig. 7. Generating AI-driven Completion from File Content.

Once the completion response is generated, the chatbot sends it back to the user as a reply via WhatsApp. Additionally, it leverages the Twilio API to send the response to the user's phone number. This multi-channel approach ensures that the user receives the completion response through WhatsApp as we can observe in Fig. 8.

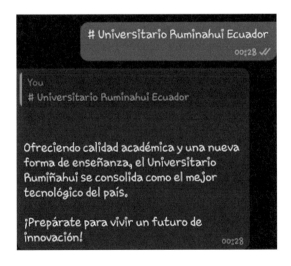

Fig. 8. Message response via Whatsapp.

5   Conclusion

In this paper, we have presented the development and implementation of an AI-powered chatbot system for educational institutions. The system utilizes a diverse dataset of conversations related to the educational domain, which includes frequently asked questions, inquiries, and specific queries about the institute's programs, admissions, courses, and facilities. The collected dataset was thoroughly pre-processed, involving cleaning, text formatting, tokenization, and other techniques to ensure accurate understanding and processing of user queries.

For the underlying model of the chatbot, we selected the GPT-2 model, a transformer-based model known for its exceptional language generation capabilities and context understanding. Additionally, we incorporated the Logic Learning Machine (LLM) algorithm, which generates intelligible rules and enhances the chatbot's decision-making process.

The implementation of the chatbot system involves a combination of technologies and libraries. We utilized the whatsapp-web.js library to establish a session with WhatsApp and handle incoming messages. The code includes functionality for displaying a QR code for session establishment and a "Client is ready" message upon successful initialization.

Furthermore, we integrated the openai library to leverage the OpenAI API for generating AI-driven completions based on user queries. The code includes functions for reading text files and PDF files, extracting their content, and passing it to the OpenAI API for completion generation. The generated completions are then sent as responses to user queries.

Additionally, we incorporated the twilio library to enable sending responses via Twilio's messaging service. The code includes a function that sends the generated completion response to the user's phone number using Twilio.

Through this implementation, we have demonstrated the capabilities of the developed chatbot system in accurately understanding and responding to user queries in the educational domain. The combination of a diverse dataset, advanced language models, and intelligible rule generation has contributed to the chatbot's effectiveness and efficiency in providing relevant and informative responses.

Overall, the implemented chatbot system serves as a valuable tool for educational institutions to enhance their communication with students, address their inquiries, and provide them with timely and accurate information about programs, admissions, courses, and facilities. The system's effectiveness and adaptability make it a promising solution for improving the overall user experience and streamlining communication processes in the educational sector.

References

1. Muselli, M.: The logic learning machine: a new efficient tool for data mining. IEEE Trans. Knowl. Data Eng. **19**(7), 917–932 (2007)
2. Vaswani, A., et al.: Attention is all you need. In: Advances in Neural Information Processing Systems, pp. 5998–6008 (2017)
3. Brown, T.B., et al.: Language models are few-shot learners. arXiv preprint arXiv:2005.14165 (2020)
4. Jurafsky, D., Martin, J.H.: Speech and Language Processing, 3rd edn. Pearson (2019)
5. Young, T., Hazarika, D., Poria, S., Cambria, E.: Recent trends in deep learning based natural language processing. IEEE Comput. Intell. Mag. **13**(3), 55–75 (2018)
6. Vasilescu, F., Hogan, B.: Artificial intelligence in education: challenges and opportunities for sustainable development. Inf. Softw. Technol. **106**, 101–108 (2019)
7. Sutskever, I., Vinyals, O., Le, Q.V.: Sequence to sequence learning with neural networks. In: Advances in Neural Information Processing Systems, pp. 3104–3112 (2014)
8. Mikolov, T., Chen, K., Corrado, G., Dean, J.: Efficient estimation of word representations in vector space. arXiv preprint arXiv:1301.3781 (2013)
9. Lipton, Z.C.: The mythos of model interpretability. Queue **16**(3), 31–57 (2018)
10. Manning, C.D., Raghavan, P., Schütze, H.: Introduction to Information Retrieval. Cambridge University Press, Cambridge (2008)
11. Pennington, J., Socher, R., Manning, C.: GloVe: global vectors for word representation. In: Proceedings of the 2014 Conference on Empirical Methods in Natural Language Processing (EMNLP), pp. 1532–1543 (2014)
12. Ruder, S., Peters, M.E., Swayamdipta, S.: Transfer learning in natural language processing. arXiv preprint arXiv:1910.01108 (2019)
13. Hochreiter, S., Schmidhuber, J.: Long short-term memory. Neural Comput. **9**(8), 1735–1780 (1997)
14. Jurafsky, D., Martin, J.H.: Speech and Language Processing: An Introduction to Natural Language Processing, Computational Linguistics, and Speech Recognition. Pearson (2020)
15. Bahdanau, D., Cho, K., Bengio, Y.: Neural machine translation by jointly learning to align and translate. arXiv preprint arXiv:1409.0473 (2014)

Economics and Management

Software and Methods Applied in the Standardization of the Grinding Process for Automobile Engines as a Business Solution

Ana Álvarez-Sánchez[✉] 🆔 and Alexis Suárez del Villar-Labastida 🆔

Grupo de Investigación en Sistemas Industriales, Software y Automatización SISAu, Facultad de Ingeniería, Industria y Producción, Universidad Indoamérica, Quito, Ecuador
anaalvarez@indoamerica.edu.ec

Abstract. When the research addresses the problem of empirical engine rectification based exclusively on the experience of the workers, which results in nonconformities on the part of the customers due to delays in delivery times. To solve this, it was proposed to standardize the operation cycles using the Stop and Observe Method. Through techniques such as direct observation, time studies and process diagrams, the grinding process was analyzed in detail. The data collected were compared with the Crystal Ball program, demonstrating consistency with 95% confidence. It was determined that the time required to rectify an engine was 27.23 days, with supplements affecting 4.8 h of an 8-h workday. The investigation proposed improvements, including the implementation of a hydraulic scissor table to reduce the physical strain on the operators and the introduction of hearing protection. This resulted in a noise reduction from 30 dB to 15 dB, exceeding the regulations of Executive Decree 2393. With these measures, the total grinding time was reduced to 26 days, representing an efficiency gain of 639.15 min or 10.65 h. This standardization not only improves the efficiency of the process, but also allows for more accurate communication with the customer on delivery times, increasing customer satisfaction.

Keywords: Standardization · Rectification · Supplementation · Standard time

1 Introduction

The annoying squeaking and rattling of the vehicle, whether it is a new or used car, is an irritating factor for the customer [1, 2]. There are three basic areas in engine grinding that work independently, these are crankshaft, engine block and cylinder head [3], crankshaft is the core part of an automobile engine, and the accuracy requirements of various shape and position errors are very high [4, 5], a series of experiments including chemical composition, microstructure, mechanical properties, hardness, toughness and fractography are performed on the failed crankshaft [6, 7]. For better engine overhaul services, it is important to have defined cycle times for the operations and in turn standardize the process [8, 9]. With the application of standardization, decisions can be made to ensure that in a better way the work will be performed with higher quality and time

M. Z. Vizuete et al. (Eds.): CI3 2023, LNNS 1041, pp. 243–250, 2024.
https://doi.org/10.1007/978-3-031-63437-6_20

savings increasing customer satisfaction [10, 11]. These findings support future studies to determine the optimal design of experiments that minimizes testing time and costs with a satisfactory and accurate estimation of engine responses [12].

2 Method

The Stop and Observe method is applied in time studies and process standardization based on the identification of the elements that are part of the operation under study; the percentage of constant and variable supplements involved in each operation for employees to recover from fatigue, noise, applied force, etc., to which they are exposed; the researcher uses it in an exhaustive and thorough manner based on the regulations of the International Labor Organization [13]. The Westinghouse Electric Corporation System was applied to calculate the speed rating, evaluating the performance of the operator considering four factors, Effort, Ability, Consistency and Conditions [14]. The results obtained were used in the calculation of the normal time, Fig. 1 shows the steps to follow with the application of the method.

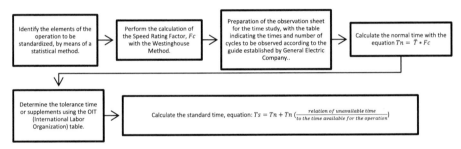

Fig. 1. Stop and Observe Methodology

2.1 Results of the Application of the Stop and Observe Methodology

The operation investigated is the Rectification of 1600 CC engines, the initial situation was diagnosed with the determination of the elements that compose it and the main activities that are carried out so that they can be fulfilled, see Fig. 2; The times for the study were taken for ten cycles applying the continuous timing, see Table 1, where it is evident that the number of lapses observed directly satisfies the sampling error; subsequently, the normal time was calculated, which was obtained by multiplying the average time of the elements by the global performance factor or speed rating, see the results detailed in Table 2.

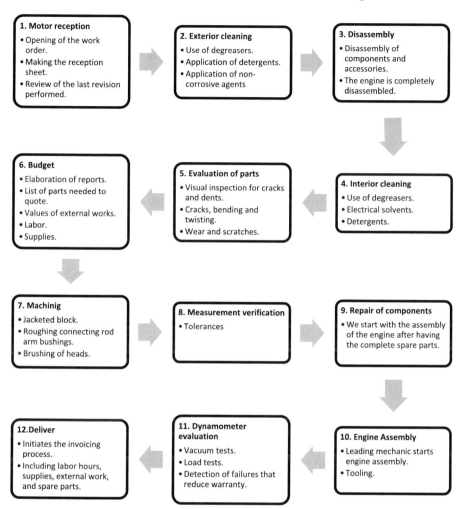

Fig. 2. Diagram of current analysis of process operations.

The value used for the calculation of the overall performance factor or speed rating (Fc) of result 1.19 was obtained with the application of the Westinghouse rating method as shown in Table 3.

To determine the standard time, the percentage of tolerance or supplement was established using the predetermined and validated tables of the International Labor Organization (ILO) with a value of 40%, see Table 4, which for 480 min of the workday represents a rest of 192 min, and with the normal time of 199.86 h, which when converted into minutes is 11.991.6; subsequently, the standard time of the initial operation was calculated by means of the following equation:

$$Ts = Tn + Tn \left(\frac{relation of unavailable time}{to the time available for the operation} \right) \tag{1}$$

Table 1. Recommended cycle times for continuous cronometraje.

Cycle time in minutes	Recommended number of cycles
0,10	200
0,25	100
0,50	60
0,75	40
1,00	30
2,00	20
2,00–5,00	15
5,00–10,00	10
10,00–20,00	8
20,00–40,00	5
40,00–en adelante	3

Source: Information taken from time Study Manual of the Erie Works at General Electric Company, developed under the guidance of Albert E. Shaw, manager of salary administration

$$Ts = 11991.6 + 11991.6(\frac{40}{440})$$

$$Ts = 11991.6 + (11991.6 * 0.090)$$

$$Ts = 13070.84 \text{ min}$$

The standard time of 13070.84 min obtained in the diagnostic stage decreased its value with the improvement proposal that positively modified the value of the percentage of variable supplements (see Table 5), as follows: intermittent and loud noise decreased from 2% to 0% and the use of muscular force/energy was reduced from 35.5 kg to 5 kg to lift, pull and push the motor, which was 22% and was shortened to 1%, these improvements occur when using a double scissor lift hydraulic table selected after the analysis of several alternatives under 5 criteria: Lowest Cost: 340 usd, Platform Size W × L(mm); (520 × 1010), Actuation and ease of handling, Capacity: 500 kg, Maximum Height: 1500 mm and Minimum Height: 440 mm, in addition to the use of the TRR 21 Db Ear Cover Type Protector used in environments with noise above 85 Db, as shown in Fig. 3.

$$Ts = Tn + Tn \left(\frac{relation\ of\ unavailable\ time}{to\ the\ time\ available\ for\ the\ operation} \right) \tag{2}$$

$$Ts = 11991.6 + 11991.6(\frac{17}{463})$$

$$Ts = 11991.6 + (11991.6 * 0.0367)$$

Table 2. Time Study Sheet (hours).

Element's	1	2	3	4	5	6	7
1. Motor reception	2,80	2,70	2,90	2,80	2,90	2,70	2,60
2. Exterior cleaning	2,30	2,20	2,10	2,20	2,40	2,30	2,30
3. Disassembling	24,60	24,50	24,30	24,40	24,50	24,60	24,60
4. Interior cleaning	6,70	6,70	6,90	6,70	6,70	6,90	6,80
5. Evaluation of parts	18,60	18,80	18,70	18,70	18,80	18,70	18,60
6. Quotation	5,90	6,50	5,80	6,60	6,00	6,20	5,90
7. Machining	20,70	20,60	20,60	20,80	20,70	20,90	20,70
8. Measurement verification	4,30	4,30	4,20	4,40	4,50	4,40	4,30
9. Component repair	35,60	35,60	35,70	35,50	35,60	35,60	35,70
10. Engine assembly	28,70	28,70	28,80	28,60	28,90	28,60	28,80
11. Dynamometer evaluation	11,70	11,80	11,70	11,90	11,70	11,80	11,70
12. Total	5,40	5,20	5,20	5,50	5,40	5,40	5,50
TOTAL, TIME BY CYCLES	167,30	167,60	166,90	168,10	168,10	168,10	167,50

Element's	8	9	10	Σ T	TPS	Tn
1. Motor reception	2,80	2,70	2,90	27,80	2,78	3,31
2. Exterior cleaning	2,40	3,40	2,20	23,80	2,38	2,83
3. Disassembling	24,30	24,40	24,60	244,80	24,48	29,13
4. Interior cleaning	6,80	6,70	6,80	67,70	6,77	8,06
5. Evaluation of parts	18,60	18,70	18,60	186,80	18,68	22,23
6, Quotation	6,30	6,20	6,30	61,70	6,17	7,34
7. Machining	20,70	20,90	20,60	207,20	20,72	24,66
8. Measurement verification	4,30	4,40	4,30	43,40	4,34	5,16
9. Component repair	35,80	36,50	35,70	357,30	35,73	42,52
10. Engine assembly	28,80	28,80	28,90	287,60	28,76	34,22
11. Dynamometer evaluation	11,80	11,80	11,90	117,80	11,78	14,02
12. Total	5,30	5,40	5,30	53,60	5,36	6,38
TOTAL, TIME BY CYCLES	167,90	169,90	168,10	168,10	167,95	199,86

Table 3. Results of the study with the application of the Westinghouse Rating Method.

Factors in the Westinghouse method	Rating awarded	
Skill	B1 – Excellent	+0.11
Effort	C1 – Good	+0.05
Conditions	D – Average	+0.00
Consistency	B – Excellent	+0.03
$Cv = 1 \pm C$		1.19

$$Ts = 12431.69\,min$$

Table 4. Breakdown of total 40% supplements.

Supplements	Percentage they represent
Constants: 9%	Gender: Male
A. Supplement for personal needs	5%
B. Base Fatigue Supplement	4%
Variables: 31%	
A. Supplement for standing work	2%
B. Supplement for awkward posture (bent over)	2%
C. Use of muscular force/energy 35.5 kg, (lifting, pulling, pushing)	22%
F. Concentration, intense for precise or fatiguing work	2%
G. Noise, intermittent and loud	2%
H. Mental tension, quite complex process	1%

a   b

Fig. 3. Post criteria analysis selection: a. Hydraulic double scissor table and b. TRR 21 Db Ear Cover Type Protector

Following the proposal, a normalized time of 12431.69 min was obtained, equivalent to 207.19 h.

3 Results

Fred E. Meyers [1] describes in his book that companies that do not have their processes standardized have a performance of 60% applying the standardization reach 85% performance in production or services; in the initial diagnosis the operators did not have any hearing protection being exposed between 86. 79 dB to 98.32 dB range of decibels that were measured with an integrating sound level meter; Executive Decree 2393 Art. 55 establishes that the noise level criterion is 85 dB, the study was conducted based on the Ecuadorian standard NTE INEN-ISO 9612, "Acoustics. Determination of occupational noise exposure. Engineering method" (ISO 9612:2009, IDT), with a measurement sampling of 8 h in full days. As it can be seen that the decibel level is exceeded, arc hearing protectors should be used to avoid occupational diseases in the operators. To reduce the percentage of supplemental use of force from 22% to 1%, a double hydraulic

Table 5. Breakdown of total supplements of 17%.

Supplements	Percentage they represent
Constants: 9%	Gender: Male
A. Supplement for personal needs	5%
B. Base Fatigue Supplement	4%
Variables: 8%	
A. Supplement for standing work	2%
B. Supplement for awkward posture (bent over)	2%
C. Use of muscular force/energy 5 kg, (lifting, pulling, pushing)	1%
F. Concentration, intense for precise or fatiguing work	2%
G. Noise, intermittent and loud	0%
H. Mental tension, quite complex process	1%

scissor lift table is currently used, which loads up to 160 kg at a maximum height of 2.50 m; it is used to move and lift the engine and its components. Below is a table with the comparison of the results of the standard times of the research (Table 6).

Table 6. Comparison of standard times of the operation: Rectification of engines.

Initial standard time				Proposed standard time			
Supplement	Minutes	Hours	Days	Supplement	Minutes	Hours	Days
40%	13070,84	217,847	27	17%	12431,69	207,19	26

4   Conclusions

In the standardization of the 1600 CC engine rectification operation, it was possible to reduce the time by 639.15 min, which represents 10.65 h and 1 day, specifically with the decrease in the percentage of total supplements from 40% to 17%. It was possible to specify through an analysis of technical data the weight of 35.5 kg being handled and the range of 86.79–98.32 decibels to which the three operators working directly in this area were exposed, exceeding the norms established in the NTE INEN-ISO 11228-1-2 for the handling of loads and in the Executive Decree 2393 in the weight and noise supplement, a double hydraulic scissor lift table and arc hearing protectors were implemented to obtain the desired results.

References

1. Meyers, F.: Estudio de Tiempos y Movimientos para la Manufactura Agil, Mexico: Camara nacional de la industria (2000). https://www.academia.edu/28556729/Meyers_Estudio_de_T iempos_y_Movimientos_para_la_Manufactura_Agil_2_ed
2. Mainkar, V., Gosavi, S., Ghadage, K., Wayal, A.: Molesto chirrido y crujido de la carrocería del automóvil: un enfoque sistemático de detección y prevención. Sociedad Acústica Turca Congreso Internacional y Exposición sobre Ingeniería de Control de Ruido 36 (2017). INTER-NOISE ESTAMBUL (2017)
3. Krysmon, S., Dorscheidt, F., Classen, J., Düzgün, M., Pischinger, S.: Real driving emissions: conception of a data-driven calibration methodology for hybrid propulsion systems combining statistical analysis and virtual calibration platforms. 14(16), 4747 (2021). https://doi.org/10. 3390/en14164747
4. Valladares, S.: Rectificado de motores, partes, proceso, Cómo y Cuándo hacerlo. Mundo del motor (2017)
5. Tingting, G., Xiaoming, Q., Peihuang, L.: Investigación sobre la evaluación de errores de redondez del diario de bielas en la medición síncrona del diario del cigüeñal. Synchron. Meas. Appl. Sci. 12 (2022). https://www.mdpi.com/2076-3417/12/4/2214
6. Aliakbari, K., Nejad, R., Toroq, S., Macek, W., Branco, R.: Análisis de fallos del cigüeñal del motor diésel de cuatro cilindros. Revista de la Sociedad Brasileña de Ciencias Mecánicas e Ingeniería 30 (2019). http://rd.springer.com/journal/40430
7. Zhao, N., Zhang, J., Ma, W., Jiang, Z., Mao, Z.: Descomposición variacional del dominio del tiempo de las señales de vibración de impacto múltiple de la máquina recíproca. Sistemas mecánicos y procesamiento de señales (2022). https://www.sciencedirect.com/science/article/abs/pii/S088832702200156X?via%3Dihub
8. González-López, M., Laureano-Anzaldo, C., Pérez-Fonseca, A.: Cálculo de eficiencias de adsorción y ciclos de reutilización recuperando el concepto de líneas operativas. Separation Science and Technology (Philadelphia) (2022)
9. Gonca, G., Sahin, B.: Investigación y evaluación del rendimiento de un motor que funciona con un ciclo dual modificado. Revista Internacional de Investigación Energética 46(3), 2454–2466 (2022)
10. Cruz-Peragón, F., Torres-Jiménez, E., Armas, O.: Metodología para simular ciclos de prueba normalizados para motores y vehículos mediante el diseño de experimentos con bajo número de tiradas. Conversión y gestión de energía, pp. 817–832 (2018)
11. Kolahchian Tabrizi, M., Cerri, T., Bonalumi, D., Lucchini, T., Brenna, M.: Modernisation of diesel engines with H_2 for a possible decarbonisation of non-electrified railways: assessment with life cycle analysis and advanced numerical modelling. 17(5), 996 (2024). https://doi.org/10.3390/en17050996
12. Cruz-Peragón, F., Torres-Jiménez, E., Lešnik, L., Armas, O.: Mejoras metodológicas para simular el rendimiento y las emisiones de los ciclos transitorios del motor de los modos de funcionamiento estacionarios: un estudio de caso aplicado a los biocombustibles. Combustible 312 (2022). https://doi.org/10.1016/j.fuel.2021.122977
13. Kanawaty, G.: Introducción al estudio del trabajo, Ginebra (1996). https://teacherke.files.wordpress.com/2010/09/introduccion-al-estudio-del-trabajo-oit.pdf
14. Villanes, L., Montero, A.: Revista Científica EPigmalión, Huacho: Universidad Nacional José Faustino Sánchez (2019). https://www.unjfsc.edu.pe/facultades/ing_indust_sistema/Epigmalion/contenidos/Vol1Num1-Articulo06.pdf

Productivity Strategy in the Peruvian Pharmaceutical Industry

Roli David Rodríguez-Palomino[1]([⊠]) [ID], Omar Bullón-Solís[2] [ID],
Fernando Alexis Nolazco-Labajos[2] [ID], and Sheylah Hoppe-Coronel[3] [ID]

[1] Laboratorio Farmacéutico Markos, Lima, Peru
roli_ro@hotmail.com
[2] Universidad Cesar Vallejo, Lima, Peru
[3] Tecnológico Universitario Espíritu Santo, Guayaquil, Ecuador

Abstract. The COVID-19 pandemic spurred an escalation in the demand for specific medications, consequently exposing pharmaceutical enterprises to a discernible productivity gap that necessitated resolution in order to effectively address the requisites and obligations of this sector. The core objective of this investigation is to scrutinize the augmentation in productivity within the realm of medication packaging processes, localized within a Peruvian pharmaceutical corporation. The research methodology deployed to attain the aforementioned objective adheres to a quasi-experimental framework, wherein observation assumes a pivotal role as the principal mechanism for capturing primary data sources. This methodology facilitates the documentation of acquired data within the ambit of manufacturing records. The outcomes unveiled innovative configurations of labor organization and equipment orchestration, which manifest as a strategic avenue for the enrichment of productivity. Thereby, this stratagem in turn expedites the amelioration of executed procedures aimed at meeting the exigencies of the pharmaceutical industry. In summation, the adoption of a novel blister-packaging paradigm, seamlessly integrated within the production line, has demonstrated conclusive impacts on the expansion of productivity, consequently bolstering the domestic and international production capacities at a national level.

Keywords: Productivity · Pharmaceutical industry · Blister Packaging · Peru

1 Introduction

The optimization of resources has been a historical concern of civilizations worldwide, from ancient to modern times [1]. The central focus has been on large enterprises, exploring ways to enhance the productivity of labor in organizations like factories. Frederick Taylor work on improving worker productivity and Henry Fayol principles for effective organizational management have been pioneering in the field of administrative thought [2].

In the realm of business management, the scientific study of organizational productivity employs primarily quantitative analyses of the working process, encompassing: 1)

the workforce and its organization, and 2) the means of production (Marx, cited in [3]). The aims of such studies have revolved around optimizing processes by diagnosing gaps between production capacity and consumption needs within specific industries, thereby leading to new productive combinations and organizational forms.

Starting from 2020, with the onset of the COVID-19 pandemic, the demand for drugs to treat COVID-19 surged [4–6] emphasize that the most sought-after medications during the pandemic included 500 mg Paracetamol, followed by the non-opioid analgesics and non-steroidal anti-inflammatory drugs (NSAIDs) category, with tablets being the most widely marketed pharmaceutical form [6].

Amidst the anti-COVID measures, the Peruvian government has recommended since 2020 that the population maintain a stock of medications, including paracetamol and others, as a preventive measure. This has triggered significant organizational changes in the pharmaceutical business, aimed at improving productivity to ensure the supply of COVID-19 treatment medications [7].

Against this backdrop, it becomes pertinent to focus on the production issues of medications within the context of COVID-19 [8], with a specific focus on a pharmaceutical company in Peru. The studied company is a pharmaceutical laboratory with extensive experience in producing medications for both public and private customers. Their product range includes anti-inflammatories, analgesics, antibiotics, antimigraine drugs, cardiotonics, neuropsychiatry drugs, and gastroenterology medications. The demand for products such as carbamazepine 200 mg tablets, ciprofloxacin 500 mg tablets, naproxen 550 mg tablets, and acetylsalicylic acid-caffeine-paracetamol 500 mg tablets has generated issues related to the presence of two identical blister lines with production discrepancies.

Furthermore, the offered quantity does not meet the projected demand due to new production goals and existing issues such as substantial waste, losses, downtime, overtime, and occasional reprocessing in blistering to reduce waste. As a consequence, there have been delivery failures to both the domestic and international markets [9].

This scenario has unveiled a productivity gap, intensifying the pressure on management to address the disparity between units produced and units demanded [10, 11]. Thus, the objective of this study has been to examine the enhancement of productivity in the medication packaging area of a Peruvian pharmaceutical industry.

2   Features of the New Medication Packaging Model: Towards Productivity Enhancement

[12] indicate that contemporary economic studies are focused on enhancing productivity in processes, which is fundamental for the advancement of any organization. In terms of productivity, this impacts efficiency, understood as practical production and the intended formal production, as well as effectiveness, which is goal attainment. Productivity thus aligns performance with efficient resource utilization. On the other hand, [13] asserts that productivity serves as a supportive strategy in process description aimed at productivity enhancement. This involves analyzing procedures and tasks through measurement.

This section analyzes the conceptual, methodological, and practical features employed to enhance the productivity of the medication packaging process. Therefore, it is necessary to address fundamental concepts that underpin the research and practice of medication packaging.

Blistering in-line is a form of medication packaging where a machine places coated tablets into protective plastic, followed by sealing by the machine. The sealed tablets are then transported via a conveyor belt to a conditioning area, where operators place them into their secondary packaging [14].

Productivity is recognized as the improvement in a process by comparing the amount of inputs utilized to the amount of manufactured products [15, 16]. The blistering model, to enhance productivity, emerges as a business strategy to improve the productivity of high-demand medication packaging under the context of COVID-19.

Methodologically, the focus is on the Blister Machine 260, where modifications will be implemented in both dimensions of productivity: work organization and means of work. As per [17], the study follows a quantitative approach, commencing with the emergence of an inquiry, establishing parameters, objectives, and questions. From these questions, verified premises are established through practical evidence using statistical tools, culminating in case-specific conclusions [17].

This business research aims for the economic benefit of the host organization by elevating productivity in in-line packaging and conditioning for the projected production. The research adopts a quasi-experimental design. According to [18], this involves manipulating one or more unconfirmed experimental variables to reveal the reasons and processes underlying a specific situation. [19] refers to making modifications to a specific variable while observing, detecting, and examining the obtained results and potential explanations.

Observation will be the primary mode of collecting primary sources, documenting data acquired in manufacturing records. Through these records, the researcher can capture positive or negative information regarding the proposed in-line blistering method. The mentioned records include: labor-hour lists, machine-hour lists, packaging and conditioning guides for each processed batch.

Next, Table 1 presents the features of the current blistering model vs. the new blistering model:

This table synthesizes the features of medication packaging or blistering models, considering the existing work organization and the technical-administrative changes in production equipment. Noteworthy changes encompass: work processes, the number of operators, and the machinery used for packaging.

The research was designed with consideration for an experimental group and a control group, each represented by a blistering machine and its associated operational staff. A total of 120 batches of pharmaceutical products were assigned to each, implemented across both workstations (Experimental Blistering and Control Blistering) in three 8-h shifts, resulting in a continuous 24-h work cycle. The study duration spanned five weeks.

The selection of the pharmaceutical company was based on the organization's request to augment production in the specific medication line mentioned in earlier sections. The organization provided authorization for the implementation of the proposed in-line blistering approach due to the researcher's track record. To adhere to organizational norms and mitigate potential brand influence risks, the identification of the studied products was deliberately withheld.

Table 1. Comparison of medication packaging models

Features	Current blistering model (Control Group)	New blistering model (Experimental Group)
Work method	• Existence of six packaging and conditioning processes • Maintenance of existing coordination with the warehouse department	• Reduction and consolidation of activities to transition from six to four packaging and conditioning processes • Alterations in coordination with the warehouse department
Number of operators	• Nine operators	• Six operators
Equipment (Blister Machine)	• Separation of packaging and conditioning booths	• Integration of packaging and conditioning booths for inline placement • Implementation of a conveyor belt in the conditioning area

Source: Own elaboration

3 Indicators and Outcomes of Implementing the New Medication Packaging Model

In this section, we will address key productivity indicators that have been considered in the implementation of this study. These indicators include: productivity, personnel utilization efficiency, packaging inputs, and batch waste.

Productivity arises from comparing the offered quantity to the expected formal quantity, determined by a percentage, while considering the department's own standard for packaging. Personnel utilization efficiency refers to useful labor hours, calculated by dividing the actual labor hours by the theoretical labor hours using a ratio [13].

Packaged products are determined by the ratio obtained from factory production to packaging, as well as the overall batch conformity, compared to the theoretical quantity using a ratio. Batch waste is the comparison of the expected formal quantity against the difference between the expected formal quantity and the actual obtained quantity, using a ratio.

The results of productivity are presented for the pre- and post-experimental stages of the two blistering machines (Table 2).

Blistering machine #1 in the experimental group exhibited a minimum productivity of 92.56% in the pretest, a maximum of 97.90%, and an average of 96.25%. In the posttest, the minimum productivity rose to 99.08%, with a maximum of 99.93% and an average of 99.45%. The variation in average productivity between the pretest and posttest for experimental blistering machine #1 was 3.2%. This increase is attributed to the implementation of the in-line blistering proposal, which enhanced productivity by increasing the quantity of medications delivered per batch.

Table 2. Productivity in a pharmaceutical industry

	Pretest			Posttest		
G. Experimental	Min	Max	Prom	Min	Max	Prom
Blistering machine 1	92.6	97.90	96.25	99.08	99.93	99.45
G. Control	Min	Max	Prom	Min	Max	Prom
Blistering machine 2	94.25	97.82	95.94	95.24	96.96	96.32

Note: Data is expressed in percentages. Legend: Min: Minimum; Max: Maximum; Avg: Average; G.: Group

Concerning control blistering machine #2, the pretest exhibited a minimum productivity of 94.25%, a maximum of 97.82%, and an average of 95.94%. In the posttest, the minimum productivity was 95.24%, with a maximum of 96.96% and an average of 96.32%. The pretest-to-posttest variation in the control group was 0.92%. While there was a productivity increase, it was substantially lower than the productivity variation observed in the experimental blistering machine.

Regarding personnel utilization efficiency, both phases of investigation for the two blistering machines are outlined (Table 3).

Table 3. Efficiency in a pharmaceutical industry

	Pretest			Posttest		
G. Experimental	Min	Max	Prom	Min	Max	Prom
Blistering machine 1	88.46	96.55	93.96	27.27	100	34.99
G. Control	Min	Max	Prom	Min	Max	Prom
Blistering machine 2	93.75	100	95.73	35.89	96.55	86.14

Note: The data is expressed in percentages. Legend: Min: Minimum; Max: Maximum; Avg: Average; G.: Group

In relation to blister machine #1 (experimental group), in the pretest, the minimum efficiency was 88.46%, the maximum was 96.55%, and the average efficiency was 93.96%. In the posttest, the minimum efficiency dropped to 27.27%, with a maximum of 100%, and an average of 34.99%. The pretest-to-posttest variation in personnel utilization efficiency (real labor hours/theoretical labor hours*100) was −58.97%. This reduction was due to the decreased personnel in the in-line blistering process, transitioning from eleven operators to six. This reduction in personnel also led to a shortened blistering duration, resulting in the negative percentage variation in the posttest. This decrease is consistent with the reduction in the use of operators required to perform the same work.

Regarding blister machine #2 (control group), in the pretest, the minimum efficiency was 93.75%, the maximum was 100.00%, and the average was 95.73%. In the posttest, the minimum efficiency dropped to 35.89%, with a maximum of 96.55%, and an average of 86.14%.

It is worth noting that the difference in work efficiency in the control group only experienced a reduction of –9.59%, placing it significantly below the efficiency achieved in the experimental group for this dimension or variable. In Table 4 presents the results regarding packaged products in the pretest and posttest of both blister machines.

Table 4. Packaged products in a pharmaceutical industry

	Pretest			Posttest		
	Min	Max	Prom	Min	Max	Prom
G. Experimental	Min	Max	Prom	Min	Max	Prom
Blistering machine 1	97.59	99.98	99.47	99.35	99.82	99.65
G. Control	Min	Max	Prom	Min	Max	Prom
Blistering machine 2	99.08	99.99	99.63	99.30	99.84	99.66

Note: The data is expressed in percentages. Legend: Min: Minimum; Max: Maximum; Avg: Average; G.: Group

Regarding blister machine #1, in the pretest, the minimum batch waste was 2.15%, the maximum was 7.44%, and the average was 3.85%. In the posttest, the minimum batch waste decreased to 0.07%, with a maximum of 0.93%, and an average of 0.54%.

As evident from the posttest results, the batch waste was notably reduced, resulting in a negative variation of –3.31% between the average posttest batch waste and the pretest batch waste of experimental blister machine #1. This indicates a recovery in the percentage of waste for finished products. For blister machine #2, in the pretest, the minimum batch waste was 2.22%, the maximum was 6.10%, and the average was 4.23%. In the posttest, the minimum batch waste increased to 3.14%, with a maximum of 4.99%, and an average of 3.77% (Table 5).

Table 5. Waste in a pharmaceutical industry

	Pretest			Posttest		
	Min	Max	Prom	Min	Max	Prom
G. Experimental	Min	Max	Prom	Min	Max	Prom
Blistering machine 1	2.15	7.14	3.85	0.07	0.93	0.54
G. Control	Min	Max	Prom	Min	Max	Prom
Blistering machine 2	2.22	6.10	4.23	3.14	4.99	3.77

Regarding blister machine #1, in the pretest, the minimum waste was 2.15%, the maximum was 7.44%, and the average was 3.85%. In the posttest, the minimum waste decreased to 0.07%, with a maximum of 0.93%, and an average of 0.54%. As observed in the posttest, the waste was significantly reduced, resulting in a negative variation of –3.31% between the average posttest waste and the pretest waste of experimental blister machine #1. This indicates a recovery in the percentage of waste for finished products. For blister machine #2, in the pretest, the minimum waste was 2.22%, the maximum was 6.10%, and the average was 4.23%. In the posttest, the minimum waste increased to 3.14%, with a maximum of 4.99%, and an average of 3.77%.

4 Comparison of Packaging Models: Elements for Productivity Analysis

In this section, it is pertinent to provide elements for the study of productivity enhancement through the implementation of the new inline blister packaging proposal in a pharmaceutical manufacturing organization in the year 2021. This study was conducted within the packaging process using two blister machines with similar technical characteristics. One of the blister machines was designated as the experimental blister machine, while the other was considered the control blister machine.

Quality standards for medications adhered to the guidelines of American, European, and British industries. These encompassed the purity of the active ingredient, pharmacological activity, uniformity of content for each pharmaceutical form, bioavailability, and product stability. All these aspects are compromised in the event of defective manufacturing or packaging issues, among other factors. The therapeutic effect of a medication can be nullified due to poor quality, resulting in adverse reactions that endanger the health of consumers. Therefore, pharmaceutical industries prioritize quality control measures.

For this research endeavor, four key productivity indicators were employed: productivity, efficiency in personnel utilization, packaging inputs, and batch waste. The preceding results demonstrated an increase in productivity with a rise in the quantity of medications delivered per batch of inputs. The efficiency in personnel utilization was positively impacted by the reconfiguration of workspace and equipment modifications, leading to a novel organizational structure for medication packaging within the company. It is evident that a reduced number of operators can achieve inline packaging and conditioning. Notably, the speed of the blister machine dictates the conditioning time in the new blistering model. This concurs with assertions by [15] that novel techniques play a pivotal role in a company's productivity by automating processes, thus potentially attaining higher production quotas [5], and adhering to designated delivery deadlines. However, the potential effects of these modifications on personnel health remain a subject for further study [15].

The introduction of the new model has led to an increase in products to be packaged. The number of approved batches for packaging has risen, owing to enhanced coordination with the quality control department. This department approves bulk materials to be used by the blister machines based on a weekly plan encompassing all approved bulk products. Increases in efficiency can be achieved through improved coordination within the packaging department for the products to be packaged, thereby reducing waiting and idle times that might adversely affect the product's shelf life.

The reduction in waste became evident in the new model due to the reduced occurrence of faulty blister packs. This decrease in product defects contributes to cost reduction, while also enhancing the productivity of the packaging and conditioning area. This aspect holds significant importance in the pharmaceutical industry. According to GMP regulations, all pharmaceutical manufacturing is strictly controlled at every step of the process [11]. A higher waste percentage warrants attention, prompting administrators to delve into the reasons behind it as it deviates from established specifications. [21] also contends that there exist numerous additional factors that can increase the cost of a product during manufacturing, which should be considered, although these aspects exceed the scope of this study.

5 Conclusions

In the realm of the pharmaceutical industry, the strategy for productivity enhancement has been directed towards the analysis of work processes and equipment, fostering continuous improvement in both administrative and production domains. This approach has successfully identified factors influencing manufacturing and administrative inefficiencies. The most substantial impacts of these endeavors were observed within the logistics domain, as evidenced by the on-time delivery of products to both domestic and international clients.

The resultant shift in production methodology, driven by the newly implemented packaging model, has facilitated smoother input and output flows in production, better integration of logistical functions, and heightened productivity achieved by optimizing process, machinery, and method variables. To further augment productivity, it is recommended to incorporate considerations for the medium and long-term effects on workers' health into the realm of productivity enhancement best practices. This would contribute to holistic strategies for productivity elevation.

In conclusion, this study has validated a novel blistering method centered on inline production, culminating in reduced medication costs and amplified productivity in medication fabrication. Another salient effect has been the improved spatial organization within the plant, infusing greater dynamism into medication manufacturing processes.

References

1. Claude, S., Alvarez, L.: Historia del pensamiento administrativo. Pearson education (2005)
2. Gonzales-Miranda, D.: La escritura de las prácticas administrativas como fundamento del corpus teórico de la disciplina: Reseña de Historia de la Administración. Escribir las prácticas. Innovar **32**(85), 205–208 (2022). https://doi.org/10.15446/innovar.v32n85.103406
3. Harnecker, M.: Los conceptos elementales del materialismo dialectico. Siglo XXI Editores (2007)
4. Tenorio-Mucha, J., Lazo-Porras, M., Hidalgo, A., Málaga, G., Cárdenas, M.: Prices of essential drugs for management and treatment of COVID-19 in public and private Peruvian pharmacies. Acta Méd. Peru **37**(3), 267–277 (2020). https://doi.org/10.35663/amp.2020.373.1560
5. Novak, H., Tadić, I., Falamić, S., Ortner Hadžiabdić, M.: Pharmacists' role, work practices, and safety measures against COVID-19: a comparative study. J. Am. Pharm. Assoc. **61**(4), 398–407 (2021). https://doi.org/10.1016/j.japh.2021.03.006
6. Quispe, F., Tocas, C.: Caracterización de los medicamentos más comercializados durante la pandemia covid-19: botica Farma Junín (2022)
7. Ministerio de Salud: Tiempos de Pandemia 2020–2021. Ministerio de salud. 191 p (2021)

8. Bahlol, M., Dewey, R.S.: Pandemic preparedness of community pharmacies for COVID-19. Res. Social Adm. Pharm. **17**(1), 1888–1896 (2021). https://doi.org/10.1016/j.sapharm.2020. 05.009
9. Khan, H., Wisner, J.D.: Supply chain integration, learning, and agility: effects on performance. Oper. Supply Chain Manag. **12**(1), 14–23 (2019). https://doi.org/10.31387/oscm0360218
10. Li, Y., Yu, W.: Exploration of the new mode of pharmaceutical care in the outpatient dispensary during epidemic of Corona Virus Disease-2019. Pharm. Care Res. **21**(3), 216–220 (2021). https://doi.org/10.5428/pcar20210312
11. Severin, T., et al.: How is the pharmaceutical industry structured to optimize pediatric drug development? Existing pediatric structure models and proposed recommendations for structural enhancement. Ther. Innov. Regul. Sci. **54**(5), 1076–1084. (2020). https://doi.org/10. 1007/s43441-020-00116-4
12. Felsinger, E., Runza, P.: UCEMA: Posgrado-download (septiembre de 2002). Homepage. http://www.ucema.edu.ar/posgradodownload/tesinas2002/Felsinger_MADE.pdf
13. Gutiérrez, H., De La Vara, R.: Statiscal Quality Control and Six Sigma, 3rd edn. Mc Graw-Hill/Interamericana Editores S.A. (2021)
14. Marco, E.: Guía de acondicionamiento y embalaje (2009). Homepage, https://recursos.exp ortemos.pe/guia-acondicionamiento-embalaje.pdf
15. Fontalvo, T., De La Hoz, E., y Morelos, J.: La productividad y sus factores: incidencia en el mejoramiento organizacional. Dimensión Empresarial **16**(1), 47–60 (2018). https://doi.org/ 10.15665/dem.v16i1.1375
16. Carro, R., González, D.: Productivity and Competitiveness. Facuty of Cs (2012)
17. Hernández, R., Fernández, C., Baptista, P.: Metodología de la Investigación (6ta. Ed) Mc Graw Hill (2014)
18. Chaves Morelli, M.S.: Los diseños cuasi-experimentales en la investigación clínica. Su utilidad y limitaciones para la inferencia causal en la práctica clínica (2021). https://www.tdx. cat/handle/10803/674481
19. Cortez, M., Iglesias, M.: Generalities on Research Methodology, México (2004)
20. AMBIT: Qué es la normativa GMP? Significado y normativa (2023). Homepage. https:// www.ambit-bst.com/blog/qu%C3%A9-es-la-normativa-gmp-significado-y-normativa
21. Kanawaty, G.: Introducción al estudio del trabajo, 4th edn. OIT, Ginebra (2014)

Efficiency as a Competitive Factor in MSMEs to Generate Profitability

Verónica Patricia Arévalo Bonilla$^{(\boxtimes)}$ ⓘ, Paúl Rodríguez Muñoz ⓘ,
Marco Verdezoto Carrillo ⓘ, Sergio Carrera Guerrero ⓘ, and Alain Quintana ⓘ

Instituto Tecnológico Universitario Rumiñahui, Sangolquí, Ecuador
veronica.arevalo@ister.edu.ec

Abstract. Competitiveness and efficiency are important aspects in business management as they generate a competitive advantage of sustainability in the market. The objective of this study is to determine the predominant factors that influence the achievement of efficiency and competitiveness results in MSMEs in the textile production sector. The analysis was based on the Data Envelopment Analysis (DEA) methodology, which describes the results of the input variables versus the incidence they have on the output variables of a system. The input variables considered were assets and equity, while the output variables were net income and total income. On the other hand, the hypotheses raised in this research were tested for their validity through the acceptance or rejection of the chi-square test by means of factor analysis. The research was based on a database of the Superintendencia de Compañías de Ecuador. The results obtained yielded important data showing and concluding that no region of the country exceeds 33% efficiency in its MSMEs. In addition, it was observed that the constructs of assets and equity generate influence in the generation of income and finally in the net profit, which is considered a key factor for the financial management and sustainability of a company.

Keywords: Efficiency · competitiveness · MSMEs · competitive advantage · profitability

1 Introduction

The data for the last five years presented by the Comisión Económica para América Latina y el Caribe, CEPAL, shows that MSMEs represent 99.5% of the region's companies, but account for only 24.6% of production, due productive diversity [1]. The description of these data counteracts the sustainability of these companies in the market, so it becomes an important problem to be studied, to understand the root causes that affect the efficiency and competitiveness of these companies for sustainability over time, through their strategies, systems and processes. In this area, Ecuador's manufacturing industry comprises a considerable portion of MSMEs, and their presence in the textile sector contributes to economic development. According to the 2021 business ranking published by the Superintendencia de Compañias y Seguros de Ecuador, there are 264 entities belonging to the segment of micro, small and medium-sized enterprises (MSMEs) in this sector.

M. Z. Vizuete et al. (Eds.): CI3 2023, LNNS 1041, pp. 260–272, 2024.
https://doi.org/10.1007/978-3-031-63437-6_22

Productive entities make profits with the resources they have in the company throughout an economic period, but not all companies do. Not all companies reach the optimal level of efficiency, as shown by net profits and total revenues. This is shown by comparing them with those companies with similar characteristics and conditions that are at the edge of efficiency, according to the application of the methodology of Data Envelopment Analysis (DEA). This research is based on the change of variables that should be considered to achieve business efficiency and competitiveness as a factor of sustainability, being assets and equity as input variables versus net income and total income as output variables. Competitiveness brings added value to companies and in this area the aim is to determine the efficiency variables that affect the sustainability of companies and their profits, understanding sustainability as an integrated system [2] of inputs and outputs, which are the financial results.

The scope of the research is to show the variation that the variables of results, such as net profit and total income can have, through the fluctuation of the variables of assets and equity in the MSMEs of a productive sector, in order to obtain information that will allow to adopt new financial, productive or commercial strategies to achieve efficiency with the optimal utilization of resources.

Research on related topics can be referred to as follows: Business innovation: Factor of competitiveness and quality of life in Popayan, Colombia [3], where it details that MSMEs have implemented quality and competitiveness systems, but these have been incipient. Another study refers to the factors that impact the sustainability of MSMEs in the food sector of the municipality of Monterrey [4], where it refers to the use of technology and quality of service as factors for sustainability, likewise, in the article: Sustainability of commercial MSMEs as a factor of competitiveness, it refers to the relationship between suppliers and customers to obtain sustainability [5]. Different articles such as: Las Pymes y factores para obtener el éxito, inicio para el marco referencial [6], La competitividad de las pequeñas y medianas empresas manufactureras de Bogotá y su percepción del apoyo que les brinda la política de competitividad [7], Evaluación de la eficiencia social en el sistema de cooperativas de ahorro y crédito del Ecuador [8], and the article: Customer service quality in financial entities, a factor for generating competitiveness in times of covid 19 pandemic, outlines the importance of customer service as a competitive factor [9]. All the publications mentioned here show competitiveness as a considerable factor for the profitability or sustainability of companies. In business sustainability, it is important to identify, processes, or vital systems, vital being understood as that which allows the model under study to be functional.

In order to continue with the study of this context, important factors that have an influence on this research, and the relationship of its variables, are detailed below.

1.1 Efficiency

According to Coll, & Blasco [10] describes the efficiency in three important areas of the evaluation of the data envelopment analysis, DEA, for the context of study "the Technical Efficiency, shows the capacity that a unit has to obtain the maximum output from a given set of Inputs, it is obtained by comparing the observed value of each unit with the optimal value determined by the estimated production edge (isoquant efficiency). (p. 12)".

Price or allocative efficiency refers to the Unit's capacity to use the different Inputs in optimal proportions given their relative prices.

Global or economic efficiency is given by the product of Technical Efficiency and Price Efficiency [10].

Technical efficiency, defined by Farell as the capacity of an entity to obtain the maximum output from a given set of inputs [11].

1.2 Sustainability

In this study, sustainability is considered as a systemic approach and for this purpose "a system is simply a set of elements (or subsystems) related to each other. The elements can be molecules, organisms, machines [1], or parts of them, social entities and even abstract concepts. Likewise, the relationships, interconnections, or "linkages" between the elements can show up in very different ways (economic transactions, flows of matter or energy, causal links, control signals, among others)".

In this way, open systems maintain energy, matter and information exchange with their environment in order to work, as well as depending on the factors, elements or variables coming from the system's environment and that influence it (input variables); on the other hand, the system produces variables that affect the environment (output variables or products) [2].

Sustainability focuses on the importance of identifying the factors that allow companies to maintain business sustainability through profitability.

1.3 Competitiveness

The concept of competitiveness has some aspects related to human and technological capital or natural resources and emphasizes the generation of a competitive advantage in the market, either by creating capabilities or skills that are developed in a product, system or process. García and Coll state that these capabilities do not constitute an advantage for all products. García and Coll [11], "The possibility that this transformation will take place depends on the product itself, the company, the sector and the environment". Under this concept, it becomes important to define the characteristics and capabilities that are appropriate and the way to transform all of them into a single position that makes the product stand out in an advantageous way, i.e. it becomes a competitive strategy [11].

According to Chiavenato [12], Collins [13] business competitiveness as a process for creating competitive advantages once the resources, skills and competencies that create value have been found and analyzed. In this aspect, competitiveness depends on the characteristics of the products of each company and its strategy to achieve a differentiation or competitive advantage considering that efficiency is the core in the context of the study.

1.4 Competitive Strategy

Becerra Bizarron shares Porter's concept, Porter says that competitive strategy consists of taking defensive or offensive actions to establish a defensible position in an industry that allows facing its competitive forces to achieve a return on investment, and for this purpose companies must find the strategy according to their particular circumstances [14].

In another context, strategy specifies the way in which an organization adjusts its own capabilities with the existing opportunities in the market to achieve its objectives [15]. Based on these notions for this research, it is concluded that each company has the power to insert a strategy that allows it to obtain profitability and sustainability in the market, under the absent factors such as management, liquidity, technology, risks and resistance to change [4].

Strategy is also seen as a pattern of objects, policies, plans to achieve goals, which allows a company to know what kind of organization it wants to be and where it stands [13].

As described above, each organization must identify the factor or factors that will allow it to obtain a competitive strategy in order to get efficiency based on the profitability or usefulness of the investment.

1.5 Variables Considered for the Analysis of Efficiency with the DEA Methodology: Assets, Equity, Performance, Income

In the DEA Data Envelopment Analysis methodology, this study reviews the input variables, which in this case are the equity and assets, as data presented by the MSMEs in the textile manufacturing sector in Ecuador, and the output variables that will be defined as the results, and are the data reflected in the yield or net profit and total income.

Asset. A resource controlled by an entity, as a result of past events, from which the entity expects to obtain future economic benefits.

Equity. The residual portion of the entity's assets, after deducting all its liabilities [16]. The International Accounting Standards Board defines assets as "present economic resources controlled by the entity as a result of past events" [17].

Income. Increases in economic benefits, produced during the accounting period in the form of inflows or increases in the value of assets, or decreases in liabilities, which result in an increase in equity and are not related to the contributions of the owners to this equity [16].

Profit (equity). The elements directly related to the measure of profit are revenue and expenses. The profit figure is often used as a measure of the entity's performance, or is the basis for other evaluations, such as return on investments or earnings per share [16].

With the published literature, the research of the efficiency of the MSMEs is carried out through the methodology of the Enveloping Analysis of data, in order to know the companies that are located in the edge of the efficiency, and on the other hand, the following hypotheses, based on the methodology of the factorial analysis are raised:

1.6 Hypothesis

The proposed conceptual model is presented in Fig. 1, which details the relationship of the factor variables and the direct relationship between them.

For the hypothesis statement, we considered the feasibility of admitting or excluding the null hypothesis that details that the input variables do not generate income and these do not generate net profit, for this purpose the following alternative hypotheses were also proposed.

H0: Assets do not generate total revenues or profits when they show efficiency.
H1: Assets generate total revenues
H2: Assets generate total revenues
H3: Total revenues generate net income in a range of efficiency.

The null hypothesis H0 is rejected under the results achieved in the Chi-Square study and the KMO Test analysis, once the results are obtained, this analysis recommends working with the alternative hypotheses H1, H2, and H3, as described in Fig. 1.

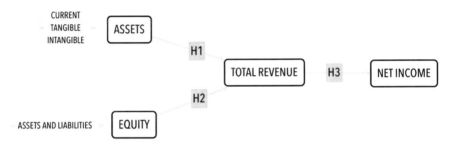

Fig. 1. Proposed conceptual model

Once the study was carried out, the results show that the companies located in the provinces of Pichincha, Guayas and Tungurahua, show a percentage that ranges between 5% and 6% of efficient companies, while the rest of the companies show unfavorable results compared to these, the comparative study is carried out in a group of companies of similar characteristics.

In this context, it can be deduced that MSMEs need to adopt new financial, commercial or productive strategies, to obtain positive results in their profit or total income variables, in order to achieve sustainability in the market, since it is obvious that there are companies that make excellent profits with lower asset values, as they manage to be on the efficiency curve.

The sustainability of companies is based on strategies, systems or processes that companies can adopt according to their needs and characteristics, allowing them to achieve development and competitiveness.

2 Methodology

This study was an exploratory, descriptive and correlational research that allowed the observation of the behavior of the object of study. The database published by the Superintendencia de Compañias y Seguros del Ecuador for the year 2021 was considered, focused the study on MSMEs in the textile manufacturing sector of the country. This research was based on the Data Envelopment Analysis, DEA, in the CCR-O model [18], which analyzes the fluctuation of the output variables, in this case, total revenues and net income. In this occasion the total income and net profit; against the constant of the input variables, which are represented in the assets and equity, which defined the efficiency of the MSMEs in the most representative provinces and in the different regions of the country, which is detailed below from Figs. 2, 3, 4, 5 and 6.

On the other hand, for the reliability analysis the confirmatory factor analysis [19, 20], was used, which allows countering a model built in advance, which allows the researcher to determine how the elements match and how the data is structured, based on a defined theory, technically it seeks to "explain the covariances or correlations between a set of observed variables through a reduced set of latent variables or factors" [21], to find the link that exists between the elements of the assets and equity variables with the total income variables and finally with the net profits, which led to describe the correlation that exists between the factors proposed through the correlation and anti-image matrix.

The Chi-square index and Bartlet's Sphericity Test were used to test the null hypothesis, the data reflected in the result provided a Chi-square value of 103.781 with a degree of freedom of 6, determining the acceptance of the alternative hypothesis.

The analysis of the KMO, Keiser-Meyer-Olkin measure, allowed observing the correlation that exists between the variables of the factors or constructs raised, as detailed in Table 1, in the same way it indicates that it is appropriate to apply the factor analysis, [20] with a result of 0.715, which is considered acceptable.

Table 1. Bartlett KMO test

Kaiser-Meyer-Olkin measure of sampling adequacy		0,715
Bartlett's test for sphericity	Approx. chi-squared	103,781
	Gl	6
	Sig.	,000

The results obtained through the DEA methodology and factor analysis, performed in SPSS, are shown below.

3 Results

The efficiency of MSMEs in the textile manufacturing sector in Ecuador was observed in the results of the Data Envelopment Analysis, DEA, which is detailed in the Sierra and Costa regions, in the most representative provinces. The Oriente region has only one company in the sector studied, so it has not been considered for comparison.

Figure 2 shows the efficiency of MSMEs in the Sierra region, in the province of Pichincha, according to the data collected, 95% of the object of study showed that they are not efficient, compared to 5% (4 companies) that were efficient, according to the results of net profits, reflected in the output variables of the DEA methodology. These results were related to those values identified with 1, indicating that they are located in the periphery of efficiency when performing an analysis and comparing them among themselves, additionally, it was observed that inefficient MSMEs should make a movement in their input variables which are assets and equity, to achieve the output variables of total income and net profits to be in the efficiency edge.

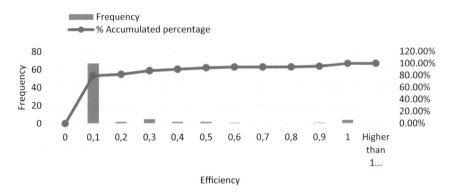

Fig. 2. Efficiency of MSMEs in Pichincha

The following results, detailed in Fig. 3, show the efficiency results obtained under the DEA methodology for MSMEs in the textile manufacturing sector in the province of Azuay in the Sierra region of Ecuador. The data obtained show an efficiency of 32% in the companies studied, while 68% of the companies are outside this range.

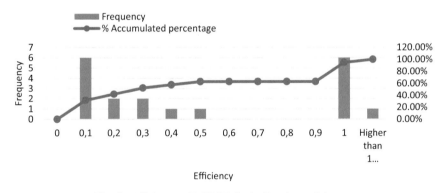

Fig. 3. Efficiency of MSMEs in the Province of Azuay

The results of the Fig. 4 show the efficiency in the province of Tungurahua, these data were obtained by entering the input and output variables in the DEA methodology, which basically shows that 25% of the MSMEs show an efficiency that is found in the net profits.

Continuing with the context of the study, it was noted that in the province of Imbabura and in the rest of the provinces of the Sierra, the MSMEs showed average efficiency values of 27% and 29%, respectively, as shown in Figs. 5 and 6, while the companies that do not show efficiency levels according to the methodology applied correspond to the difference of these levels.

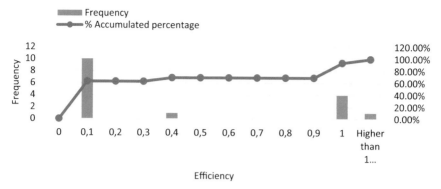

Fig. 4. Efficiency of MSMEs in the Province of Tungurahua

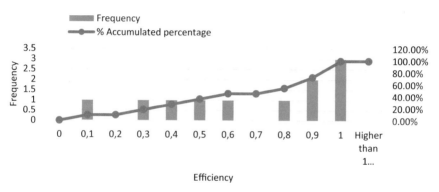

Fig. 5. Efficiency of MSMEs in the Province of Imbabura

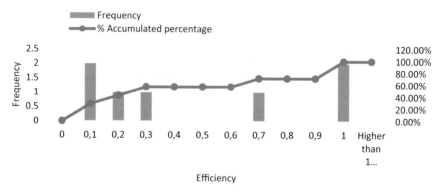

Fig. 6. Efficiency of MSMEs in other provinces of the Sierra.

In the Costa region, the province of Guayas was analyzed according to the number of MSMEs in the sector, in this case 99 companies; other provinces were analyzed as a single group because the number is not representative for individual analysis. Figure 7 shows that 4% of MSMEs in Guayas are efficient, while 96% are not at or near the efficiency edge.

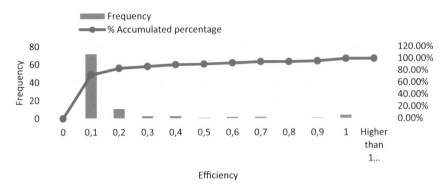

Fig. 7. Efficiency of MSMEs in the Province of Guayas

Finally, Fig. 8 shows the efficiency values represented by the other Costa region provinces, where 33% were found to be at the efficiency edge.

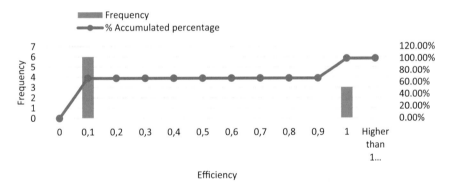

Fig. 8. Efficiency of MSMEs in other provinces of the Costa region.

In another context, through the data obtained from the factor analysis in SPSS IBM, it was found that there is correlation in the variables studied and a linear relationship as detailed in Table 2 of the Correlation Matrix, with a determinant close to zero, which indicates the suitability of the analysis.

In Table 3 of the Total Variance Explained, the analysis showed the direct relation of the variables of the assets construct and the equity construct, understanding that in the first factor are the current, tangible, and intangible assets, while in the second factor or equity construct the variable of the assets minus the liabilities has been considered.

This factor in turn has a direct impact on total income and will subsequently influence the net income variable. This result is shown in Table 2 of the Correlation Matrix.

The results of the proposed model showed that with high levels of significance (p-value < 05), the hypotheses H1, H2, H3 were fulfilled, as described in Table 4, showing

Table 2. Correlation matrix[a]

		Asset values	Equity values	Net income values	Total revenue values
Correlation	Asset values	1	0,739	0,542	0,924
	Equity values	0,739	1	0,843	0,787
	Net income values	0,542	0,843	1	0,596
	Total revenue values	0,924	0,787	0,596	1

a. Determinant = ,015

Table 3. Total variance explained

Component	Initial eigenvalues			Sums of the squared saturations of extraction			Sum of the saturations squared by rotation		
	Total	% of variance	Accumulated %	Total	% of variance	Accumulated %	Total	% of variance	Accumulated %
1	3,223	80,566	80,566	3,223	80,566	80,566	2,064	51,604	51,604
2	,595	14,864	95,430	,595	14,864	95,430	1,753	43,827	95,430
3	,110	2,760	98,190						
4	,072	1,810	100,000						

Extraction method: Principal Component Analysis.

that the assets construct and the equity construct influenced total income and, in turn, this had an impact on net income. This scenario allows to understand the efficiency of the companies focused on the perceived profit as a factor of sustainability, additionally it allows to know the impact that a change in the input variables will have on the output variable, which in this case is the net profit.

Table 4. Chi-square tests

	Value	gl	Bilateral asymptotic significance
Pearson's Chi-square H1	143,003	1	0,000
Pearson's Chi-square H2	24,578	1	0,000
Pearson's Chi-square H3	11,193	1	0,001

In another context, the companies that were profitable on their assets were surveyed through the ROI, both in the year 2021 and 2022, and shows values greater than zero, which are shown in Table 5, however, the companies that have remained profitable and efficient, in the period studied, reach 8.89%, for a better comparison, three decimal places have been taken in the profitability index.

In this way, it was possible to show that the companies that remained efficient and profitable, adopted strategies in their input variables to get better results in their output variables or in their efficiency and profitability.

Table 5. Rentabilidad 2021, 2022

Profitable 2021 & 2022		2021	2022
Province	Company	ROI = profitability net income/assets	ROI = profitability net income/assets
Pichincha	Reyes industria textil cia. Ltda	0,0002%	0,1803%
Pichincha	Silkproducts cia.ltda.	1,7389%	0,4761%
Pichincha	Sportslocker sociedad anónima	0,0045%	11,386%
Azuay	Dicovs, diseños, confecciones y ventas c. Ltda.	3,3396%	58,682%
Azuay	Industria de la moda inmoda cia.ltda.	0,8097%	5,150%
Azuay	Maki-fairtrade s.a.	0,0287%	48,598%
Azuay	Procostura cia. Ltda.	0,0003%	2,316%
Azuay	Soluciones industriales y medicas soinmed cia. Ltda.	1,5832%	3,033%
Tungurahua	Arfatexcia c.l.	0,0008%	8,327%
Tungurahua	Incalza s.a.	0,0003%	21,404%
Tungurahua	Industrias y textiles pequeñin cia. Ltda.	5,794%	0,093%
Imbabura	Importgroup cia.ltda.	5,413%	10,399%
Guayas	Cristex creaciones textiles y merchandising cristexcreaciones s.a.	66,807%	57,465%
Guayas	Salvadanio s.a.	0,0001%	0,069%

With the study carried out and under the hypotheses proposed, the conclusions of the present investigation are defined below.

4 Conclusions

The analysis showed that the MSMEs in Ecuador show a variation in the efficiency limit according to the Data Envelopment Analysis, the characteristics of the manufacturing sector show a similarity because no province shows an efficiency of more than 33% and for the most representative provinces such as Pichincha and Guayas, its index is 5%, representing a minority in its scope.

The sustainability of the companies is based on the capacity of these entities to adapt to new changes and to adopt strategies in their productive, commercial and economic areas, strategies that will be in accordance with the needs and characteristics of each of the MSMEs.

The constructs of assets and equity influence the obtaining of total income and these generate net profits in the companies, which is why the companies must carry out a financial management that allows variations in the variables of assets and equity. It is concluded that the low efficiency indexes found in the results of the DEA analysis are affected by the lack of management in the correct use of the assets and equity variables.

Finally, the profitability and efficiency indexes show a low percentage of 8.89% of the total number of companies studied, evidencing the importance of business management for the creation of competitiveness and sustainability in the market, through the efficient management of tangible or intangible resources and systems such as organizational productive capacity.

References

1. Dini, M., Stumpo, G.: Mipymes en América Latina: un frágil desempeño y nuevos desafíos para las políticas de fomento, pp. 439–471. Documentos de proyectos (LC/TS.2018/75) CEPAL (Comisión Económica para América latina y el Caribe), Santiago, Chile (2020). ISBN: 978-92-1-058625-2 (pdf). Cap. IX
2. Hugo, F., Flores, C., Peralta, A., Lara, P.: Sostenibilidad empresarial en relación a los objetivos del desarrollo sostenible en el Ecuador. RECIAMUC 3(1), 670–699 (2019)
3. Saldarriaga, M., Guzmán, M., Concha, E.: Innovación Empresarial: Factor de competitividad y calidad de vida en Popayán, Colombia. Revista Venezolana De Gerencia 24(2), 151–166 (2020)
4. Torres, E., López-Lira, A.: Factores que influyen en la sostenibilidad de las PyMES del sector restaurantero del municipio de Monterrey. InnOvaciOnes De NegOciOs 18(35), 1–19 (2021)
5. Villalba, A.: La Sustentabilidad de las Pymes comerciales como factor de competitividad. Ciencia Latina Revista Científica Multidisciplinar 3(1), 134–155 (2019)
6. Santamaría, R.: Las PYMES y factores para obtener el éxito, inicio para el marco referencial. Ingeniería Industrial. Actualidad y Nuevas Tendencias 6(21), 131–144 (2018)
7. López, L., Ríos, M., López, C.: La competitividad del as pequeñas y medianas empresas manufactureras de Bogotá y su percepción del apoyo que les brinda la política de competitividad. Pistas Educativas 39(129), 92–118 (2018)
8. Rodriguez, P., Villareal, L., Arévalo, V.: Evaluación de la eficiencia social en el sistema de coopertativas de ahorro y crédito del Ecuador. Conectividad 2(1), 1–21 (2021)
9. Arévalo, P.: Customer service quality in financial entities a factor for generating competitiveness in times of Covid 19 pandemic. In: Zambrano Vizuete, M., Botto-Tobar, M., Diaz Cadena, A., Durakovic, B. (eds.) CI3 2021. LNNS, vol. 511, pp. 516–527. Springer, Cham (2022). https://doi.org/10.1007/978-3-031-11438-0_41

10. Riaño, C., Larrea, O.: Análisis envolvente de datos y sus aplicaciones en sostenibilidad. Ingeniare **17**(31), 11–19 (2021)
11. García, E., CollSerrano, V.: Competitividad y eficiencia. Estudios de Economía Aplicada **21**(3), 423–450 (2003)
12. Chiavenato, I., Sapiro, A.: Planeación estratégica, fundamentos y aplicaciones, 3rd edn. McGraw Hill, México (2010)
13. Collins, N., Rosales, F., Villao, J.: Competitividad sostenible: una herramienta clave en la gestión administrativa, 1st edn. Compas (2017)
14. Becerra, M., Rivera, P.: Dimensiones de la competitividad: Factores internos y externos a las empresas, 1st edn. Plaza y Valdés Editores, Madrid (2018)
15. Belleflamme, P., Peitz, M.: Organizacón industrial Mercados y estrategias. Universidad del Rosario, Bogotá (2021)
16. Superintendiencia de Bancos y Seguros Marco Conceptual Catálogo de cuentas y fondos complementarios previosionales cerrados. https://www.superbancos.gob.ec/bancos/wp-content/uploads/downloads/2019/04/marco_conceptual_CUC_seg_social.pdf. Accessed 10 Mar 2023
17. CONSEJO DE NORMAS I Consejo de Normas Internacionales de Contabilidad, Marco conceptual para la información financiera. https://www2.deloitte.com/content/dam/Deloitte/cr/Documents/audit/documentos/niif-2019/El%20Marco%20Conceptual%20para%20la%20Informaci%C3%B3n%20Financiera.pdf. Accessed 10 Mar 2023
18. Farrell, M.: The measurement of productive efficiency. J. R. Stat. Soc. Ser. A (General) **120**(3), 253–290 (1957)
19. De la Garza, J., Morales, B.: Análisis estadístico multivariante un enfoque teórico y práctico, 1st edn. McGraw Hill, Madrid (2012)
20. Meza, C.: Econometría, 1st edn. Ediciones Unisalle, Bogotá (2022)
21. Bollen, K.: Structural Equations with Latent Variables. Wiley, New York (1989)

ICT for Educations

The Impact of Virtual Reality on Reading Comprehension in Rural Areas

Jorge Beltrán$^{(\boxtimes)}$, Carolina Basantes, and Maritza Quinzo

Instituto Superior Universitario Vida Nueva, Quito, Ecuador
{jorge.beltran,carolina.basantes,
maritza.quinzo}@istvidanueva.edu.ec

Abstract. Reading comprehension, according to Bransford, Brown, and Cocking; Shuell, mentions that contemporary learning theories place students at the center of learning activities. In order to enhance the reading comprehension skills of fourth-grade students in general basic education, an innovative reading system has been developed. This system combines conventional reading activities with the use of OCULUS QUEST 2 glasses in conjunction with the coespace platform. In this virtual world, different scenarios can be observed, allowing for navigation. The purpose of the study is to evaluate the effectiveness of glasses in improving students' reading comprehension. To achieve this objective, literature related to reading theory and comprehension will be reviewed. The obtained results and the implications of the study are presented as elements for analyzing the research landscape. From the perspective of these authors, these contributions to the study's object are demonstrated.

Keywords: Reading · Literature · Education · Comprehension

1 Introduction

According to UNESCO, in its report "Ecuador in ERCE 2019," the country ranked below the regional average, along with Guatemala and Honduras, in the area of reading. This is a clear demonstration of the deficiency in reading competence within the education system.

In this context, educators face new challenges in attempting to improve the reading process. Children in rural areas live in a different environment, and teachers must be ready to promote significant changes using innovative strategies with the use of TAC (Assistive Technology and Communication).

Virtual reality in the educational context has gained importance and benefits students, as stated by Panerai, Catania, Rundo, and Ferri cited in [1]. Hence, there arises a need to investigate the impact of virtual reality on comprehensive reading in rural settings. The research objective lies in utilizing new technologies to develop the skill of reading comprehension in elementary school children [2].

The information was gathered from the "Pichincha" educational unit located in the Lloa parish, involving fourth-grade students in the study. After the survey application, the results of this research, which employed an analytical method with a quantitative approach, are encouraging. Among the students who participated in the experiment using virtual reality glasses, 46.67% exhibited a favorable reaction, demonstrating an increased interest in reading.

1.1 Reading

Reading is a daily routine, but critical reading doesn't seem to be. Students invest a lot of time in comprehending the information contained in various resources (books, monographs, articles, brochures, visual presentations, and more), but sometimes they lack the ability to identify what they are reading. They read to understand and understand to learn, but they don't read to think critically. "Understanding requires formulating content, but also discovering underlying viewpoints or values according to one's ideology" [3]. Through reading and writing, boys and girls learn by grasping the world and its organization, expressing ideas, feelings, and emotions, and socializing through both oral and written interaction and comprehension.

Reading promotes many skills that educators strengthen in boys and girls from early education. Reading is associated with expanding vocabulary and becoming familiar with the syntactic and grammatical functions of the language. It's designed as a personal mission. The decoding technique is the same for everyone; texts convey the same content, and it doesn't make sense to comment on what's understood, except for texts that develop the emotional function of language, such as literature [3].

Reading is considered a process that everyone should engage in, partly because it's about shaping individuals for life and for relationships not only based on respect and tolerance but also as a tool for critical analysis and reflection about the world. It forms a link between understanding and action mediated by knowledge to achieve transformation.

The habit of reading is a fundamental link for the development of logical thinking, lifelong self-education, a constant interest in knowledge, enrichment of the intellect, expansion of vocabulary, and the ability to enhance human relationships and attitudes toward life. Reading is one of the primary sources of knowledge acquisition and is one of the most significant activities in the process of cultural formation [4].

1.2 Reading Skills

The skills developed through the educational curriculum and now a competency-based curriculum are the four language skills (speaking, reading, writing, listening) [2]. These language skills must be mastered to communicate effectively in all situations [5]. There's no other way but to use language for communicative purposes. For this reason, there are also four skills that should be developed in the classroom with a communicative approach: language skills.

The use of reading aloud links written comprehension with oral expression, developing competencies such as understanding the entire text, intuiting details, settings, characters, and actions. Additionally, the ability to deduce the meaning of an unknown term based on these micro-skills, which are part of reading comprehension [2]. This is why reading is considered a strategy [6].

1.3 Reading Comprehension Skills

Reading presents to the individual a possible world, both pleasant and unpleasant. It is imprinted with a relationship called animator and scene structure, offering a broad perspective on what life is. Through dissemination and collaborative work, scenarios can be created with children, involving important events they were part of. This enhances the ability to comprehend context and content, thereby boosting processes of socialization and cognitive development.

This is a situation that can be effectively addressed through engagement with reading. This approach fosters the development of technical, visual, strategic, and cognitive skills in children within General Basic Education.

1.4 Levels of Reading Comprehension

Reading comprehension is the ability to perceive, use, reflect upon, and take an interest in written texts in order to achieve personal goals, develop knowledge, and potentials, consequently participating in society [6]. This is why texts are organized in various ways. Depending on the type of text—narrative, descriptive, explanatory, among others—and how the content is presented—continuous, discontinuous, or mixed [7].

Continuous texts exhibit three characteristics. Firstly, the sense is a set of semantic relationships that exist among characters and other languages, resulting from interdependencies between contextual and linguistic elements. Secondly, cohesion involves linking sentences together, where the transition from one idea to another doesn't signify a rupture but rather a logical continuity. Various mechanisms enable text to be coherent: repetition, connectors or textual markers, anaphoras, punctuation. And thirdly, coherence is a property of the text related to thematic unity; in other words, the entire content revolves around a central theme or reference. Furthermore, consistency refers to the relationship between each part and all other parts, the organization's progression of pieces and themes [8].

Reading comprehension is important because it enables students to understand what they are reading, expand their knowledge to form solid opinions about different types of texts and situations, and it aims to automate learning. Therefore, it is crucial for educational institutions to enhance their methodological competencies [9].

Within the educational curriculum of Ecuador, as conceptualized by the Ministry of Education (MinEduc), the degrees, scales, or levels of text comprehension are understood as the possession of strategies or capacities. In the literal or basic degree or scale, students should obtain explicit information from the text with a specific objective. Next is the level or scale of interpretation or inference of the text. At this point, the student lays the foundation for and shapes the reading. The third level is the critical reflective level, where the student expresses their opinion or judgment of the text they read, taking into account the effects that occur, combining it with other readings, the social and cultural context of the text, and their own context as the reader [8].

There are three levels of reading comprehension according to Smith. The first level is literal, referring to the student's understanding to clarify what is stated in the text. The second level is inferential, which involves reasoning, demonstrating the students' ability to make conjectures and hypotheses about the text. The third level of reading comprehension is critical-evaluative, related to the capacity to judge the quality of the text and to issue reasonable judgments about it [9, 10].

1.5 Pedagogy and Didactics of Reading Comprehension

Reading strategies encompass the nature of higher-order mental, metacognitive, and learning processes. These occur in children from an early level, both consciously and unconsciously. This unconsciousness is a part of the previously mentioned general law of inheritance regarding the cultural development method from Vygotsky.

Working with children through metacognition to reach the Zone of Proximal Development signifies that the teacher plays a crucial role as a fundamental axis in the teaching and learning process, developing children's awareness. As educators, we are a part of their consciousness and their evolution towards metacognition [11].

The required skills identified in reading comprehension are related to the cognitive and metacognitive strategies that elementary school children need in order to construct meaning and self-regulate the reading process. It starts by identifying the skills that allow readers to process the text. Different levels of depth are involved: literal value, inferential value, critical value. Moreover, skills that aid readers in problem-solving and understanding when it occurs are identified, influenced by various factors depending on the text type and the reader themselves. The construction of meaning supports students in strengthening their reading skills [12].

It's important to emphasize the significance of using Information and Communication Technologies (ICT), as defined in the National Education Plan for the 2006–2016 decade, to enhance learning in a dynamic and productive way, thereby developing the competencies of children. The Ministry of Education has indicated this as one of its challenges. The use of ICT in pedagogy and education significantly improves connectivity with quality and equity standards to support the educational process [13]. It recognizes the curricular transversality of ICT use based on pedagogical research. The curriculum underscores the need to strengthen literacy processes as an essential condition for human development, eradicating illiteracy, and promoting social and civic participation.

1.6 Virtual Reality in Education

Using interactive scenarios to complement classroom learning allows for a more playful understanding of concepts that might initially seem distant. While the technology is just beginning to be widely adopted, developers are already making efforts to incorporate it into the classroom. Sony's glasses for their PlayStation 4 and HTC's Vive viewer also contribute to this trend. If things continue as planned, 2016 could be the year when virtual reality solidifies as an increasingly widespread tool, spanning from entertainment to education [14].

It is a technology particularly suitable for education due to its ability to capture students' attention by immersing them in virtual worlds related to various areas of knowledge. This can aid in learning the content of any subject [15, 16].

It is still difficult to imagine that virtual reality learning experiences could be widely adopted. However, there is no doubt that we are at the beginning of that widespread adoption," believes Fernando Valenzuela, President of Cengage Learning for Latin America, an educational solutions company. "The biggest challenge lies not only in having access to these platforms and devices, which are becoming increasingly accessible every day, but also in the proper design and implementation of a learning experience that allows students and teachers to extend and enhance the impact of what is learned" [17].

1.7   Contribution of the Metaverse to Education

The metaverse is an acronym formed by the prefix "meta" (beyond) and the contraction of the term "universe" – "verse" - which refers to a 3D virtual world that is immersive, interactive, and collaborative. It constitutes a new dimension on the Internet, with a significant impact on the education of the future [17]. It was American novelist Neal Stephenson (1992) who first used this term in his novel "Snow Crash" to refer to a new version of the Internet. This allows different members of a group to work together in a virtual space using avatars that interact face-to-face and create three-dimensional objects that represent ideas, values, or emotions [18]. The metaverse encompasses several characteristics that surpass extended reality [19]:

- It is persistent. It doesn't restart or pause but continues indefinitely.
- It promotes users' proactivity, giving them control, enabling decision-making, and anticipating potential events.
- It is synchronous and exists in real-time.
- There are no limits to the simultaneity of users in an activity.
- It provides an experience that spans the digital and real world, blurring the line between private and public networks.
- It offers data and digital element interoperability. For instance, a virtual object from a video game can be gifted to a friend through Facebook.

1.8   Challenges of the Metaverse in Education

The main challenge lies in transitioning from a sophisticated yet standalone virtual reality environment to an integrated network of 3D virtual worlds or metaverse. This highlights the need to advance knowledge in four aspects:

1. Realism (making users feel immersed in an alternative dimension)
2. Ubiquity (establishing access to the system through digital devices and maintaining users' virtual identities within the system)
3. Interoperability (allowing the creation and movement of 3D objects outside the system)
4. Scalability (enabling efficient usage of the system with massive numbers of users connected simultaneously).

From a practical standpoint, the initial experiences that have approached the metaverse have focused on the written and oral comprehension of languages [17]. The metaverse offers numerous educational possibilities related to the acquisition of cultural knowledge [18]. In this context, a key concept in metaverse development is culture, which refers to the backgrounds and experiences of the individuals forming part of this dimension [19], encompassing two related concepts: who we are and how we live [20]. Castells [18] asserts that a culture in the knowledge society is necessary, based on the establishment of certain norms. The author suggests that this culture is part of a system of real virtuality, where reality itself (the existence of people) is submerged in a scenario of virtual images, where appearances are not only on the screen through which the experience is communicated but become the experience. In this context, it is essential to construct a culture or set of foundational norms and values to communicate, work, and exist in the metaverse – an alternate reality where the boundaries between the physical and virtual remain hazy and need to be precisely defined to safeguard users' identities [22].

In this context, Smithson [23] suggests the need for an ethical code to develop metaverse culture, which is grounded in five aspects:

1. Transparency of knowledge and opinions between consumers and creators of products based on the metaverse contribute to improving its functioning.
2. A value system that promotes coexistence and relationships among users.
3. The principle of inclusion is fundamental. The metaverse should be accessible and easy to use for all individuals.
4. Avatars (or alter egos) represent real people, so harassment and fraudulent behaviors will result in the temporary removal of users from this space.
5. Ethics is the cornerstone, which translates into designing programs and learning experiences that ensure consumer confidentiality. The metaverse is an acronym formed by the prefix "meta" (beyond) and the contract.

2   Methodology

The current research project is descriptive in nature and follows a quantitative approach. Through this methodology, data collection was carried out with the objective of determining whether virtual reality enhances reading comprehension or not.

2.1   Instrument Design

For data collection, a Likert scale questionnaire consisting of eleven items was designed. The first six items are related to the independent variable, reading comprehension, three items correspond to the dependent variable regarding virtual reality, and two items are for hypothesis validation. Specifically, whether virtual reality enhances and broadens the ranges of comprehension along with the micro skills of reading, listening, and writing. The instrument was administered to 30 students, and all of them responded to it.

According to DeVellis, RF and Thorpe [24], the Likert scale is used to measure attitudes and opinions, allowing respondents to express their levels of agreement or disagreement with specific statements. The name comes from Rensis Likert, an American psychologist and educator, whose contribution lies in the formulation and application of this method in social research, although the term "Likert scale" itself was not his direct creation.

The following structure presents an evaluation matrix based on the Likert scale, which is used to rate various aspects related to specific skills and characteristics. The evaluated criteria include text comprehension, vocabulary, reading fluency, analysis and reflection, written expression, and interaction with virtual reality in an educational context.

In the Likert scale used, the numerical values 1, 2, 3, and 4 have the following meanings:

(VERY GOOD): Indicates a very positive performance in the evaluated criterion.
(GOOD): Indicates a positive performance, with areas that may be improved.
(AVERAGE): Indicates an acceptable performance, with opportunities for improvement.
(POOR): Indicates an unsatisfactory performance that requires improvement.

For example, if you rate "Item 6: Virtual reality sparks students' interest" with a value of 1, it indicates a very positive impact of virtual reality on students' interest.

On the other hand, if you rate it with a value of 4, it suggests that virtual reality does not effectively capture students' interest.

This table as a whole provides a detailed evaluation of how various aspects related to text comprehension, virtual reality, and other skills are perceived and rated in the educational context.

2.2 Reliability Validation

The reliability analysis of the scale from the designed study materials was conducted using the Cronbach's alpha coefficient. This coefficient was calculated using the scale assessment function in SPSS software version 2. The obtained results of the assessment for both scales are presented below (Table 1):

Table 1. Reliability value.

Cronbach's Alpha	Item Number
0.780	11

After applying Cronbach's alpha, it is verified that the reliability is high, so the questionnaire is then administered to the students.

2.3 Sample

The sample that participated in the research consisted of thirty fourth-grade students from the "Unidad Educativa Pichincha" enrolled in the 2022–2023 school year. Of these, 14% are girls and 16% are boys. This sample adheres to the selection of a non-probabilistic or purposive sampling criterion.

3 Results

3.1 Descriptive Study

The results obtained from the survey conducted with thirty fourth-grade students from the Unidad Educativa Pichincha, when presenting the story using virtual reality through goggles, revealed the potential to enhance reading comprehension by providing a more immersive and engaging experience. This could assist readers in visualizing and experiencing the concepts and scenarios described in the text more vividly. Readers could explore virtual environments related to the story they are reading, interact with virtual characters, or solve puzzles and challenges within the virtual environment.

In Item 1, 53.33% of the students showed an improvement in reading comprehension by being able to recognize the title of the story, in contrast to the 20% of students who were not proficient in identifying the title. In Item 6, 100% of the students indicated that virtual reality sparked their interest in reading. Finally, in Item 11, which pertains to hypothesis validation, it is considered that 51% of students improved their reading comprehension compared to traditional reading methods.

These results suggest that virtual reality can have a positive impact on reading comprehension by generating greater interest and engagement among students.

However, it's important to note that these results are based on the perceptions of the participants and may be influenced by individual and contextual factors (Table 2).

4   Results Analysis

In Table 3, it is evident that 53.33% showed improvement in terms of reading comprehension, in contrast to 3.34% of students who were not able to recognize the title of the story.

In Table 4, it is evident that 46.67% determined that virtual reality sparks interest in reading, in contrast to 6.67% regarding the handling of TAC technology.

5   Discussion

The text discusses various perspectives on reading comprehension and its relationship to education and emerging technologies like augmented and virtual reality.

[1] highlight the significance of reading in early education, stressing how the habit of reading contributes to language development, imagination, creativity, and critical thinking in children. They argue that reading enriches vocabulary, textual comprehension, and knowledge acquisition, crucial for long-term academic success.

[2] and [3] emphasize the close connection between reading and writing in basic education. They both assert that reading practice enhances students' writing skills by familiarizing them with grammatical structures, literary styles, and written communication strategies.

On the other hand, [4] focus on early childhood reading's importance, arguing that early exposure to books and shared reading with parents or caregivers influences young children's cognitive and affective development. They underline that reading in early childhood establishes the foundation for a lifelong love of reading and continuous learning.

In contrast, [7] establish a significant connection by exploring reading comprehension levels in elementary students. They highlight the importance of promoting literal, inferential, and reflective critical comprehension in the reading process, asserting that developing these skills allows students to interpret and analyze texts more effectively.

Table 2. Results of the assessment applied to fourth-grade students of the Unidad Educativa Pichincha.

Student	Text comprehension	Virtual reality	Virtual reality enhances reading comprehension	Observation
N01	5	5	5	Nothing new
N02	5	5	5	Nothing new
N03	5	5	5	Nothing new
N04	5	5	5	Nothing new
N05	5	5	5	Nothing new
N06	5	5	5	Nothing new
N07	5	5	5	Nothing new
N08	5	5	5	Nothing new
N09	5	5	5	Nothing new
N10	5	5	5	Nothing new
N11	5	5	5	Nothing new
N12	5	5	5	Nothing new
N13	5	5	5	Nothing new
N14	5	5	5	Nothing new
N15	5	5	2	Nothing new
N16	5	5	2	Nothing new
N17	3	5	3	Nothing new
N18	2	5	1	Nothing new
N19	3	5	3	Nothing new
N20	1	5	2	Nothing new
N21	2	5	1	Nothing new
N22	3	5	3	Nothing new
N23	4	5	3	Nothing new
N24	3	5	3	Nothing new
N25	1	5	3	Nothing new
N26	3	5	2	Nothing new
N27	3	5	1	Does not get better
N28	3	5	1	Does not get better
N29	3	5	1	Does not get better
N30	4	5	1	Does not get better

Table 3. Diagnosis of reading comprehension.

Assessment	Frequency	Percentage
Ítem 1 Text comprehension.	14	53.33
Ítem 2 Vocabulary.	6	20.33
Ítem 3 Reading fluency	3	10.00
Ítem 4 Análisis y reflexión	6	20.00
Ítem 5 Written expression	1	3.34
Total	**30**	**100,0**

Table 4. Virtual Reality.

Assessment	Frequency	Percentage
Ítem 6 Virtual reality sparks the interest of students	14	46.67
Ítem 7 Identify the main ideas of the text	6	20.00
Ítem 8 "Virtual reality requires computer knowledge."	4	13.33
Ítem 9 "Virtual reality is integrated with the curriculum content."	4	13.33
Ítem 10 "Handles TAC technology."	2	6.67
Total	**30**	**100,0**

The analysis then shifts to augmented and virtual reality, emerging technologies that have garnered interest in education due to their potential to transform teaching and learning methods. The text compares authors' ideas regarding these technologies' use in education, particularly in the context of reading and comprehension.

R. M. Sainz in "Realidad Aumentada: Una nueva lente para ver el mundo" explores augmented reality's possibilities as a tool that provides an enriched and expanded view of the surroundings. Sainz believes this technology can open new perspectives in education by facilitating greater content interaction and promoting active learning.

In contrast, Margherita emphasizes the importance of virtual reality as a tool for highly immersive educational experiences. It's worth noting that while virtual reality shares similarities with augmented reality, it is based on a fully immersive experience, where users are in a digitally generated environment. Margherita argues that education can greatly benefit from this technology, as it allows simulation of real-world situations and contexts to enhance learning.

Given the current context, reading comprehension is a fundamental skill in education, involving the ability to interpret and understand the meaning of a text. The significance of reading comprehension lies in various aspects, including knowledge acquisition. Reading different types of texts expands vocabulary, internalizes text structure and organization, and teaches new forms of communication.

Virtual reality provides students with an immersive experience that enables deeper and personalized interaction with content. By visualizing scenarios and characters described in texts, students can establish emotional and sensory connections that enrich their understanding and retention of information. Additionally, the visual and experiential nature of virtual reality can be especially beneficial for visually or kinesthetically inclined learners.

The experimental group demonstrated significant improvements in various aspects of reading comprehension, from identifying fundamental details to inferring implicit meanings and retaining relevant information. These results are encouraging, as they indicate that a notable 46.67% of students involved in the experiment, when using virtual reality glasses, responded positively, showing a renewed and revitalized interest in reading activities. Table 3, it is evident that 53.33% showed improvement in terms of reading comprehension, in contrast to 3.34% of students who were not able to recognize the title of the story.

6   Conclusions

The methodology used adopts a quantitative approach with a descriptive level, collecting objective and measurable data, providing an overall view of the relationship between reading comprehension and virtual reality. The results from the questionnaire administered to the 30 participating students indicated a positive correlation between virtual reality and improved reading comprehension, as well as reading, listening, and writing skills.

Virtual reality is an emerging technology with significant potential to transform education by changing how teaching and learning take place. These technologies have garnered substantial interest in the educational realm due to the opportunities they offer to enrich the learning experience. It's essential to continue exploring and harnessing the potential of virtual reality in education. This technology can create new possibilities in the educational process, especially in rural areas.

The proposal of this research involves utilizing technological resources oriented toward virtual reality, specifically employing Oculus Meta Quest 2 glasses to present interactive and dynamic scenarios that enhance reading comprehension and improve students' learning experience. He text discusses various.

References

1. Domínguez, A.I., Rodríguez Delgado, L., Torres Ávila, Y., Ruiz, M.: Importancia de la lectura y la formación del hábito de leer en la formación inicial. Estudios del Desarrollo Social: Cuba y América Latina **3**(1), 94–102 (2015)
2. Sanz, G., Luna, M., Cassany Comas, D.: Enseñar lengua. Dialnet **1**(12), 83–99 (2007)
3. Arnáez Muga, P.: La lectura y la escritura en educación básica. Educere: La Revista Venezola de Educación **13**(45), 289–298 (2009)
4. Noreña, C.A.R., de Castro Daza, D.P.: La lectura en la primera infancia. Dialnet **1**(20), 7–21 (2013)
5. Hermosillo, M.M.M., González, J.R.V.: Leer más allá de las líneas. Análisis de los procesos de lectura digital desde la perspectiva de la literacidad. Scielo, no. 50 (2018)

6. Juárez, M.I.G.: Los textos continuos: ¿cómo se leen? La competencia lectora desde Pisa. In: INNE, México, pp. 2–27 (2012)
7. Felipe, Z.B.L.A., Gilberto, C.-B.: Comprensión lectora en los niveles literal, inferencial y crítico reflexivo de los estudiantes de educación primaria. Revista de Investigación, Formación y Desarrollo **9**(2), 6–14 (2021)
8. MinEduc: Currículo de los Niveles de Educación Obligatoria subnivel elemental, Segunda ed., Ministerio de Educación del Ecuador, Quito (2019)
9. Ortega, J.L.G., Sepúlveda, S.F., Fuentes, A.R.: La comprensión lectora de escolares de educación básica. Scielo **1**(40), 187–208 (2019)
10. Alastre, D.M.: Comprensión de la lectura inicial. Consideraciones actitudinales acerca de la lectura y la escritura en el nivel de educación inicial. Educare: La Revista Venezolana de Educación **9**(28), 83–86 (2005)
11. Cruz, J.C., Carrillo, A.T.M., Carrillo, A.T.M.: Potenciar la comprensión lectora desde la tecnología de la información. Dialnet **9**(2), 26–36 (2011)
12. Palomares Marín, M.: La realidad aumentada en la comunicación literaria. El caso de los libros interactivos. Revista de la Facultad de Educación de Albacete **2**(29), 79–94 (2014)
13. Sainz, R.M.: Realidad Aumentada: Una nueva lente para ver el mundo. Ariel, Madrid (2011)
14. Reinoso, R.: Posibilidades de la Realidad Aumentada en Educación. In: Dialnet, pp. 175–195 (2012)
15. Álvarez, D.S., Ferrer, D.H., Bohorquez, E.P.: Uso de la realidad aumentada para fomentar la lectura. Teknos Revista Científica **2**(19), 29–34 (2019)
16. Margherita, C.F.: La educación también quiere estar inmersa en el mundo de la realidad virtual. El Mercurio. https://www.proquest.com/newspapers/la-educación-también-quiere-estar-inm ersa-en-el/docview/1770584318/se-2. Accessed 06 Mar 2016
17. Garrido-Íñigo, P., Rodríguez-Moreno, F.: The reality of virtual worlds: pros and cons of their application to foreign language teaching. Interact. Learn. Environ. **23**(4), 1–18 (2013). https://doi.org/10.1080/10494820.2013.788034
18. Dionisio, J.D., Burns, W.G., Gilbert, R.: 3D virtual worlds and the metaverse: current status and future possibilities. AMC Comput. Surv. **45**(3), 1–38 (2013). https://doi.org/10.1145/248 0741.2480751
19. Davis, A., Khazanchi, D., Murphy, J., Zigurs, I., Owens, D.: Avatars, people, and virtual worlds: foundations for research in metaverses. J. Assoc. Inf. Syst. **10**(2), 90–116 (2009). https://doi.org/10.17705/1jais.00183
20. Martín-Ramallal, P., Merchán-Murillo, A.: Realidad virtual. Metaversos como herramienta para la teleformación. En: Casas-Moreno, P., Paramio-Pérez, G., Gómez Pablos, V.B. (eds.) Realidades educativas en la esfera digital: Sistemas, modelos y paradigmas de aprendizaje, pp. 15–38. Egregius Ediciones (2019)
21. Zheng, R.Z.: Cognitive and affective perspectives on immersive technology in education. IGI Global (2020)
22. Han, H.C.: Teaching visual learning through virtual world experiences: why do we need a virtual world for art education? Art Educ. **68**(6), 22–27 (2015). https://doi.org/10.1080/000 43125.2015.11519344
23. Smithson, A.: The Metaverse Manifesto. https://bit.ly/3tdHfyu. Accessed 5 de enero de 2022
24. DeVellis, R.F., Thorpe, C.T.: Desarrollo de escalas: Teoría y aplicaciones. Publicaciones de salvia (2021)

ARTRI: A Gamified Solution for the Motor Stimulation of Older Adults with Osteoarthritis of the Hands

Galo Patricio Hurtado Crespo[1]([✉]) [iD], Ana C. Umaquinga-Criollo[2] [iD],
Anddy Sebastián Silva Chabla[1], and Nelson David Cárdenas Peñaranda[1]

[1] Instituto Tecnológico Particular Sudamericano Cuenca, Cuenca, Ecuador
gphurtado@sudamericano.edu.ec
[2] Universidad Técnica del Norte, Ibarra, Ecuador

Abstract. The objective of this study is to develop a controller capable of communicating with a web platform for the implementation of gamification in physiotherapeutic scenarios aimed at older adults diagnosed with osteoarthritis of the hands, in order to migrate towards the use of technological resources and ICT. According to data from the National Institute of Statistics and Censuses (INEC), Azuay is positioned as one of the provinces with the highest life expectancy, projecting a life expectancy of 79 years in 2020. This situation is related to a high incidence rate of osteoarthritis in the elderly, which makes this project relevant for this population. The proposal combines the categories of software, music, video games and electronics in a web platform that serves as a complementary therapy, using a controller as a manipulation tool. For the theoretical foundation, data was obtained from different academic repositories, following a qualitative, transcendental, experimental, explanatory approach. The evaluation of the project was carried out through the application of a questionnaire based on the LIKERT scale, in which the teachers evaluated the project and gave their opinion as judges. This study represents a significant contribution to the field of non-pharmacological therapy and the implementation of technology in the care of older adults with osteoarthritis of the hands.

Keywords: Gamification · Osteoarthritis · Therapy · Motor stimulation

1 Introduction

Musculoskeletal disorders represent one of the main health problems worldwide. They affect the musculoskeletal system causing pain, and limiting mobility, dexterity and the general level of functioning of the person. They are the main factor that contributes to the need for rehabilitation worldwide [1]. Rehabilitation treatments aimed at dealing with the symptoms caused by musculoskeletal disorders represent a negative economic impact on the patient's finances [2]. The particular case of a musculoskeletal disorder focused on the joints, enters the field of osteoarthritis, rheumatoid arthritis, gout, among others [1].

M. Z. Vizuete et al. (Eds.): CI3 2023, LNNS 1041, pp. 287–300, 2024.
https://doi.org/10.1007/978-3-031-63437-6_24

According to the article [3] osteoarthritis causes alterations in the cartilage and negatively affects the functioning of mobile joints, causing pain, stiffness and loss of joint function.

Aging, together with muscle weakness, are risk factors in the appearance and development of osteoarthritis, according to [4] it is estimated that worldwide 10% of men and 18% of women over 60 have moderate and severe osteoarthritis symptoms. This pathology belongs to the most common type of arthritis that affects an estimated 302 million people worldwide, being most frequently seen in the joints of the knees, hips, and hands [5]. In addition, it is positioned as the third condition of rapid risk associated with disability. After diabetes and dementia, its impact is compared to rheumatoid arthritis in the context of severe arthritis [6].

Osteoarthritis of the hands is characterized by a symmetrical distribution that affects on a larger scale the distal interphalangeal (Distal Interphalangeal (IFD)), proximal interphalangeal (Proximal Interphalangeal (IFP)) and trapeziometacarpal (Trapeziometacarpal (TMC)) joints. In the cases where the condition is between the interphalangeal joints, it usually begins at the DIP joint, causing pain, tissue growth, and the appearance of nodules [7].

Treatment options for osteoarthritis of the hands fall into three categories: (i) pharmacological, (ii) non-pharmacological, (iii) surgical. Within non-pharmacological treatments, it has been shown that exercise therapies show slightly positive benefits for people suffering from osteoarthritis; at the level of functionality and pain management, at the psychological level; a decrease in depression levels is recorded [8]. The beneficial effects of exercise therapy combined with manual therapy or splints decreases the pain caused by the pathology in patients with osteoarthritis of the hands [9].

Music therapy is a complementary therapy that is accredited by numerous studies that support the benefit of its use for pain management in different diseases. Its definition according to [10] is the clinical use, based on the evidence of musical interventions to achieve individualized objectives within a therapeutic relationship by a professional. Among its benefits, stress control, memory improvement and communication stand out. The results exhibited by the study [11], in reference to the implementation of music therapy in the treatment of osteoarthritis symptoms, evidence the beneficial effect of therapy in reducing pain.

Gamification is a term that has gained popularity in recent years, especially at an educational level, being called a learning technique. Gamification refers to the use of game design elements in non-game contexts. According to [12] this type of technique transfers the dynamics of the games to educational and professional spaces, seeking to improve results and skills.

Gamification as a resource applied to the rehabilitation of musculoskeletal disorders evidences the need for stimulation and motivation in young people, adults and the elderly, and highlights advantages such as intuitive dynamics and rapid adherence to treatment compared to traditional therapies; crucial factors for the efficacy of treatment in pain management [13]. According to [14] rehabilitation-oriented gamification is closely related to the use of external devices such as motion sensors or mixed reality devices intended to support gamified activities such as exercise and physiotherapy to facilitate the performance of individuals in the rehabilitation process.

The world of technology is in constant change; progress must be related to social reality, in order to seek to satisfy collective needs [15].

The gamification proposal as a therapy for the management and reduction of symptoms caused by osteoarthritis of the hands in the elderly through motor stimulation of the joints affected by the disease, is oriented towards the use of Information and Communication Technologies (ICT) and technological resources, which allow greater accessibility with an intuitive dynamic of use, which is not very difficult to understand. It is intended to reach a considerable number of beneficiaries, starting with older adults diagnosed with osteoarthritis of the hands.

While research suggests the potential of games and feedback to improve physical rehabilitation [13, 21–24], there are targeted efforts in this area of treatment adding the use of ICT found in research such as [25], which applies the use of virtual reality glasses to provide enjoyable participation with the inclusion of interactive experiences, while the study by [26] called "MOVE-OK" incorporates gamification as an incentive to promote participation in physical activities and exercises in patients with knee osteoarthritis, and additionally, support in decision-making in the medical area [27] and care and remote home monitoring [22].

This research is being conducted to detail the variables or techniques carried out by the partial application of music therapy and exercise therapy in conjunction with electronics and computing. The proposition that seeks to be verified in carrying out the research is the effectiveness of gamification as a viable tool in complementary therapy for the symptomatic management of osteoarthritis of the hands.

2 Materials and Methods

The type of research carried out is of a qualitative, experimental, cross-sectional, explanatory approach. In [16], they state that qualitative research seeks learning through the experience of individuals, and in turn generates knowledge that is based on the perspectives of individuals. The applied research methodology (see Fig. 1) is based on the stages according to what was proposed by [16].

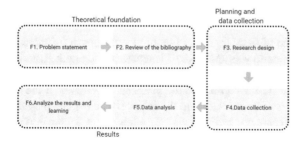

Fig. 1. Research methodology

2.1   Problem Statement

To carry out the investigation, the diagnosis is delimited by the technique of observation in different nursing homes for the elderly (Hogar Miguel León, Hogar Cristo Rey) in Cuenca, Ecuador. The absence of implementation of recreational activities that have the purpose of gradual helping improve the health difficulties of individuals suffering from osteoarthritis of the hands through the use of ICT and activities with a therapeutic objective is evident. In addition, the province of Azuay has a high level of life expectancy, which is associated as a causal effect of the incidence of osteoarthritis in people due to the risk factor of advancing age, according to data from the National Institute of Statistics and Censuses [17].

2.2   Literature Review

According to the literature review carried out, the impact of osteoarthritis as a disabling disease worldwide is known. In general, it is possible to control it; however, the damage to the joints is irreversible [18]. A study by [19] evidences the relationship between rheumatic diseases and physical disabilities; among the most prevalent is osteoarthritis of the hand.

2.3   Research Methodology

The study is of a qualitative, experimental type applied through the implementation of gamification in the field of therapies for motor stimulation of the affected joints in people from the city of Cuenca who are over 60 years of age. The experimental cross-sectional research design consists of a test session of less than 20 min, in which the participants manipulate the electronic device in conjunction with the software, seeking to compare the reaction of the different groups of test subjects to the study environment. An experimental investigation has as its main objective to explain the causal relationships between variables through controlled manipulation and the measurement of the resulting effects, which makes it an explanatory approach in scientific research.

2.4   Data Collection

As the main source of data, the Delphi method was chosen for the collection of information. A questionnaire was applied to 7 teachers of the South American Private Technological Institute of Cuenca. The response format is of the LIKERT type. The group of experts was selected according to the years of experience in the area and the type of activity they carry out in relation to the subject. The experts play the role of judges; they give their criticism and point of view through scores for the controller and the web platform.

The controller design consists of a rectangular box as a base, to which are attached flexible and elongated structures that culminate in polystyrene balls that contain buttons, a key element in therapy. The web platform is intended for gamified activity; it presents fast interaction game characteristics. The communication between the two components allows registration of the scores, which is stored in a database for later analysis.

In the context of the field tests, the selection of participants was random from a population of 60 years and older, with a total of 7 participants selected. A LIKERT type questionnaire was applied to evaluate the level of approval for the project and finally to evaluate the effectiveness using the simple verbal scale [20] that classifies pain into six categories from 0 (no pain) to 5 (unbearable pain), which depends on individual perception. The secondary information source for the theoretical foundation was the literature review, conducted in academic repositories such as Google Scholar, Scielo and Scopus.

2.5 Data Analysis

For the analysis of the data coming from the opinions of the experts and users, statistics of frequency are used in order to indicate the degree of acceptance obtained by the project and the effectiveness it presented.

2.6 Analysis of Results and Learning

Conclusions were drawn from the results obtained. They are presented in the results section by means of tables that indicate the total frequencies with which each indicator present in the LIKERT scale was valued and a graph that represents them.

2.7 Design and Elaboration of the Prototype

Next, Tables 1 and 2 describe the components used for the development of the proposed software with their proper description.

Table 1. Components used for the software.

Component	Description
Data base	ElephantSQL as database administrator PostgreSQL
Backend	SpringBoot development framework with Kotlin and the MVC design pattern
Frontend	Vite development environment in React with Typescript

Table 2 shows the components used for the controller, such as the ESPWROOM32 microcontroller that, together with the exposed modules, works wirelessly via bluetooth using a rechargeable battery.

Figure 2 shows the ESPWROOM32 microcontroller, pulsators, the MT3608 Dc-dc module and a battery It is a basic representation of the controller's operation.

Figure 3 shows the schematic design of the prototype in the design phase for subsequent simulation and implementation.

Figure 4 shows the collection of information in the platform's database.

The score that the user obtains is recorded in the database and the number of attempts made. These data will serve to measure the progress of the participants in a period of time, allowing the evaluation of the usability of the electronic device.

Table 2. Components used for the hardware.

Components	Description
Microcontroller	ESPWROOM32
Electronic components	Microswitches
Charging module	Micro-usb charging module (Tp4056) and battery indicator module
Battery	4.2v Lithium battery (rechargeable)
Voltage booster module	Module Dc-dc MT3608

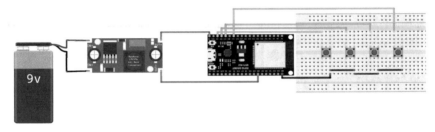

Fig. 2. Circuit design

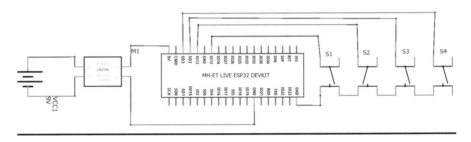

Fig. 3. Prototype schematic

Figure 5 shows the start user interface of the ARTRI application.

In this section the user validates the registration with the username and password credentials.

Figure 6 shows the interface of the dynamics of the video game developed for the user.

The user has 4 options that will be activated randomly depending on the selected topic. It should be noted that each topic has a different degree of difficulty.

Figure 7 shows the scores obtained by the participants.

You can see the score obtained, which is the reflection of the successful attempts by each user. These results are the metric that is taken into consideration when making decisions in the progress of each user.

In Fig. 8 the application and use of the controller is evidenced. Picture A of the image shows the final product; platform together with the controller. In picture B the

Fig. 4. Database in ElephantSQL

Fig. 5. Screenshot of the start screen of ARTRI

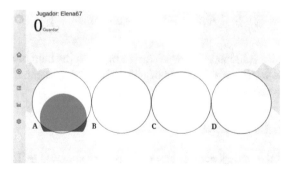

Fig. 6. Application video game interface

participant can be seen making use of the prototype and platform. In picture C the guide is shown indicating to the participant how to interact with the project. And, in picture D the validation granted by the experts is presented.

Fig.7. Interface for the points obtained by the users

Fig. 8. Tests of the prototype

The testing and validation phase was carried out with the help of expert teachers in the study´s area, as well as with the help of older adults who are the direct beneficiaries of the electronic device.

3   Results and Discussion

Below is a total of the ratings awarded by the experts, expressed in Table 3.

Based on the results obtained from the validation of the experts, a graph in Table 3 is presented, evidencing the percentage of experts who indicated that they were in total agreement with the indicators.

Figure 9 shows the results of the experts in terms of the different variables that were proposed for the development of the electronic device. 45.44% fully agreed with the technological foundation, feasibility, applicability and user interface. 27.28% fully agreed with the aspects of the controller's novelty and design. Finally, 15.91% fully agreed with the indicators for its use.

Table 3. Results of validation by the experts

Indicators	Completely agree	Agree	Neither agree nor disagree	Disagree	Completely disagree
Applicability	5	2	0	0	0
Feasibility	5	2	0	0	0
Novelty	6	1	0	0	0
Technological foundation	5	2	0	0	0
User experience	5	2	0	0	0
Controller design	6	1	0	0	0
Indicators for use	7	0	0	0	0
User interface	5	2	0	0	0
Total	44	12	0	0	0

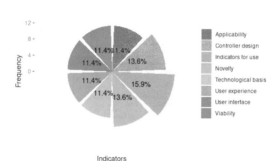

Fig. 9. Analysis of the experts that agreed completely

In Fig. 10 it can be seen that 83.35% agreed with the applicability, user interface, user experience, feasibility and technological foundation and 16.66% agreed with the controller design and novelty.

In relation to the opinions granted by the target participants, the qualifications granted are displayed, expressed in table 4.

Figure 11 shows a frequency graph that indicates that 42.86% fully agreed with the dynamics of the game, ease of use and the user interface. 28.57% of the participants fully agreed. 57.14% of the participants fully agreed with the structure of the device. Finally, 71.43% of the participants fully agreed on aspects of functionality and general level of satisfaction.

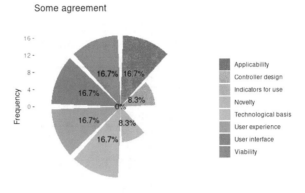

Fig. 10. Analysis of the experts that are in agreement

Table 4. Results of validation by users

Indicators	Completely agree	Agree	Neither agree nor disagree	Disagree	Completely disagree
User interface	2	4	1	0	0
Game dynamics	3	3	1	0	0
Device structure	4	3	0	0	0
Ease of use	2	4	1	0	0
Funcionality	5	2	0	0	0
Level of general satisfaction	5	2	0	0	0
Total	21	18	3	0	0

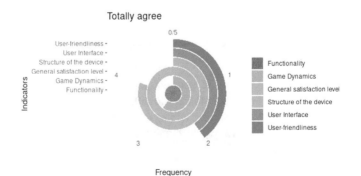

Fig. 11. Analysis of the users that are in complete agreement

In Fig. 12 it is evident that 42.86% agreed with the dynamics of the game and the structure of the device. Concerning the level of general satisfaction and functionality 28.57% of the participants agreed. Finally, 57.14% of the participants agreed concerning the aspects of ease of use and user interface.

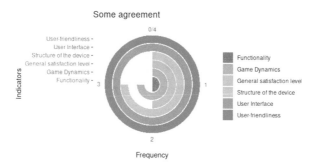

Fig. 12. Analysis of the users that are in agreement

In Fig. 13 it is evident that 14.28% neither agreed nor disagreed with the dynamics of the game, the user interface and the ease of the game.

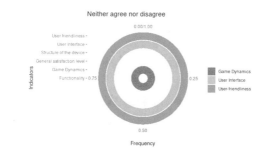

Fig. 13. Analysis of the users that neither agree nor disagree

According to the simple verbal scale, the data shown in Table 5 were tabulated in order to know the effectiveness of the project based on the perceptions of the participants.

Figure 14 shows the data obtained before the implementation of the device, and indicates that three of the seven participants, representing approximately 42.86% of the population, perceived moderate pain in the hands, followed by 28.57% of the participants indicating that they suffered from intense pain, and the remaining 28.57% indicated that the pain they perceived was light.

Figure 15 shows the data obtained after the implementation of the device, and indicates that three of the seven participants, representing approximately 42.86%, perceived slight pain in the hands. 42.86% of the participants indicated that they felt discomfort. And, 14.28% stated that they felt moderate pain.

Table 5. Simple verbal scale

Categories	Before the study	After the study
None	0	0
Discomfort	0	3
Very little, light pain	2	3
Moderate	3	1
A lot, intense, strong	2	0
Unbearable	0	0

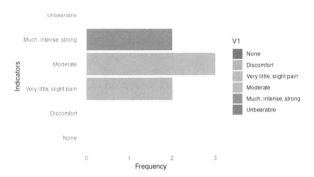

Fig. 14. Pain scale measurement before the study

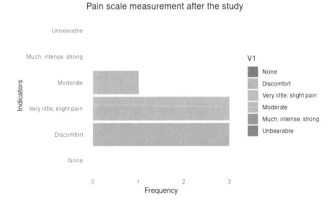

Fig. 15. Pain scale measurement after the study

4 Conclusions

The study shows that technological innovation in the field of health, through the implementation of web tools and electronic devices that integrate gamification, causes an increase in the attention and motivation of patients. This allows the elderly to get involved and participate in recreational activities that help cognitive and motor stimulation, allowing a significant improvement in various aspects of health.

Complementary therapies demonstrate that treatment integrates with the recovery process in the participants at a physical and psychological level. They are effective in the aspect of pain management and increase the sensation of control and treatment of the disease, as well as the general well-being of the patient.

In accordance with the defined and structured framework, the research and development of the prototype was completed. The evaluations of the experts evidenced the high level of impact, allowing the integration of the projects of the academy with the elderly subjects who have osteoarthritis of the hands.

References

1. OMS Trastornos musculoesqueléticos, https://www.who.int/es/news-room/fact-sheets/detail/musculoskeletal-conditions, last accessed 2023/10/04
2. Seoane-Mato, D., et al.: Prevalencia de enfermedades reumáticas en población adulta en España (estudio EPISER 2016). Objetivos y metodología. Reumatol. Clin. **15**(2), 90–96 (2019)
3. Sanchez-Lopez, E., et al.: Synovial inflammation in osteoarthritis progression. Nat. Rev. Rheumatol. **18**(5), 258–275 (2022)
4. Vidal Fuentes, J.: Artrosis y dolor: la complejidad e impacto de un síntoma. Rev. Soc. Esp. Dolor. **28**, 1–3 (2021)
5. Kolasinski, S.L., et al.: 2019 American College of Rheumatology/Arthritis Foundation guideline for the management of osteoarthritis of the Hand, hip, and knee. Arthritis Rheumatol. **72**(2), 220–233 (2020)
6. Hawker, G.A.: Osteoarthritis is a serious disease. Clin. Exp. Rheumatol. 37 Suppl 120, 5, (2019)
7. Programa de intervención de terapia ocupacional en pacientes mayores de 60 años con artrosis degenerativa de la mano–Repositorio Institucional de Documentos
8. Wang, S.-T., Ni, G.-X.: Depression in osteoarthritis: Current understanding. Neuropsychiatr. Dis. Treat. **18**, 375–389 (2022)
9. VillafaÑe, J.H., et al.: Physical activity and osteoarthritis: Update and perspectives. Pain Med. **20**(8), 1461–1463 (2019)
10. What is AMTA?, https://www.musictherapy.org/about/whatis/, last accessed 2023/10/04
11. Mukherjee, O., Suma, L.: Management of Chronic Illness through Music Therapy: a Review. Journalofhealthstudies.in. last accessd: https://journalofhealthstudies.in/uploads/229/14004_pdf.pdf
12. Vega, Z., Magdalena, Z.: Estrategias metodológicas de la gamificacion en el aprendizaje. Universidad de Guayaquil. Facultad de Filosofía, Letras y Ciencias de la Educación (2019)
13. Alfieri, F.M., et al.: Gamification in musculoskeletal rehabilitation. Curr. Rev. Musculoskelet. Med. **15**(6), 629–636 (2022)
14. Tuah, N.M., et al.: A survey on gamification for health rehabilitation care: Applications, opportunities, and open challenges. Information (Basel). **12**(2), 91 (2021)

15. Lata, E.D.B., et al.: Gamificación en la estimulación cognitiva de niños entre 5–7 años con síndrome de down en la UNAE. Ciencia Latina. **6**(1), 3676–3692 (2022)
16. Hernández Sampieri, R., Fernández Collado, C., Baptista Lucio, P.: Metodología de la Investigación (2010)
17. Osteoartritis.:   https://www.mayoclinic.org/es-es/diseases-conditions/osteoarthritis/symptoms-causes/syc-20351925, last accessed 2023/10/04
18. INEC.: Programa Nacional de Estadística. Instituto Nacional de Estadística y Censos, Quito-Ecuador (2021)
19. Guevara-Pacheco, S.V., et al.: Prevalence of disability in patients with musculoskeletal pain and rheumatic diseases in a population from Cuenca. Ecuador. J. Clin. Rheumatol. **23**(6), 324–329 (2017)
20. Ordóñez Mora, L.T., Sánchez, D.P.: editoras científicas. Evaluación de la función neuromuscular. Cali, Colombia: Editorial Universidad Santiago de Cali (2020)
21. Vaidya, M., Armshaw, B.: Surface electromyography and gamification: Translational research to advance physical rehabilitation. J. Appl. Behav. Anal. **54**(4), 1608–1624 (2021)
22. Mehta, S.J., et al.: Effect of remote monitoring on discharge to home, return to activity, and rehospitalization after hip and knee arthroplasty: A randomized clinical trial. JAMA Netw. Open. 3, 12, e2028328 (2020)
23. Zhao, X., et al.: Gamified Rehabilitation for Pain Distraction in Total-Knee-Replacement Patients. In: Extended Abstracts of the 2018 CHI Conference on Human Factors in Computing Systems. ACM, New York, NY, (2018)
24. Wijaya, R., et al.: Design application to lose weight of overweight person (Steppy Application). In: 2015 5th IEEE International Conference on System Engineering and Technology (ICSET). IEEE (2015)
25. Özlü, A., et al.: The effect of a virtual reality-mediated gamified rehabilitation program on pain, disability, function, and balance in knee osteoarthritis: A prospective randomized controlled study. Games Health. **12**(2), 118–124 (2023)
26. Leach, W., et al.: Protocol for a multi-center randomized controlled trial to evaluate the benefits of exercise incentives and corticosteroid injections in osteoarthritis of the knee (MOVE-OK). Trials **23**, 1 (2022)
27. Burgon, T. et al.: Measuring and improving evidence-based patient care using a web-based gamified approach in primary care (QualityIQ): Randomized controlled trial. J. Med. Internet Res. **23**(12), e31042 (2021)

Criteria for Creative Pedagogical Practices in Writing an Essay on National Reality Using the Logical Framework Approach

Lizzie Pazmiño-Guevara[1]([⊠]) [iD], Asdrúbal Ayala-Mendoza[1] [iD],
Jorge Álvarez-Tello[2] [iD], and Marco Duque-Romero[3] [iD]

[1] Facultad de Ciencias de La Educación, Universidad Indoamérica, Quito, Ecuador
lizziepazmino@indoamerica.edu.ec
[2] Centro de Innovación Social y Desarrollo (CISDE), Quito, Ecuador
[3] Centro Universitario (CIFE), Quito, Ecuador

Abstract. The manuscript proposes, in the current educational context, pedagogical practices influenced by technology, promoting innovative approaches in teaching using the logical framework approach (LFA) to enhance essay writing skills. To propose a methodology for the development of an academic essay on the national reality of Ecuador in higher education. A comprehensive review of 1554 publications in SCOPUS between 2020–2022 were conducted, selecting 20 relevant ones to characterize 4 evaluation criteria. As well, there is a didactic proposal that includes research in theoretical and methodological phases, utilizing the logical framework approach as the main tool. An exploratory and participatory research on the Ecuadorian reality is promoted, incorporating criteria of educational applications related to gamification, collaborative work, flipped classroom, and creative thinking. As result of the review, four evaluation criteria are identified: gamification, creative pedagogical practice, flipped classroom, and collaborative learning. Creative pedagogical practice stands out with 42% of the experiences, focusing on the use of technology and innovation. The proposal's phases and the screens of the national reality subject with which students begin the teaching-learning process for each mentioned pedagogical practice are presented. Finally, a didactic methodology based on the logical framework approach that integrates pedagogical practices and the use of Information Computing Technology for the development of academic essays on the Ecuadorian reality has been designed. The implementation of these educational applications in the proposed didactic methodology is suggested to enrich the students' learning process and encourage their participation and reflection in the elaboration of academic essays about the Ecuadorian reality.

Keywords: Logical Framework Approach · Academic Essays · Pedagogical Practices · Ecuadorian National Reality

M. Z. Vizuete et al. (Eds.): CI3 2023, LNNS 1041, pp. 301–313, 2024.
https://doi.org/10.1007/978-3-031-63437-6_25

1 Introduction

In the current educational context, pedagogical practices are being influenced by technologies, leading to new approaches in the teaching and learning process [1]. Gamification stands out as a playful approach that seeks to improve teaching through risk assessment, promotion of physical activity, and development of social skills [2, 3]. Furthermore, sociocultural approaches are integrated to promote creativity and learning [4]. The teacher's creativity is crucial in applying various tools that facilitate the teaching process [5, 6, 7]. Narrative techniques allow us to propose an educational ideology that addresses new challenges [8]. The intervention of the teacher, with the support of technology such as robotics, enhances interaction with students [9]. Developing critical and creative thinking through literature enables teachers to generate interactive strategies in the classroom [10, 11]. Collaborative learning and team physical activities have been shown to increase emotional integration among students [12, 13]. In this context, it is important to explore how the Logical Framework Approach (LFA) can be a complementary tool to establish a cause-and-effect logic [14]. In this learning context, there is a need to explore how the Logical Framework Approach (LFA), particularly in its initial phase, can be a complementary tool to establish a cause-and-effect logic. This approach allows for the identification of problems, understanding the current situation, establishing objectives, selecting alternative solutions, and measuring results [15]. While it has limitations, the Logical Framework Approach is recognized for its utility in planning, effectiveness, and adaptability across different contexts. The Logical Framework Approach has been successfully used in project planning and evaluation, and its application in academic essay writing has also been studied. It provides a solid foundation for creating logical arguments and strong academic essays [16]. Furthermore, it is highlighted that the use of a logical framework enhances writing skills by aiding in presenting ideas coherently. Considering the background, the aim of this paper is to propose a methodology for developing an academic essay on the national reality educational project in higher education, integrating creative pedagogical practices. The key stages will be analyzed, and the logical framework will be adapted to the specificities of academic essays, aiming to promote reflective and critical learning among university students.

2 Method

This research adopts a mixed exploratory-descriptive approach with the purpose of developing an evaluation framework and conducting a pilot study of a methodological proposal that promotes the learning and assimilation of both soft and specific skills. This proposal is structured into two main components:

– Development of an evaluation framework that will allow for the analysis and assessment of the effectiveness of the implemented didactic proposal.
– Design of the teaching proposal for the formative project on national reality, aiming to achieve a final product, such as the academic essay.

The method used in this research involved a comprehensive search in the Web of Science (WOS) and SCOPUS databases. Specific filters were applied to select articles published between 2020 and 2022 that focused on pedagogical practices. Initially, 1554 potential articles were identified. Subsequently, the terms "pedagogical practices" were combined with "gamming," "metacognition," "creative pedagogical practices," "collaborative learning," and "inverted classroom." Reference was made to the developed criteria for evaluating the contributions of the authors' experiences. Additional criteria were applied, such as document type and the category of Social Sciences and Education.

Articles without at least one citation were excluded to select relevant information. The titles and abstracts of the articles were then reviewed to identify research outside the scope of the present study, such as those related to foreign language development, early education, physical education, and learning difficulties or disabilities. As a result of the final evaluation process, 20 publications from both databases were selected out to be used as research sources in the results. These articles provide valuable information and support the theoretical foundation of the research, providing a solid basis for the analysis and discussion of the findings.

Finally, the results obtained were presented in a clear and organized manner (see Fig. 1). With this methodological approach, we aim to obtain a broad and detailed understanding of the pedagogical practices addressed in the scientific literature, as well as identify trends and advancements in this field. Boolean equations using AND and OR connectors were employed to conduct the article search.

In SCOPUS, the following English language search configuration is used:

TITLE-ABS-KEY (pedagogical AND practices) and (limit-to (pubyear, 2022) or limit-to (pubyear, 2021) or limit-to (pubyear, 2020)) and (limit-to (doctype, article)) and (limit-to (subjarea, "soci")) and (limit-to (pubstage, "final")) and (limit-to (srctype, "journal")).

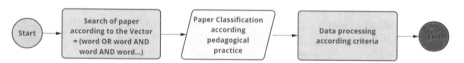

Fig. 1. Evaluation research framework diagram

2.1 Didactic Proposal for Learning About the Ecuadorian National Reality Based on Essay Writing, Applying the Logical Framework, and Analyzing the Effects

Participants. The proposal involves conducting training activi-ties with groups of approximately 205 students from various majors. This allows for comparing the outcomes between ex-perimental and control groups, providing a more robust per-spective on the effectiveness of the instructional proposal.

Training activities. Students will participate in a variety of educational activities aimed at promoting critical reflection and enhancing their research, analysis, and writing skills. Field trips to non-formal educational sites such as museums, historical sites, or government institutions are some of these activities. Classroom discussion and analysis sessions will focus on the main problems or challenges faced by Ecuador in various areas. The research work aims to gather information on the selected topic, identify relevant issues, and analyze the effects and causes. Students will conduct analysis of academic and scientific sources to support their essay ideas and reflections. Writing and essay revision sessions will be conducted, with feedback provided by the instructor and possibly among peers.

Validation of the didactic proposal.   The validation of the didactic proposal involves the participation of experts in the pedagogical field and educational projects. These experts will follow established evaluation protocols to analyze the effectiveness of the proposal in terms of students' learning outcomes, the quality of the essays produced, and participation in the educational activities. It is important to consider that the implementation of the proposal includes detailed planning, adapted to the needs and available resources in the context of higher education.

Phases of the proposal.   To gain a clear understanding of the phases and operationalization of the didactic proposal, it is essential to analyze and comprehend the main problems and challenges that Ecuador faces in various fields. In this context, the teacher aims to foster critical reflection and develop research, analysis, and writing skills. To achieve this, the following phases will be followed (see Fig. 2):

Phase 1. Sensitization based on ancestral and cultural knowledge.   Students and teachers organize a field trip to a museum to establish a framework of awareness based on historical events and social phenomena that provide different perspectives and an initial analysis of the future central topic of interest. Prior to the trip, a collaborative forum is conducted to explore relevant historical events and social phenomena.

Phase 2. Determination of the main topic related to the national situation.   Students and the teacher select a central topic of interest and relevance to Ecuador's national reality, covering social, economic, cultural, political, environmental, and institutional aspects. The chosen topic should be related to the students' professional profile.

Phase 3. Analyzing and creating a problem tree.   Students gather information by researching the topic. The logical framework is used to create a problem tree, identifying relevant problems related to the topic. Problems that have results are considered passive problems and are grouped into effects that emerge from brainstorming and culminate in further research for the corresponding support. These helps consolidate a broader problem. Then, problems that have consistent results and continue throughout the project establishment are identified.

Phase 4. Finding solutions.   Students conduct exploratory research on different solution proposals for each previously identified cause, which will lead to actions to be expressed in the essay. It is important to note that the analysis is at a design level, and the conclusions can be incorporated as questions for further research.

Phase 5. Writing an essay and recording a video.   Students write a reflective essay on the previously selected topic while reflecting on the problem identified in the effects and the general problem. The preliminary structure of the essay is provided, including a clear and concise introduction, an argumentative development, and a conclusion that synthesizes the main ideas. All content must come from scientific sources, with appropriate APA citations to support their arguments and reflections. A 2–3-min video of the essay will be submitted.

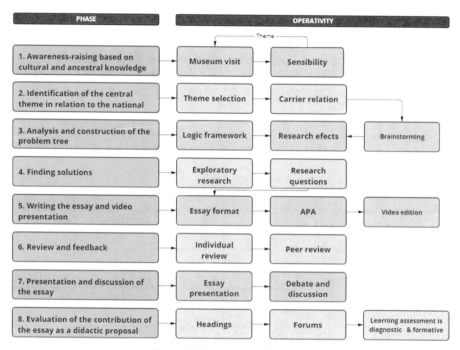

Fig. 2. Didactic proposal diagram

Phase 6. Review and feedback.   The essays are reviewed by the teacher, who provides individual feedback for improving writing, argumentation, and writing quality. Peer review is encouraged to promote collaboration and learning among students.

Phase 7. Presentation and discussion of the essays.   The essays are presented in 3 min in class, followed by group discussion and debate on the problems and solutions presented, fostering respect and communication. The essays are clustered according to the respective domains for the debate.

Phase 8. Evaluation of learning from the essay as a didactic proposal.   The teacher uses rubrics to assess the development of the problem tree, essay elaboration, and essay defense through a video.

Formative learning evaluation takes place throughout the process and through debates, where each team member self-evaluates, as well as engages in peer evaluation and teacher evaluation. At the end of each partial and semester, the evaluation is quantitative, including multiple-choice questions, case studies, and open-ended questions. Active participation in asynchronous forums is important for discussing various topics related to attitudes towards the adversities of our country's reality. Essay similarity is checked during the assessment.

3 Results

3.1 Bibliographic Review Characterization

The Table 1 presents the breakdown of publications based on the criteria from the previous table. It also provides the percentages of the most applied criteria (in the columns), along with the density of usage per author (in the rows), for the techniques or strategies to enhance learning.

Table 1. Characterization of the paper review.

N°	Author's/*Criteria*	M	N	Q	S	Total
1	[17]	♦				1
2	[18]		♦		♦	2
3	[19]				♦	1
4	[13]			♦	♦	2
5	[6]	♦	♦		♦	3
6	[20]	♦	♦		♦	3
7	[21]			♦		1
8	[9]	♦	♦		♦	3
9	[22]	♦	♦		♦	3
10	[23]		♦		♦	2
11	[24]		♦			1
12	[25]		♦			1
13	[26]	♦				1
14	[27]	♦	♦	♦		3
15	[28]		♦			1
16	[26]	♦	♦			2
17	[29]		♦		♦	2
18	[30]		♦		♦	2
19	[31]		♦			1
20	[32]		♦			1

Source: Prepared by the authors

The following Table 2, displays the density of criteria usage, allowing for the observation of trends, complementarities, and new fields of exploration.

Table 2. Criteria ranking.

Criteria	Quantity	%	Ranking
M	08	22	3°
N	15	42	1°
Q	03	08	4°
S	10	28	2°

Source: Prepared by the authors

The results not only provide us with the ability to visualize the concentrations of criteria but also to detect the fields where authors in each scenario can exploratorily implement new proposals that can aid teaching practices in virtual learning environments.

3.2 Criteria of Other Pedagogical Practices

As part of the literature review, the construction of criteria from other pedagogical practices is taken as a reference to discriminate the authors' proposals and structure information with the following references (see Table 3): (M) Gamification, (N) Creativity, (Q) Flipped classroom, and (S) Collaborative learning, as shown in the following table.

3.3 Proposal Design

For the design of the proposal, the presentation of the course material on national reality is considered, which includes the following screens (see Figs. 3 and 4). These screens outline the course sections that align with each phase of the proposal:

Through this information configuration in the virtual platform of the Higher Education Institution (Fig. 3), there are pedagogical elements for the development of participatory dynamics, as well as the production of deliverables such as the logical framework through the problem tree, and the original essay following the content (see Table 4).

In the dynamic field of education and research, new methodologies and tools are constantly being explored to enhance the effectiveness of learning and evaluation processes. The use of the Logical Framework Approach provides a strategic approach in the planning and execution of experimental courses, while the creation of an image mosaic serves as an innovative way to synthesize and visually represent the multimedia teaching resources generated throughout the course (see Fig. 5). This publication aims not only to present the implementation and outcomes achieved but also to underscore the synergy between tools that enhance the understanding, presentation, and evaluation of accomplishments.

Table 3. Criteria description

Criteria	Teaching didactic strategies	Description of the criterion
M	Gamification	Game-based learning motivates dynamic interaction among students, focusing activities on learning objectives
N	Project-Based Learning, Challenges, and Problem-Based Learning	Regular education has underestimated the potential of creativity and innovation. Teachers should harness students' innate abilities in idea generation, autonomy, and collaborative problem-solving through Problem-based learning, Project-based learning, and Challenge-based learning
Q	Flipped classroom	Tool that promotes values and creates a learning environment that is individual, shared, and participatory. It emphasizes learning from what is learned, done, and mistakes. It encourages sharing experiences, analyzing different perspectives, and fostering challenges and information. Students come prepared with answers to provide solutions through reading, analyzing reality, comparison, and constructing new ideas
S	Collaborative learning	Team-based learning is an instructional approach that promotes leadership and communication skills as alternatives for developing values, positive attitudes, and social skills. It encourages students to work collaboratively in groups to solve problems and address challenges within an educational setting. By engaging in team-based activities, students learn how to effectively communicate, negotiate, and contribute to group decision-making. This approach fosters a supportive and inclusive environment where students can develop problem-solving strategies and enhance their interpersonal skills while working towards shared goals

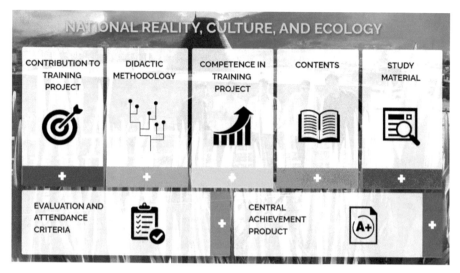

Fig. 3. Course sections

Table 4. Content section of the Logical Framework Approach subject.

CONTENT
WORKSHOP 1: Republican and democratic historical Ecuador
Training map and prior knowledge
Pre-colonial era
Colonial era
Independence era
Republican era
Contemporary era/partial evaluation
WORKSHOP 2: current Ecuador: socio-cultural, political, economic, geographical, and environmental aspects
Socio-cultural approach
Political approach
Economic approach
Institutional approach
Geographical approach
Environmental ecological approach
Product socialization/final evaluation

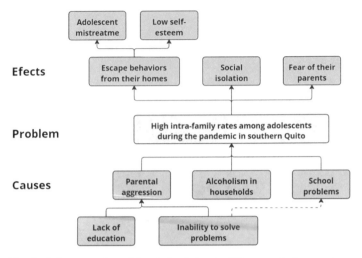

Fig. 4. A figure caption of the student´s Logical Framework Approach use.

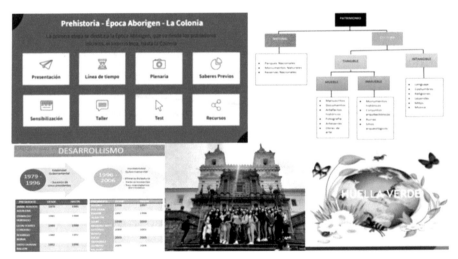

Fig. 5. A figure caption of multimedia teaching resources and visit to Fray Pedro Gocial "San Francisco" museum as a mosaic.

4 Discussion

Technology is generating new approaches in the teaching and learning process in the current educational context. Gamification is an approach that seeks to improve teaching through risk assessment, promoting physical activity, and developing social skills. Socio-cultural approaches that promote learning and creativity are also integrated. Project-based, challenge-based, and problem-based learning have been widely developed, fostering comprehension, communication, teamwork, and organization. Another strategy used in education is the flipped classroom, where students seek and share data in class.

Collaborative learning, both in person and online, has been shown to enhance students' social and emotional development. The Logical Framework Approach (LFA) is an additional tool for establishing a cause-and-effect logic in education planning and evaluation. Teachers' creativity and the use of educational technologies, such as collaborative platforms and search tools, support these educational practices. These practices aim to enhance student interaction, motivation, and learning by promoting their engagement and developing their cognitive and social skills. The proposed learning approach using the logical framework for writing an essay on the Ecuadorian national reality presents a collaborative dynamic in which students develop skills and competencies throughout the learning process, divided into phases. It begins with exposure to historical and current facts to establish a knowledge base on which students are motivated to explore the topic of interest, related to their field of study. The essay writing, in a reflective format starting with the use of the logical framework, provides students with the knowledge and ability to take a critical stance in researching and analyzing historical, present, and future events related to the chosen topic. This capacity, achieved through collaborative work, contrasts with the initial understanding of the topic, as students immerse themselves from different perspectives to develop a better judgment and discernment during the essay's development. However, limitations may arise regarding mobility and interaction with tasks assigned by other courses during the field trip.

5   Conclusion

The proposed learning in this article is achieved through the integration of field and theoretical research information and discussion in a reflective essay that serves as a bridge for the development of knowledge and experiences related to the students' career. In the field of national reality, access to a type of information that students can interpret through their own experience and perception of the analysis variables for the construction of the problem tree is unique. The educational proposal to enhance knowledge of the national reality leverages and fosters students' creative skills in problem formulation using brainstorming techniques, resulting in an original essay. The didactic proposal will be validated with the participation of pedagogy specialists. These specialists will use evaluation protocols to assess the effectiveness of the proposal in terms of students' learning outcomes, the quality of the created essays, and participation in the formative activities. An eight-step proposal is included, integrating technology, cognitive and communicative skills, and collaborative learning experiences with the goal of creating an essay about the country's reality. The use of the logical framework for writing an essay on the Ecuadorian national reality presents a collaborative dynamic in which students develop skills and competencies during the learning process. The essay is divided into phases, starting with the exposure of historical and current facts to build a knowledge base that motivates students to investigate the topic of interest and its relation to their career. The reflective format of essay writing, starting with the use of the logical framework, allows students to acquire knowledge and adopt a critical perspective to investigate and analyze current and future events related to a topic of interest. Teamwork allows contrasting the reality observed when exploring the topic from different perspectives, aiding in developing better judgment and understanding during the essay's development.

Mobility and interaction with tasks assigned by other courses during the field trip can be promoted within the limitations.

Finally, the proposal for an academic essay in the formative project of national reality represents a valuable opportunity to develop critical and reflective skills in university students. The integration of creative pedagogical practices, along with the use of the logical framework as a methodological tool, will strengthen the learning process and contribute to the development of relevant and contextualized higher education in open and self-governing collaborative scenarios, providing experiential conditions for students and ensuring that learning is not just an activity but a space for integration, inclusion, and conceptualization of new formative proposals. The simultaneous evaluation of the application of criteria for optimal learning is left open for further consideration.

References

1. Akram, H., Aslam, S., Saleem, A.: The Challenges of Online Teaching in COVID-19 Pandemic: A Case Study of Public Universities in Karachi, Pakistán. Journal of Information Technology Education: Research **20**, 263–282 (2021)
2. Fleer, M.: Digital pop-ups: Studying digital pop-ups and theorising digital pop-up pedagogies for preschools. Eur. Early Child. Educ. Res. J. **28**(2), 214–230 (2020)
3. Harper, N., Obee, P.: Articulating outdoor risky play in early childhood education: voices of forsest and nature school practitioners. Journal of Adventure Education and Outdoor Learning **21**(2), 184–194 (2021)
4. Alves, A., Hostins, R., Magagnin, N.: Autoria de jogos digitais por crianças com e sem deficiências na sala de aula regular. Revista Brasileira de Educação Especial **27**, 79 (2021)
5. Zhang, A., Olelewe, C., Orji, c., Ibezi, N., Sunay, N., Obichukwu, P., Okanazu, O.: Effects of Innovative and Traditional Teaching Methods on Technical College Students. Achievement in Computer Craft Practices. Sage Open, 10(4) (2020)
6. Shkitina, N., Grevtseva, G., Kasatkina, N., Nemudraya, E., Tsiulina, M.: Training pedagogical universities students in the field of school learning administration. Perspektivy Nauki i Obrazovania **45**(3), 140–157 (2020)
7. Niemi, A., Jahnukainen, M.: Educating self-governing learners and employees: studying, learning and pedagogical practices in the context of vocational education and its reform 23(9), 1143–1160 (2019)
8. Yang, Y., Long, Y., Sun, D., Van Aalst, J., Cheng, S.: Fostering students' creativity via educational robotics: An investigation of teachers' pedagogical practices based on teacher interviews. Br. J. Edu. Technol. **51**(5), 1826–1842 (2020)
9. Chen, S., Zhang, S., Qi, G., Yang, J.: Games Literacy for Teacher Education: Towards the Implementation of Game-based Learning (2020)
10. Katz-Buonincontro, J., Anderson, R.: A Review of Articles Using Observation Methods to Study Creativity in Education (1980–2018). Journal of Creative Behavior **54**(3), 508–524 (2020)
11. Tai, K.: Translanguaging as Inclusive Pedagogical Practices in English-Medium Instruction Science and Mathematics Classrooms for Linguistically and Culturally Diverse Students. Res. Sci. Educ. **52**(3), 975–1012 (2022)
12. Abrantes, J., Seabra, C., Lages, L.: Pedagogical affect, student interest, and learning performance. J. Bus. Res. **60**(9), 960–964 (2007)
13. Molina, A., Martí, I., Martínez, A.: Percepción del profesorado de Educación Física sobre el Aprendizaje Cooperativo y su relación con la Inteligencia Emocional. Retos: nuevas tendencias en educación física, deporte y recreación 41, 735–745 (2021)

14. Betancourt-Odio, M., Sartor-Harada, A., Ulloa-Guerra, O.: Azevedo-Gomes, J: Self-perceptions on digital competences for M-learning and education sustainability: A study with teachers from different countries. Sustainability (Switzerland) **13**(1), 1–12 (2021)
15. Cárdenas, L., Cruz, N., Álvarez, N.: Revisión del marco lógico: conceptualización, metodología, variaciones y aplicabilidad en la gerencia de proyectos y programas. Inquietud Empresarial **22**(1), 117–133 (2022)
16. Smith, J., Li, H., Rafferty, M.: The Implementation Research Logic Model: A method for planning, executing, reporting, and synthesizing implementation projects. Implement. Sci. **15**(1), 1–12 (2020)
17. Fleer, M.: Digital pop-ups: studying digital pop-ups and theorizing digital pop-up pedagogies for preschools. Eur. Early Child. Educ. Res. J. **28**(2), 214–230 (2020)
18. Flôr, A., Fernandes, F., Lima, J., Braga, V., Cruz, J.: PhysioArt: a teaching tool to motivate students to learn physiology. Adv. Physiol. Educ. **44**(4), 564–569 (2020)
19. Gemmel, P., Goetz, M., James, N., Jesse, K., Ratliff, B.: Collaborative Learning in Chemistry: Impact of COVID-19. J. Chem. Educ. **97**(9), 2899–2904 (2020)
20. Zhang, A., Olelewe, C., Orji, C.: Effects of Innovative and Traditional Teaching Methods on Technical College Students. Achievement in Computer Craft Practices. Sage Open 10(4), (2020)
21. Torres, S., Casillas, F., Cabezas, M.: University digital transformation: the challenge of meeting the educational and emotional needs of the academic community. In: Miguelánez, S., Frutos, F. (eds) 2nd International Conference of Research in Education Retos de la educación post-pandemia 2, 1–175 (2021)
22. Alves, A., Hostins, R., Magagnin, N.: Authorship of digital games by children with and without disabilities in the general education classroom. Revista Brasileira de Educacao Especial **27**, 971–990 (2021)
23. Bereczki, E., Kárpáti, A.: Technology-enhanced creativity: A multiple case study of digital technology-integration expert teachers' beliefs and practices', Think Skills Create 39, (2021)
24. Cremin, T., Chappell, K.: Creative pedagogies: a systematic review. Res. Pap. Educ. **36**(3), 299–331 (2021)
25. Rankin, J., Garrett, R., MacGill, B.: Critical encounters: enacting social justice through creative and body-based learning. Aust. Educ. Res. **48**(2), 281–302 (2021)
26. Harper, N., Obee, P.: Articulating outdoor risky play in early childhood education: voices of forest and nature school practitioners. Journal of Adventure Education and Outdoor Learning **21**(2), 184–194 (2021)
27. Hernández, Y., Díaz, L.: Learning mathematics from philosophy for/with children | A aprendizagem das matemáticas desde filosofia para/com crianças | El aprendizaje de las matemáticas desde filosofía para/con niños. Childhood and Philosophy 17, (2021)
28. Piršl, D., Piršl, T.: Using literary texts in norwegian language teaching. Journal of Teaching English for Specific and Academic Purposes **9**(3), 505–516 (2021)
29. Tai, K.: Wei, L: Constructing playful talk through translanguaging in English medium instruction mathematics classrooms. Appl Linguist **42**(4), 607–640 (2021)
30. Guo, L., Wang, C.: Enabling automatic retrieval of schemas from long-term memory in English grammar practice. Asia Pac. Educ. Rev. **23**(2), 361–373 (2022)
31. Silva, E., Lino-Neto, T., Ribeiro, E.: Going virtual and going wide: comparing Team-Based Learning in-class versus online and across disciplines. Educ Inf Technol (Dordr) **27**(2), 2311–2329 (2022)
32. Vigna, J., Michael, R., Russon, P: My COVID teacher – pedagogy and technology: Frontiers of online teaching in the creative writing classroom. Text (Australia) 26(1), (2022)

Collaborative Learning in Higher Education in Distance Mode

Karina Salomé Ayala Jaramillo[1]([email]) and Luis Fermando Ávila-Ascanio[2]

[1] Universidad Indoamerica Quito, Quito, Ecuador
karinasayala@uti.edu.ec
[2] Escuela Normal Superior Cristo Rey, Barrancabermeja, Colombia

Abstract. In higher education in the distance mode, there is a need to strengthen collaborative learning techniques. The aim of this research is to identify the problem of working in groups in higher education. The methodology has a mixed approach nested in the quantitative with a survey applied to 138 students and the process carried out is through an action research approach due to the interpretation of the social phenomenon mentioned. In this research, the shortcomings of group work are identified, obtaining recommendations to promote this learning in the distance modality. Results of the survey were analyzed to conclude that students confuse disorganized group work with collaborative work, recognize the benefits of this work, but point out as a great challenge the participation and responsibility of classmates. The lack of leadership in the groups is also highlighted and 53% of the students surveyed state that they prefer individual activities. It is important that students recognize the true characteristics and implications of collaborative work and be trained to take more leadership roles in this type of work.

Keywords: Collaborative learning · Higher education · Distance education

1 Introduction

Currently, it has been observed that higher education students in the distance mode perform autonomous work because they focus on reading important academic documents, preparing, reviewing, and submitting assignments. This collaborative work allows an organization to accomplish a certain task assigned, which involves the leadership, participation, and cooperation of the members that the teacher has guided for the development of learning and obtaining knowledge.

[1] It should be understood that it is not only the presentation of the tasks or projects assigned by the teacher, but it also goes much further, that is, committing to collaborative learning and that they can get something more than a grade.

Considering what has been analyzed above, it should be noted that [2] refers to knowledge as part of each person and for this it is necessary to have mental processes that allow the understanding of everything that surrounds him/her and something important is collaborative work that allows social interaction.

In higher education in the distance mode, teachers are support guides to promote learning, but there has been evidence of a lack of commitment on the part of university

© The Author(s), under exclusive license to Springer Nature Switzerland AG 2024
M. Z. Vizuete et al. (Eds.): CI3 2023, LNNS 1041, pp. 314–323, 2024.
https://doi.org/10.1007/978-3-031-63437-6_26

students in the fulfillment of tasks, expositions in collaborative work, which causes the participation of few members and a lack of commitment on their part.

[3] They conducted a mixed-cohort study applied to students of a master's degree in teacher training in youth education and a bachelor's degree in educational guidance. The aim of the research was to develop an innovative approach based on collaborative learning methods and the use of ICT. A Jigsaw-based approach to activity planning was introduced, based on comprehension tasks, complex decision-making tasks, peer essay writing, problem solving, structured academic discussions and group quizzes. The results obtained show the high price students pay for cooperation. Moreover, the contribution of ICT to such training is also significant. Therefore, it is important to develop these two topics among teachers from initial training and to keep them updated through continuous training.

On the other hand, [4] in his article refers to the strategies to promote collaborative learning in distance education and proposes several activities which are: Working on a common project or proposal, review or critique of work among peers and role play, when working with students, interaction and interest in the expected work and total participation can be observed it is important to mention that technological resources such as discussion or debate forums, blogs, wikis, virtual groups, email, among others can be used.

Similarly, [5] conducted a study on the study chains in small groups and the construction of knowledge taking into account the socio-cognitive chains that are formed within the asynchronous interaction and is based on 4 phases: clarification and organization of the task, synthesis and final agreements, elaboration of meanings and the exchange of information based on the above, it is shown that students participate adequately in the first two phases and the other two are not fully complied with.

On the other hand, [2] mention that collaborative work is characterized by the organization in teams, in which members work together to achieve a common goal. In a collaborative environment, meaningful learning is fostered since the intervention of team members can demonstrate their skills and knowledge to work on specific tasks and share ideas. Considering the aforementioned, it can be stated that students combine different skills and knowledge to achieve a common goal.

For his part, [6] mentions that distance education refers to the creation of a space that, through the meaningful use of available resources, provides authentic and meaningful learning in learning moments where Higher Education Students perform autonomous work because they focus on reading important academic documents, elaboration, revision, and submission of assignments.

[1] They emphasize online collaborative learning, the level of interdependence among members is high and is determined by the objectives of the task, the distribution of responsibility, leadership, acquired communication skills, the development of functions to fulfill certain assigned tasks, in which leadership, participation and cooperation of the members that the teacher has guided for the development of learning and obtaining knowledge is immersed.

They emphasize collaborative learning as an interaction among group members in which each one makes his or her contribution and social knowledge is fostered, allowing for reflection and participation among individuals who are interested in learning more about a given topic.

He states that there are several tools that allow collaborative work, and it should be noted that, at present, there are countless programs easily accessible to work in groups and this facilitates learning. A conscious work that allows reflection and participation of each member of the group is to achieve a social interaction in which there is peer learning, it should be considered that in higher education who wants to acquire knowledge is the student and for this they have access to the use of technologies as a tool to facilitate their academic life.

Therefore, this article aims to diagnose the current state of online collaborative learning, especially in higher education, and to identify the success factors associated with this approach. The aim of this project is to obtain information that evidences the lack of cooperation in collaborative work among the members of the group taking as background the lack of time, lack of concern, conformism, lack of leadership among others [2] That is why a strategy of online collaboration, that is, using technologies for work is when collaborative discussions occur in order to implement practical solutions for learning development.

2 Methodology

This research project is framed in the application of surveys to 138 students of the career of initial distance education, systematically taking a sample of students of third, fourth, fifth and sixth semester through the *Google forms* tool, in order to know the weaknesses and strengths of working in collaborative groups for the development of learning, showing the concerns that each of the participants have.

2.1 Participants

The population selected to carry out this research are teachers and students of higher education in early childhood education. The participants are 138 university students between the third and sixth semesters.

2.2 Research Method

The research method used in this study has pragmatic epistemological support and is of a mixed nature, since it makes inferences from the handling of several quantitative variables and qualitative analysis categories in both surveys applied, as illustrated in Tables 1 and 2.

On the other hand, the units of analysis presented in Table 2 are extracted from the "Survey directed at teachers".

In this study, neither approach is privileged, but both are given equal status [9].

Likewise, it is important to mention that the type of research is descriptive considering that data were collected and analyzed on a particular phenomenon in order to

Table 1. Variables and categories of the instrument "Survey directed to students".

Variable	Category
Level of education completed	Attitude towards collaborative work
Knowledge of collaborative work	Attitude towards peer participation
Frequency of participation in group activities Student input to group activities in educational planning Perceived active participation of peers in group activities Frequency of group activities in education planning	Attitude towards leadership in collaborative activities

Table 2. Variables and categories of the instrument "Survey directed to teachers".

Variable	Category
Range of years of teaching experience	Perception of student performance during collaborative work
Frequency of planning in group activities	Perception of students' active participation in group activities
Willingness to take part in training on collaborative learning	Use of digital resources for collaborative learning
Perception of learning when performing collaborative tasks	Perception of student leadership during collaborative work Non-use of digital resources for collaborative learning Perception of the importance of collaborative work for learning Taste for collaborative activities in the classroom

describe it [10]. With it, the manifestation of the perception of collaborative work in the participants was analyzed, identifying patterns, and relating units of analysis in order to better understand the behavior of the participants towards collaborative learning.

2.3 Procedure for the Analysis of Information

Once the data were collected, the information was organized according to its classification as quantitative or qualitative and an analysis was made with descriptive statistics of the data obtained. The authors discussed the results in light of their knowledge of the research context. The actions carried out were framed in the phases shown in Fig. 1.

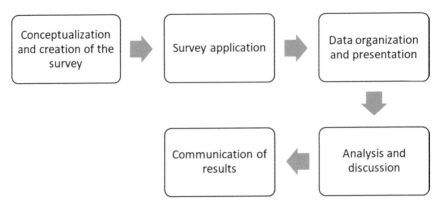

Fig. 1. Research phases.

3   Results and Discussion

The objective of this research was to analyze students' attitudes towards collaborative learning in the distance mode. First of all, it was found that of the 138 students surveyed, 93% stated that they knew what collaborative work is and 64% of them expressed liking for this way of working. The students recognize the benefits of collaborative work as the sharing of ideas, mutual help, the use of individual strengths and social relationships. Students who do not like collaborative work mention factors such as unequal workloads, unfairness in the final grade of the work, difficulty in finding common spaces in the agenda, and the irresponsibility of some people when working in teams. Thus, it can be said that students confuse group work with collaborative work and more than a third of them have had previous negative experiences that make them uneasy about non-individual work.

When asked in particular about peer participation in group work, 73 students (53% of the sample) acknowledge such participation compared to 65 students who do not. This indicates a divided opinion on the commitment of peers to collaborative work, which is influenced by their previous experiences. Among the most common arguments of those who perceive a lack of commitment are the lack of responsibility and discipline of their peers when it comes to this type of work. Those who like it point out that, if all students are committed, it is possible to achieve results that are more ambitious and broader in scope. 98% of the students affirm that they actively contribute their knowledge when they are doing collaborative work (Fig. 2).

As shown in the figure, the majority of students emphasize their individual commitment to group activities. Likewise, 93% of the students recognize that their teachers assign tasks that are solved collaboratively.

Regarding student leadership in collaborative activities, only 37% of those surveyed were willing to assume leadership of group activities. The students perceive that the leader is "overloaded" with work and is responsible for micro-managing tasks by pressuring other classmates to do their part of the work. Students also stand out from this group who mention not being able to be leaders due to lack of time, work outside the university and lack of internet.

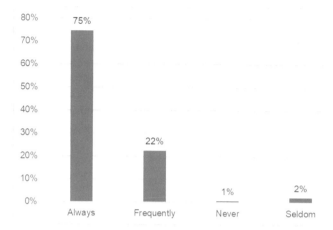

Fig. 2. Shows the students' personal disposition towards participation in group activities.

Figure 3 presents the last question asked to the students, which defines what type of activities the students prefer:

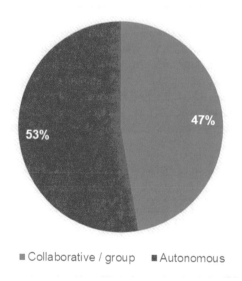

■ Collaborative / group   ■ Autonomous

Fig. 3. Responses to the question "What type of activities do you prefer when developing your learning?".

Note. 73 students prefer autonomous activities, and 65 students prefer collaborative ones.

It is found that slightly more than half of the students prefer activities that are developed individually. This tendency was present throughout the survey. Students recognize the importance and value of collaborative work, and the frequency with which their teachers assign such work, but they are also unwilling to do it, based on their previous negative experiences, where they confuse disorganized group work with true collaborative work, in which roles, responsibilities and time management must be established.

Statistical analysis of measures of central tendency was performed on the responses obtained from the test, separating them into two main types, according to the original question. The statistical values for yes/no questions are grouped in Table 3 and those with Likert scale in Table 4. For yes/no questions, yes was assigned a value of 1 (Yes = 1) and no a value of 2 (No = 2). In the case of Likert scale questions, it was assigned as follows: Always = 4, Almost always = 3, Almost never = 2, Never = 1.

Table 3. Analysis of statistical measures of yes/no questions.

	Mean (ϵ)	Median (Me)	Standard deviation (σ)
Knowledge of collaborative work	1.1	1.0	0.3
Interest in group activities	1.4	1.0	0.5
Perception of peer participation	1.5	1.0	0.5
Attitude towards leadership	1.6	2.0	0.5

The statistical analysis of the variables shows that the data are not very dispersed since the standard deviation of each variable is equal to or less than 0.5. On average, the students show that they have knowledge of what collaborative work consists of. However, a mean of 1.5 indicates a divided opinion in the students' perception of the participation of their peers; with a mean of 1.4, it reflects the manifestation of a weak liking for group activities. Finally, with a mean of 1.6, there is a weakness in the leadership of the sample; less than half of the sample claims to be willing to lead collaborative activities.

Table 4. Analysis of statistical measures of Likert-scale questions.

Variable analyzed	Mean (ϵ)	Median (Me)	Standard Deviation (σ)
Participation in collaborative activities	3.7	4.0	0.6
Perception of participation	2.7	3.0	1.0
Frequency of knowledge contribution	3.6	4.0	0.5
Perception of active participation	3.0	3.0	1.0
Frequency of collaborative activities in the classroom	3.4	3.0	0.8

On the other hand, from the statistical analysis of the questions posed with Likert scale, it can be extracted that the students present more dispersed opinions, mainly in the perception of the participation of their peers, where a standard deviation of 1.0 is given, with which it can be affirmed that on average, the students perceive that they do participate in the collaborative activities and do it actively while their peers do not do it in the same way. This phenomenon is also evident when analyzing the variables in a box-and-whisker diagram, as shown in Fig. 4.

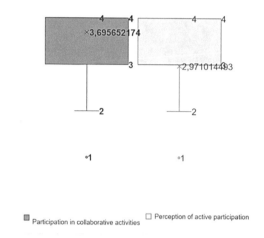

Participation in collaborative activities   Perception of active participation

Fig. 4. Box-and-whisker plot for questions that assessed participation.

Note. The box-and-whisker plot shows that students on average, feel that they actively participate in collaborative activities $(\epsilon = 3.7)$ while their perception of their peers' active participation is low $(\epsilon = 3.0,$ right on the line of the lower quartile). Students also consider on average, that they contribute a lot of their knowledge to such activities and the frequency with which they perceive that their teachers assign collaborative work is high, as seen in Fig. 5.

Note. The box-and-whisker plot shows that students on average, feel that they frequently contribute their knowledge in collaborative activities $(\epsilon = 3.6)$ and perceive that collaborative activities in the classroom occur frequently $(\epsilon = 3.3)$.

From the above analysis it is possible to infer that the students' self-concept in collaborative work is high but their perception of the benefits of collaborative activities or the work of their peers is low, that is, there is a negative predisposition towards collaborative work.

In this study it was found that the participants have a misconception of the characteristics of collaborative work and do not use it to their advantage. Although they express liking, interest, and active participation in it, they also recognize that in the group activities they have been involved in so far, they feel overloaded with responsibilities and with very little support from their group mates.

One of the precursors of this situation is the lack of leadership expressed by more than 60% of the respondents and sometimes technical difficulties or difficulties in the

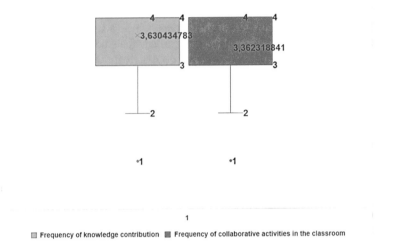

Fig. 5. Box-and-whisker plot for questions that assessed the frequency of knowledge contribution and collaborative classroom activities.

students' agenda, considering that some of them also work. The students' self- concept is high when asked about their contribution to collaborative work, but there is a negative perception of the participation and contribution of other students.

It is urgent to propose an initiative that reinforces students' leadership skills and teaches them the true characteristics and conditions of collaborative work, as opposed to mere disorganized group work.

Although many authors highlight the benefits of collaborative work [1]; this study highlights the challenges presented by this type of work in distance higher education, given the extra-university occupations of students and the differences in learning styles and rhythms. Everything indicates that having a distorted view of collaborative work and not assigning roles properly, there is an imbalance of roles among students and negative experiences as a result of this type of work.

4 Conclusions

For the participants of this research, it was possible to verify the current state of collaborative learning, showing that there is very little participation on the part of some members of the groups and that the students in question confuse the concepts of collaborative and group work. The students have a high self-concept of their participation and influence in collaborative work, but at the same time they perceive low benefit in collaborative activities and low participation of their peers, that is, they have a negative predisposition towards collaborative work.

[11] Another phenomenon described in this study has to do with the low leadership attitude that students show towards collaborative work, precisely because of the low use that is made of it. Students show a preference for autonomous and individual activities.

Due to these phenomena, it is necessary to provide teachers with innovative strategies to promote collaborative work, leadership and active participation that will benefit the students. This will better train students as professionals and as individuals, considering that their field of action will require them to lead actions within the educational community.

Future research in this context will try to determine the positions of different teachers regarding collaborative work and the interest in discovering new strategies that contribute to the development of learning and thus meet the needs presented here.

References

1. Castillo Ceballos, G.: Editorial Nº12. Estudios Sobre Educación, 12. https://doi.org/10.15581/004.12.24327 (2007)
2. García Hernández, C., Espinosa Meneses, M., y Peñalosa Castro, E.: Interacción discursiva y representaciones sociales de jóvenes universitarios en torno al uso de las TIC en la educación. Reencuentro. 1(62). 46–54 (2011)
3. Diaz, F., Morales, L.: Collaborative learning in virtual environments: an instructional design model for continuing professional development. Educational Technology and Communication, 22–23(47–48), 1–8 (2008)
4. Scagnoli, Mag. N.I.: El Aprendizaje Colaborativo en Cursos a Distancia. Investigación y Ciencia, 14(36), 39–47. https://www.redalyc.org/articulo.oa?id=67403608 (2006)
5. Castellanos Ramírez, J. C., y Niño, S.A.: Aprendizaje colaborativo y fases de construcción compartida del conocimiento en entornos tecnológicos de comunicación asíncrona. Innovación Educativa (México, DF), 18(76), 69–88. http://www.scielo.org.mx/scielo.php?script=sci_arttext&pid=S1665-26732018000100069&lng=es&nrm=iso&tlng=es (2018)
6. Mendoza Castillo, L.: Lo que la pandemia nos enseñó sobre la educación a distancia. Revista Latinoamericana de Estudios Educativos, 50 (ESPECIAL), 343–352. https://doi.org/10.48102/RLEE.2020.50.ESPECIAL.119 (2020)
7. Guerra Santana, M., Rodríguez Pulido, J., Artiles Rodríguez, J.: Aprendizaje colaborativo: experiencia innovadora en el alumnado universitario, Revista de Estudios y Experiencias en Educación 18(36), 269–281 (2019)
8. Cuetos Revuelta, M.: Valoración de una experiencia con Linoit para el aprendizaje colaborativo en línea. Campus Virtuales. http://uajournals.com/ojs/index.php/campusvirtuales/article/view/894 (2021)
9. Johnson, R. B., Onwuegbuzie, A.J.: Toward a Definition of Mixed Methods Research, Journal of Mixed Methods Research, 1(2), 112–133 (2007)
10. Felder, R.M., Brent, R.: Cooperative learning, ACS Symposium Series, 970, 34–53, (2007)
11. Hernández-Sellés, N.: Self-directed learning in collaborative work processes in higher education, Educar, 58(2), 389–403 (2022)
12. Quesada, A.: Aprendizaje colaborativo e interuniversiario en línea: una experiencia asíncrona y síncrona. Revista de Lenguas Modernas (2010)
13. Scagnoli, N.I.: Estrategias para Motivar el Aprendizaje Colaborativo en Cursos a Distancia, pp. 1–15 (2005)
14. Castro, E.P., García Hernández, C., Meneses, M.E.: La argumentación como recurso para el aprendizaje colaborativo en internet: una propuesta, pp. 1–19. http://ilitia.cua.uam.mx:8080/jspui/handle/123456789/69 (2011)
15. Collazos, C.: How to take advantage of cooperative learning in the classroom. Educación y Educadores 9, 61–76 (2006)

Psychosocial Risks in the Work Environment of the ITCA Higher Technological Institute

Patricia Lisbeth Esparza Almeida$^{(\boxtimes)}$ ⓘ, Marlon Fabricio Hidalgo Méndezⓘ,
Mónica Monserrath Chorlango Garcíaⓘ, Byron Sebastián Trujillo Montenegroⓘ,
Juan Carlos Jaramillo Galárragaⓘ, and Luis Shayan Maigua Moralesⓘ

Instituto Superior Tecnológico ITCA, Ibarra EC100150, Ecuador
`plesparza@itca.edu.ec`

Abstract. This research studies the level of occupational psychosocial risk in 2021 of the teaching and administrative staff at a Higher Technological Institute in the city of Ibarra-Ecuador. With a qualitative approach, this field research compares the results obtained from a study on the level of psychosocial risk in 2019 and the results of this same evaluation during the SARS-CoV-2 pandemic in 2021. For the data collection, in both cases, the Psychosocial Risk Assessment Questionnaire established by the Ministry of Labor of Ecuador was applied and validated with the Cronbach's Alpha values of the instrument by dimension, and the Pearson coefficient was calculated. The results showed that between 2019 and 2021, the level of risk did not increase. The indicators in 2021 show 93% low risk and 7% medium risk. In this case, the pandemic did not contribute to the exponential increase in psychosocial risk due to the corrective actions proposed and executed in the prior evaluation carried out in 2019, the adaptability of the staff, and the measures taken by the institute during the pandemic. These measures effectively mitigate psychosocial risks in the work environment, corroborating the importance of evaluations and continuous improvement plans in Higher Education institutions.

Keywords: Psychosocial risks · Psychosocial Risk Factors · Evaluation Manual

1 Introduction

From the perspective of psychosocial risk, the inquiry about the work environment implies that the lack of certainty in the workplace produces insecurity and discomfort regarding the present and future in individuals. The risk of being fired, subject to a salary cut or reduced benefits, is a reality that is part of the world of work. Depending on the social context, individuals' perception of this risk may be minimal or increase. This increase or decrease in the perception of risk directly impacts work performance and the mental health of employees.

In 2020, the International Labor Organization (ILO) warned that both the reality of the risk and its perception tended to be aggravated by the pandemic, thereby increasing the dangers of suffering from physical and mental disorders. For this reason, the ILO

M. Z. Vizuete et al. (Eds.): CI3 2023, LNNS 1041, pp. 324–338, 2024.
https://doi.org/10.1007/978-3-031-63437-6_27

advocated the urgent evaluation and adequate management of psychosocial risks during the pandemic to appease their psychological effects (decreased mood, anger, low motivation to work, mental exhaustion, anxiety, depression, and suicidal thoughts), as well as its physical effects (physical exhaustion, digestive problems, changes in appetite and weight variations, dermatological reactions, cardiovascular diseases, musculoskeletal disorders, headaches, and unexplained pain). In addition, the ILO pointed out that these psycho-physical effects on the individual are prone to manifest themselves in the workplace, directly affecting companies' economic life and daily life [1].

Along the same lines, other investigations have observed the effects of COVID-19 at different latitudes. Lozano Vargas (2020) [2], points out that studies in China show that the psychological state of workers infected by the virus was seriously affected. Work pressure, frustration, fear of being discriminated against, family detachment, and physical and mental exhaustion affect work performance. Another investigation carried out with health personnel in Brazil carried out by Soares e Silva et al. (2020) showed that psychosocial factors such as discrimination due to contagion, conflicts derived from the hierarchy, difficulties related to the nature of work, emotional exhaustion, increased workload, lack of recreational activities and lack of adequate work implements to avoid contagion, increased uncertainty and its effects on physical and mental health [3].

Psychosocial risk factors in an emergency caused by COVID-19 increase occupational psychosocial risks. From the theory of psychosocial risk in the workplace, a difference is postulated between occupational psychosocial risks and psychosocial risk factors. Moreno and Baez (2010) point out eleven dimensions related to psychosocial risks that affect the psychophysiological responses of workers and their health [4]. These dimensions are job content, overload and pace, schedules, control, environment and equipment, organizational culture and functions, interpersonal relationships, role in the organization, career development, work-family relationships, and contractual security.

Exposure to psychosocial risks that affect health in the workplace is related to the organizational structure and environmental conditions such as climate and organizational culture. Risk factors, therefore, are potentially harmful to the health and well-being of workers since they are social factors that alter the staff of an institution and have different effects on each individual. Psychosocial risks are work circumstances that sometimes have a high probability of seriously damaging workers' health physically, socially, or mentally [5].

Psychosocial risks occur in the work context, while psychosocial risk factors respond to the social context. Both are related to occupational health and the global environment; therefore, they directly affect the core of all human organizations. The psychosocial risks present in the work environment are related to environmental conditions and workplace design. Regarding the organization and management of work, the risk can be associated with inadequate conditions in breaks and breaks, work schedules, decision-making, monotony, autonomy, mental load, responsibility, and interpersonal relationships, among others. These conditions can manifest through work stress, Burnout syndrome, and workplace bullying [6].

Through the Psychosocial Risk Questionnaire of the Ministry of Labor of Ecuador (2018), the levels of risk to which workers of organizations in Ecuador may be subjected are determined. This work compares the psychosocial risk factors affecting workers' health, generating actions to prevent or reduce psychosocial risk at the ITCA Higher Technological Institute before and during the SARS-CoV-2 pandemic [7].

2 Methodology

The present investigation addresses the psychosocial risks in the work environment in the context of the SARS-CoV-2 pandemic during 2021. The unit of analysis in which this study was carried out is the ITCA Higher Technological Institute staff in the city of Ibarra-Ecuador.

The research is an explanatory field, and the approach is qualitative. The population used corresponds to 100% of the ITCA staff, made up of 45 people, of which 11 are administrative staff and 34 teachers. The collection of information was carried out through the Psychosocial Risk Assessment Questionnaire produced by the Ministry of Labor of Ecuador, which has been validated and presents the following values: Cronbach's Alpha of the instrument: 0.967 (High-reliability level); Cronbach's alpha by dimension: values from 0.806 to 0.904 (High-reliability level for the eight dimensions); R Pearson: all items are above 0.30 (all items are Valid). (Ministry of Labour, 2018) [7].

The questionnaire consists of fifty-eight questions that measure the existence and levels of psychosocial risk in workers. Regarding the existence of risk these are measured based on eight dimensions: Workload and rhythm, Skills development, Leadership, Margin for action and control, Organization of work, Recovery, Support, and support. These dimensions have been evaluated based on a Likert scale, which provides results of risk levels. The levels are measured by three parameters: High, Medium, and Low as shown in Table 1.

Table 1. Risk levels according to the Ministry of Labor.

Risk level	Qualification	Description
High	58 to 116	Risk of high potential impact on the safety and health of people. The danger levels are intolerable and can immediately harm people's health and physical integrity. Safety and prevention measures must be applied continuously and follow the specific need identified to avoid an increase in the probability and frequency of risk
Half	117 to 174	Risk of moderate potential impact on health and safety. However, it can compromise safety and health in the medium term. If security and prevention measures are not applied continuously and according to the specific need identified, the impacts can be generated with greater probability and frequency
Low	175 to 232	Risk of minimal potential impact on health and safety. It does not generate harmful effects in the short term. The effects can be avoided through periodic monitoring of the frequency and probability of occurrence and occurrence of an occupational disease. Preventive actions should focus on ensuring that the level is maintained

Source: Ministry of Labor (2018)

The Ministry of Labor of Ecuador establishes three stages for applying the Questionnaire detailed in Table 2. This protocol allows the evaluation of risk and its levels to determine the level of occupational psychosocial risk produced by the SARSCoV-2 pandemic 2021 on ITCA staff.

Table 2. Psychosocial Risk Assessment Questionnaire of the Ministry of Labor

Activities
1. Form the work team Occupational Health and Safety Technician or Manager occupational physician Head of Human Talent
2. Prepare, disseminate, and socialize information Inform the highest authority Search for communication channels for the application of the questionnaire Inform all staff
3. Instrument application phase
Have trained staff Comfortable space free of distractions Delivery of questionnaires to participants Provide the necessary information to the participants who require it Collect information for further analysis

Source: Ministry of Labor (2018)

The evaluation process of existing psychosocial risks for ITCA staff was carried out using descriptive statistics, exposing the research data through tables and figures in a concise, clear, and summarized way [8]. In addition, through the preliminary analysis of variables, descriptive statistics use statistical inference to address the conditions under which the inferences drawn from a sample are valid, to conclude the population of interest [9]. Data processing and analysis are used to highlight the relevance and importance of the results obtained in this research.

3   Results and Discussion

In 2019, the occupational health and safety personnel of the ITCA institute carried out a study of psychosocial risks by applying the Questionnaire designed for such purposes by the Ministry of Labor of Ecuador [7]. According to the parameters established by this instrument, the overall results showed 33% medium risk and 67% low risk.

Two years later, a new study was carried out using the same evaluation instrument, and it was believed that a pandemic scenario would create favorable conditions for an increase in psychosocial risks, so negative results were expected in the new evaluation. On the contrary, the results showed a considerable decrease in risk levels. The data showed a decrease in risk during 2021 when ITCA staff presented 93% low psychosocial risk and 7% medium risk.

The results of 2019 set a direction in terms of the need to contain risk levels and seek their reduction. Therefore, the decrease in risk levels in 2021 could be attributed to permanent monitoring and the application of preventive measures based on previous evaluations' results.

These data are significant compared to other research on psychosocial risks in higher education institutions. A study that analyzes the socio-emotional well-being and psychosocial risk in higher education teachers at a Colombian university in 2020 showed that the pandemic negatively affected the well-being of teachers at a social, labor, and economic level. Despite having a positive level of resilience, this was not enough to counteract the adverse effects of the crisis, which is why teachers experienced high levels of stress and their mental health was significantly affected [10].

In the field of health, it is evident that the pandemic increased the possibility of the existence of psychosocial risks by potentially increasing the risk of suffering from conditions such as stress, anxiety, mental fatigue, and depression; 65% of the personnel of the General Teaching Provincial Hospital of Ambato presented high risks related to psychological demands, social support and social capital for patient care [11]. These data reflected an upward trend during the pandemic, indicating the high occupational risk experienced by health personnel.

Orejuela (2021) analyzed a higher education institution, reporting that 6% of the staff presented a high-risk level, 37% medium risk, and 57% low risk [12]. The minimal impact responds to the fact that the psychosocial risk factors generated by SARS-CoV-2 did not harm workers' safety and health in the medium term, and its periodic monitoring maintains a low risk, compared to the probability of an occupational disease occurring. The difference between the risk levels in both studies is due to the nature of the line of work in each institution. During the pandemic, the workers of health institutions were directly exposed as front-line personnel, facing higher levels of exposure than the workers of educational institutions, who, in many cases, carried out their activities electronically.

Table 3 presents the results of each dimension related to psychosocial risks concerning the low, medium, and some risks in the ITCA staff.

The dimensions related to psychosocial risks reveal that, in terms of high risk, the personnel affected by Load and Pace of work and Leadership present a value of 2%; the Recovery dimension presents a value of 7%, and the rest of the dimensions present a value of 0. These values indicate that in the work environment of the institution, most of the dimensions studied did not present a high risk of acquiring adverse psychological or physical effects. These results contributed to the announcement of the COVID-19 pandemic, promoting and demanding an urgent evaluation and adequate management of psychosocial risks during the pandemic to appease the adverse psychosocial effects.

At the medium risk level, the values of the three types of bullying presented a value of 0, and the Workload and Pace dimension presented a value of 42%. The rest of the dimensions presented values lower than 50%. These values are significant if one considers that the psychosocial risk factors produced in the health emergency by COVID-19 increased the psychosocial risks in Ecuador.

At the Low-Risk level, the values of the dimensions were between 56% and 100%. The three types of harassment presented values of 100%. Of the fifteen levels studied,

Table 3. Assessment of Psychosocial Risks in ITCA workers

Questionnaire Dimensions	Low risk	Medium risk	High risk
1. Workload and pace	56%	42%	2%
2. Development of skills	89%	11%	0%
3. Leadership	87%	11%	2%
4. Margin of action and control	84%	16%	0%
5. Organization of work	93%	7%	0%
6. Recovery	58%	36%	7%
7. Support and support	91%	9%	0%
8. Other important points:			
8.1. Discriminatory harassment	100%	0%	0%
8.2. Workplace Harassment	100%	0%	0%
8.3. Sexual harassment	100%	0%	0%
8.4. Addiction to work	67%	33%	0%
8.5. Working conditions	87%	13%	0%
8.6. Double presence (work – family)	76%	24%	0%
8.7. Work and emotional stability	93%	7%	0%
8.8. Self-perceived health	76%	24%	0%

ten presented values higher than 80%. Thus, the low-risk levels presented high values, indicating a low probability of being potentially harmful to occupational health and well-being.

The work of Sarabia et al. (2020) shows that the pandemic exposed the workers of the Multifunctional Logistics Operator Group R to high psychosocial risks in terms of the Influence of Social Support and Leadership Factors with a high percentage of 4.07% according to the CoPsoQ-istas21 Method Test. These risks are focused on dimensions that present high levels of risk exposure, such as Sense of Work 4.71%, Development Possibilities 4.64%, Group Feelings 4.42%, and Role Clarity 4.44%. In addition, workers were mainly exposed to unfavorable health situations concerning the pace of work, which presented an average of 4.04%. This value is considered a high-level concerning exposure to psychosocial risks within the parameters of the test used by the authors. The high exposure directly affected the workers regarding their Development Possibilities and Sense of Work. Psychosocial risk factors also affected their perception of Justice and Vertical Trust within the company, which presented percentages of 4.22% and 4.25%, respectively [13].

The comparison between the results obtained in the evaluation of the year 2019 [14] before the pandemic and the one carried out during it in 2021 in the institution is shown in Fig. 1.

Regarding high risk in the Workload and Rhythm dimension, the level was 2% and presented a value of 7% for recovery. In 2019, in Workload and Pace, the ITCA

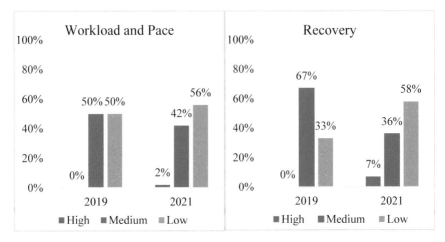

Fig. 1. Risk levels in Workload and Pace and Recovery.

presented data of 50% at the Medium and Low levels. In 2021, 2% were at High Risk in this category. However, the Medium Risk decreases from 50% to 42%. It is observed that the affectation in this dimension is low in both years. High of these percentages are concentrated in the medium level with a total of 22% in six dimensions: Workload and Pace 51% and Recovery 39%. However, it is worth mentioning that if the results obtained by applying the same evaluation instrument in 2019 are considered, there is evidence of an increase in high and medium risk levels in the dimensions of Load and Rhythm, Leadership and Recovery that must be considered.

When the staff was already exposed to a new reality, in 2021, an 8% decrease in Workload and Rhythm was obtained at the medium risk level, but an increase of 2% in risk was noted. High indicates that a part of the population was affected by the new environment and its demands.

The Workload and Pace dimension, in 2019, had a value of 50% in terms of medium and low risk; for the year 2021, the high risk appeared at 2%; however, the average risk decreased from 50% to 42%, which shows that the condition in this dimension is relatively low and when analyzing the characteristics of the worker, he is a person who fulfills several functions within the institution, for which it is this increase is understandable.

Figure 2 presents the values obtained for the dimensions corresponding to leadership and the development of competencies.

The leadership dimension in 2019 presents percentages of Medium Risk and Low Risk with 33% and 67% respectively compared with the values of 2021, it is observed that the average risk decreases from 33% to 11%, and 2% manifests itself in High Risk; while the average risk decreases by 22%.

In the Skills Development dimension, high risk remains at 0% and medium risk decreases significantly, presenting only 11% in 2021 compared to 50% obtained in 2019; therefore, the low risk in 2021 increases to 89%. Regarding the average risk, the percentage and dimension is, due to the development of skills and leadership, 11%.

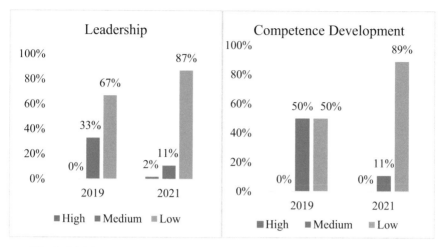

Fig. 2. Risk levels in the dimension of Competence Development and Leadership.

Based on the comparison of different works that revolve around the study of occupational psychosocial risk in Peru, Colombia and Ecuador, Orozco and Rincón (2023) report that leadership is one of the main psychosocial risk factors in the three countries [15]. For this reason, the authors argue that the attributes of leaders added to adequate communication, motivation and good relations with the staff are key elements of a preventive business culture against occupational psychosocial risks, to minimize them or to contain their increase. Even in contexts in which psychosocial risk factors significantly influence occupational psychosocial risk, leadership can remain standing and be used to appease the appearance of fatigue levels due to overwork, as Córdoba et al. have shown. (2022) in their work on psychosocial risk factors and fatigue in workers of an Ecuadorian pharmaceutical company in the context of a pandemic in 2021 [16].

In the case of ITCA, despite the decrease from 33% to 11%, the appearance of 2% in the leadership dimension in ITCA is observed in relation to 2019. This does not necessarily respond to the fact that ITCA senior managers lost in view of the importance of their leadership, but rather because the staff entered the teleworking modality during the COVID-19 pandemic, which generated inconveniences in communication, information management and adaptation to the teleworking modality.

In this sense, what happened at ITCA is not an isolated case; other Higher Education institutes in Ecuador presented adaptability problems to teleworking.

An investigation on the psychosocial risks in public higher education in Ecuador with a sample of 600 participants showed that 4 out of 10 respondents considered that the preparation to assume the new modality of work and studies needed more time and resources than they did exist, giving rise to essential inconveniences of adaptation [17].

Figure 3 presents the changes that have occurred in the dimensions of Margin of Action and Control and Work Organization.

The Margin of Action and Control dimension presents low risk values of 50% in 2029 and this dimension increased by 34% in 2021. The risk related to the Action and Control Margin in terms of the high level remains at 0%, most interviewees agreed that

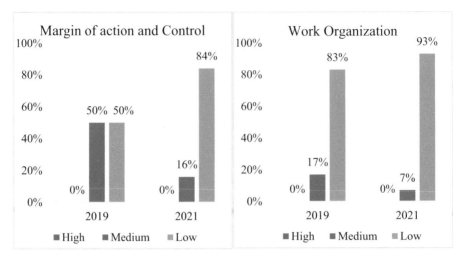

Fig. 3. Risk levels of Margin of Action and Control and Work Organization.

there was no high risk in these dimensions during that period. The medium risk values decrease reaching 16%, which generates an increase in the low risk that rises to 84%.

The Work Organization dimension indicates that the average risk has decreased by 7% by 2021, which generated an increase in the low risk level to 93%, contrary to the 83% obtained in 2019. This dimension did not present large variations during the study period and there was no high risk in these dimensions during that period.

Figure 4 compares the levels of risk in the dimensions of recovery and support and support.

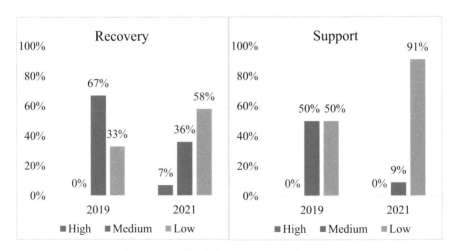

Fig. 4. Risk levels in Recovery and Support.

It can be seen that the Recovery dimension shows an increase of up to 7% in 2021 in high-risk values, a non-worrying value that is offset by medium- and high-risk values for the same year. The average level decreased significantly by 31% in the period 2019–2021. The appearance of 7% in the Recovery dimension is significant data, since it accounts for an increase in the increase in occupational psychosocial risks. This is shown by the work of Córdoba et al. (2022) on the relationship between risk factors and fatigue in workers. In this study, the authors show that the difficulty to recover in the time that elapses between the end of working hours and the start of a new day influences institutional life [16].

Despite the fact that the correlations are weak, the authors maintain that the lack of recovery in the workers of a pharmaceutical company in Ecuador is concatenated with the appearance of various types of fatigue and its increase in 36% of the participating population. The investigation revealed that the lack of adequate rest increases the feeling of tiredness in workers, which influenced their work performance and their mental health in the context of the COVID-19 pandemic.

A Mexican higher education institute found that 57.7% of workers reported high and very high levels of psychosocial risks after 15 months of confinement related to work-family life interference and violence related to fatigue and work. Fatigue increases the risk of mental health conditions among staff [18].

In the case of the ITCA Institute, the Recovery and Support dimension shows a notable decrease in average risk, considering that by 2021, it presents a decrease of 41%.

The Support and Support dimension shows a notable decrease in the average risk, taking into account that by 2021 it presents a decrease of 41%; which is offset by a 41% increase in low risk.

The work of Jhayya (2020) on risk assessment in the Financial and Economic Analysis Unit (UAFE), Quito-Ecuador is based on the risk assessment questionnaire that is also used for this research and, unlike the UAFE, ITCA presents low values since 2019, and in the context of the pandemic, the risk continues to increase. These values reveal that the UAFE was already an institution prone to psychosocial risks before the pandemic; which suggests keeping risks low through prevention, reduces the chances that risk factors have a negative impact on occupational psychosocial risks [19].

The manifestation of 2% of the risk at a high level is related to workers who fulfill several functions within the ITCA. This increase in work is understandable if one takes into account that at the peak of the pandemic there was a cut in personnel, and as a result, several workers assumed functions that corresponded to the dismissed employees.

Figure 5 presents the results of Discriminatory Harassment and Workplace Harassment.

The results show that there is no difference in Discriminative Harassment during the period 2019–2021; however, in the Workplace Harassment dimension, the average risk level decreased to such an extent that in 2021 it completely disappeared, and 100% low risk was obtained.

In the Workplace Harassment dimension, an increase of 17% in average risk can be observed in the same period. In the workplace bullying dimension, the average risk level

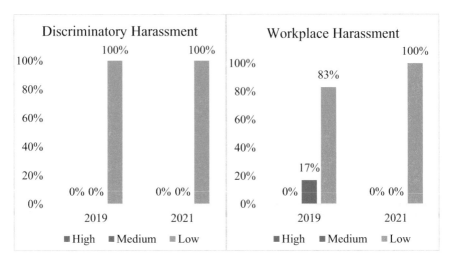

Fig. 5. Risk levels of Discriminative Harassment and Workplace Harassment.

decreased to such an extent that in 2021 it completely disappeared and 100% low risk was obtained.

The results of Sexual Harassment and Addiction to Work are presented in Fig. 6.

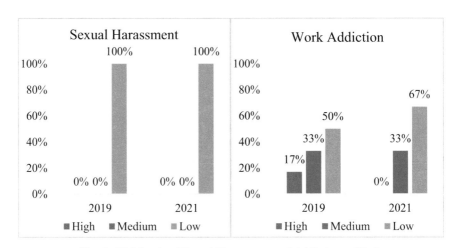

Fig. 6. Risk levels of Sexual Harassment and Addiction to Work.

It can be noted that there were no changes in the dimension of Sexual Harassment during the period 2019–2021. The Workaholic dimension presents, in the 2019–2021 period, a 17% decrease in high risk, similar values are presented in medium risk and an increase of 17% in low risk.

Next, Fig. 7 is presented, detailing the values for the Working Conditions and Double Presence.

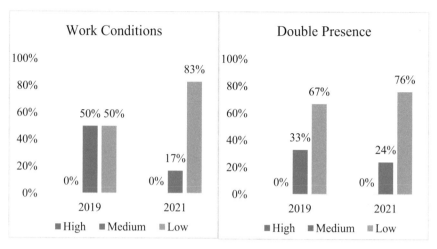

Fig. 7. Risk levels in the dimensions of Working Conditions and Double Presence.

The data show an absence in high risk and an increase in low risk of 33%, in 2021. The Double Presence dimension shows a maintenance of values in high risk, 0%, in the period 2019–2021 and significant values of the Double Presence in medium and high risks with a difference of 9% for medium and low risks.

Figure 8 presents the values obtained for the dimensions Labor and Emotional Stability and Self-Received Health.

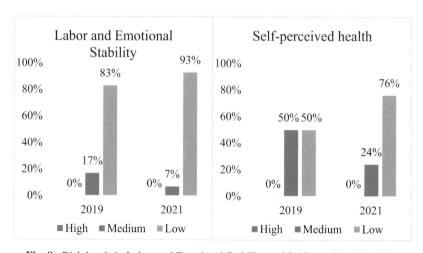

Fig. 8. Risk levels in Labor and Emotional Stability and Self-perceived Greeting.

The results show a low risk an increase of 10% and a decrease to medium risk of 10%; while at high risk the value was maintained at 0%. The self-perceived health dimension presented a decrease in value, at medium risk, of 26% and at low risk it presented an increase of 26%.

The dimensions Discriminatory, labor and sexual harassment, Work addiction, Work conditions, Double presence, Labor and Emotional Stability and Self-Perceived Health have decreased to such an extent that the high-risk level has been completely eliminated, while the level of average risk in some cases has decreased to 0% and in others it has decreased considerably. This fact is based on the new modality of work (teleworking) since it must be considered that the appearance of a new work environment in which the physical interaction between colleagues is almost null but not non-existent, the schedules have been adapted, although in many cases functions have been increased, but the existing flexibility and work based on goals have decreased several dimensions of risk.

4 Conclusions

The application of the Psychosocial Risk Assessment Questionnaire from the Ministry of Labor of Ecuador to ITCA personnel in 2019 and 2021 shows that the social risk factors for SARSCoV-2 did not increase the percentages of occupational psychosocial risks; on the contrary, it is evident that the risk levels decreased considerably. The instrument used to collect the information was very useful since it allowed a comparison between the information obtained in 2019 and 2021. This finding is significant if other studies agree that the pandemic generally harmed the psychosocial risks of other companies and higher education institutions.

Despite the fact that the results found were mostly positive, in some dimensions the level of risk increased from medium to high, as is the case of Workload and Pace, which shows that the condition in this dimension is quite low and at the same time analyze the characteristics of the worker, who fulfills various functions within the institution, for which this increase is understandable; In addition, taking into account that at the peak of the pandemic there was a cut in staff and some workers had to take on new assignments. The leadership dimension has also been affected by the same dynamics, in 2019 the staff was under a medium and low level of risk. Another dimension that generated controversy was the recovery that, on the one hand, the average risk level decreased in 2021, taking into account that there was a staff cut, some workers assumed new responsibilities in addition to those previously imposed, the difficulty of adapting to the new environment, as well as the complexity of assuming new schedules had an impact on the appearance of this risk. Another dimension that generated controversy was the recovery that, on the one hand, the medium risk level decreased from 67% in 2019 to 36% in 2021, the high risk level appears again this time with 7% in 2021, which is related to what was previously described, taking into account that there was a personnel cut, some workers assumed new responsibilities in addition to those previously imposed, the difficulty of adapting to the new environment as well as the complexity of assuming new schedules had an impact on the appearance of this risk.

During the pandemic period, the Institute has directed its resources to train staff on issues related to the pandemic, improvement of labor skills, and others; they have taken into account the opinion of the workers, to a certain point the functions have been made more flexible in search of the person being able to adapt to the new reality, always communicating the objectives to be met and reporting on the results obtained; The Institute has ensured the safety and health of the workers, for which it has personnel that provide team support, as well as medical and psychological personnel that are freely accessible to all employees.

A series of actions and instruments have been proposed that allow the risk levels to decrease within the Institute, these are contained in the Comprehensive Operational Control and are detailed in a better way in the Occupational and Health Environmental Surveillance Program focused on the psychosocial factor; In the case of personnel who are inevitably exposed to psychosocial risk, or who do not know how to identify it, a Psychosocial Risk Prevention Manual was prepared through which personnel can identify the dimension to which they are exposed and the actions to be taken, in case of If an adverse event had occurred that affects the psychological part, a section containing Psychological First Aid was also included, which can be applied by anyone, not necessarily by a mental health professional.

The instrument used for the evaluation of existing psychosocial risks in the *Instituto Superior Tecnológico* ITCA, facilitated the collection of data as well as the comparison between the data obtained in 2019 and it was demonstrated that there was a minimal increase in the appearance of a high risk level, being the case of the dimensions: Workload and Pace, Leadership and Recovery; while the risks that were previously at a medium level tended to decrease in the last year; therefore, the pandemic that hit the entire world has not been a factor that exponentially increases the psychosocial risk within the institution's staff. Based on the evaluation of psychosocial risks, a series of corrective actions contained in the Comprehensive Operational Control were recommended, which in the medium and long term can be evidenced through a new evaluation of psychosocial risks.

With the results of this investigation, a Labor Environmental and Health Surveillance Program focused on the psychosocial factor was proposed, which contains relevant information to avoid a deterioration of the worker's mental health, as well as the institutional work environment and a program was designed. Manual for the Prevention of Psychosocial Risks that contains measures and actions to be taken in the event of an adverse event that may generate or trigger the appearance of a psychosocial risk within the Institute's personnel.

The results present a significant improvement in terms of the situation prior to the pandemic, due to the fact that the Institute in this period has focused its resources on training staff in issues related to the pandemic, improvement of labor skills and others; taking into account the opinion of the workers, to a certain point the functions have been made more flexible in search of the person being able to adapt to the new reality, always communicating the objectives to be met and reporting on the results obtained; The Institute has ensured the safety and health of the workers, for which it has personnel that provide team support, as well as medical and psychological personnel that are freely accessible to all employees.

References

1. Organización Internacional del Trabajo: Gestión de los riesgos psicosociales relacionados con el trabajo durante la pandemia de COVID-19 (2020)

2. Lozano Vargas, A.: Impacto de la epidemia del Coronavirus (COVID-19) en la salud mental del personal de salud y en la población general de China. Revista de Neuro Psiquiatría **83**(1), 51–56 (2020)

3. Soares e Silva, J., Batista de Carvalho, A., Carvalho Santos Leite, H., Neves de Oliveira, E.: Reflexiones sobre los riesgos ocupacionales en trabajadores de salud en tiempos pandémicos por COVID 19. Revista Cubana de Enfermería **36**(2), 1–11, (2020)

4. Moreno Jimenez, B., Baez León, C.: Factores y riesgos psicosociales, formas, consecuencias, medidas y buenas prácticas, 1nd edn. Universidad Autónoma de Madrid, Madrid (2010)

5. Brown, L.W.: worker and wellbeing. An overview. In: AA.VV, A healthier work environment. Basic concepts and methods of measurement. Proceedings of an international meeting. Stockholm, 27–30 mayo (1991)

6. Mansilla Izquierdo, F.: Manual de Riesgos Psicosociales en el trabajo: Teoría y Práctica (2016)

7. Ministerio del Trabajo: Guía para la aplicación del cuestionario de evaluación de riesgos psicosociales. Dirección de seguridad, Salud en el trabajo y Gestión integral de riesgos, Ecuador (2018)

8. Rendón-Macías, M.E., Villasís-Keever, M.A., Miranda-Novales, M.G.: Estadística descriptiva. Revista Alergia México **63**(4), 397–407 (2016)

9. Faraldo, P., Pateiro, B.: Estadística y Metodología de la investigación. Curso 2012–2013 grado en enfermería, Tema 1. Estadística descriptiva. Universidad de Santo Domingo de Compostela, España (2012–2013)

10. Hernández-Suárez, C.A., Prada-Núñez, R., Solano-Pinto, N., Fernández-Cezar, R.: Factores de riesgo y resiliencia durante el aislamiento obligatorio de la pandemia de Covid-19: Una experiencia en docentes de Educación Superior. Mundo FESC **11**(S1), 27–37 (2021)

11. Valle, S.: Plan de prevención de riesgos psicosociales en una institución pública de la provincia de Tungurahua. Tesis de 4to nivel. Pontificia Universidad Católica del Ecuador, Ambato (2023)

12. Orejuela, M.: Diagnóstico y propuesta de intervención de riesgos psicosociales en el personal administrativo de la Escuela Politécnica Nacional en el año 2020. Tesis de 3er Nivel. Pontificia Universidad Católica del Ecuador, Quito (2021)

13. Sarabia, L., Suárez, C., Tipán, D., Sarabia, D.: Factores de riesgo psicosocial en los trabajadores del operador logístico multinacional Grupo R. Quito-Ecuador, Periodo Noviembre-Diciembre **4**(12), 37–51 (2020)

14. Instituto Superior Tecnológico José Chiriboga Grijalva: Informe evaluación riesgos psicosociales. Dirección Talento Humano ITCA, Ibarra (2019)

15. Orozco, S., Rincón, S.: Semejanzas de los Factores del Riesgo Psicosociales Laborales entre Colombia, Ecuador y Perú. (Monografía). Corporación universitaria Minuto de Dios, Pereira – Colombia (2023)

16. Córdoba, D., Chávez, A., Villacrés, A., Colunga, C., Barrera de León, C.: Factores de Riesgo Psicosocial y Fatiga en trabajadores de una empresa farmacéuttica ecuatoriana en el contexto de pandemia en el 2021 **24**(1), 1–17 (2022)

17. Vargas, L., Vargas, V., Cedeño, L., Piloso, D.: Riesgos psicosociales y la educación pública en la Pandemia. Caso Ecuador. Centro Sur. Social Science Journal **5**(1) (2020)

18. Díaz Patiño, D.G., Anaya Velasco, A., Santoyo Telles, F.: Factores de riesgo psicosocial y calidad de vida durante el confinamiento por covid-19 en universidades. RIDE. Revista Iberoamericana para la Investigación y el Desarrollo Educativo **12**(24), e028 (2022). https://doi.org/10.23913/ride.v12i24.1168

19. Jhayya, A.: Evaluación de riesgos Psicosociales en los servidores de la Unidad de Análisis Financiero Y Económico (UAFE). Tesis de cuarto nivel. Universidad Internacional SEK, Quito Ecuador (2020)

Low-Cost Didactic Models for the Teaching-Learning of Automated Systems in Academic Environments

Marco V. Pilco[1]([✉]) [ID], Jerry Medina Schaffry[1] [ID], Paul S. Espinoza[1] [ID],
Liliana M. Calero[2], and E. Fabian Rivera[1] [ID]

[1] Instituto Tecnológico Superior Universitario Oriente, Joya de los Sachas 220101, Ecuador
mpilco@itsoriente.edu.ec
[2] Ministerio de Educación, Joya de los Sachas 220101, Ecuador

Abstract. This article consists of creating low-cost didactic models to be used in practical activities in the area of Industrial Automation applied to Electromechanics students. In this sense, we analyze the prices and activities that can be performed in the existing models in the market according to the type of production present in the northern Amazon of Ecuador. Subsequently, the training stations are proposed by analyzing the mechanical and electrical materials that exist in the area, the mechanical and electrical designs are made to assemble the systems by coupling sensors, actuators, programmable automatons, pneumatic systems, etc., to the existing ones. Finally, the manufacturing price of the constructed model is compared with the standards, resulting in an affordable and economic value for educational institutions. The applicability of the didactic modules to 80 students of the Electromechanics career is evaluated, presenting motivating results regarding improvements in the acquisition and development of learning skills and abilities in practical activities.

Keywords: Didactic Model · Industrial Automation · Low Cost · Teaching and Learning Process

1 Introduction

Over the last few decades, there has been a steady increase in the automation of various production processes. This progress has been facilitated by the development and cost reduction of the required technology, which has allowed this trend to remain in force [1]. Likewise, industrial automation is a field that most companies have chosen to implement within the production lines, as it includes several fields such as electronics, electricity, mechanics, software and information technology [2, 3] that over time has facilitated human tasks in the generation of products and services.

Higher education institutions, according to the needs of each sector, offer young people study careers that prepare them to develop their skills and abilities in order to contribute to society and solve social problems [4]. These include the areas of automation, electromechanics, electronics and mechatronics, which, through the application of theoretical knowledge, allow students to learn how to handle elements of automated systems [5]. One of the tools that help in the teaching-learning process are the didactic models (DM) that enable the representation of ideas on a reduced scale, creating an object that is easy to visualize and understand from the students' perspective [6].

One of the problems facing higher education in Ecuador is the lack of DM to strengthen the knowledge acquired in the classroom [7] because they are not manufactured in the country, or because the cost is high and they must be imported from abroad. In the field of automation, training equipment is expensive, which is why we choose to build low-cost DM with as many materials as can be acquired within the country [8]. Analyzing the work of Cabré, who developed a low-cost didactic platform to teach discrete control theory concepts, he found that this tool helps teachers to effectively guide the learning process of students [9].

On the other hand, the use of virtual tools in training processes are widely used to analyze industrial environments, as mentioned by Proaño in the creation of a three-phase horizontal separator applied to the oil industry [10]. The application of augmented reality (AR) has a great boom in educational processes, since it produces drastic changes due to the fact that it allows to generate significant learning [11, 12]; including this subject matter with MD 3D provides endless applications where the student benefits from the acquisition and application of knowledge by developing spatial perception skills [13]. Likewise, automation in augmented reality is an innovation that integrates several technologies, with the objective of developing mixed real and virtual experiments, making use of mobile devices, which facilitate the comprehension of the student environment [14].

In particular, one of the strategies used in the development of low-cost didactic automation models is the use of open-source prototyping and programming platforms such as Arduino [15], that allow students to interact directly with automation components. The use of Arduino can enhance students' understanding and promote active, hands-on learning of automated systems [16].

In the practical experience at the Instituto Tecnológico Superior Universitario Oriente, located in the canton of La Joya de los Sachas, province of Orellana, the link between theory and practice in the Electromechanics course, in the Industrial Automation I and II module, is developed through the use of DM; these simulate the processes that are developed in the area such as: oil industry, coffee and cocoa food processing, agriculture and services. Having said that, DM have a high value in the market, reaching values of thousands of dollars, consequently, they are not easy to acquire; therefore, through the proposal of projects within the institution, prototypes have been built whose main material is wood, which, due to climatic conditions, tend to be damaged.

It was observed how the quality of the materials and components used can affect the durability and accuracy of the prototypes, which in turn can impact the learning experience. In addition, the lack of clear standards and guidelines for the development of certain DM can hinder replication and adoption in different educational contexts. Therefore, it is important to address these challenges and seek solutions that ensure the quality and effectiveness of automation DM.

In this context, the present work proposes the development of DM of industrial processes that are used in the fields described above. It begins by analyzing the processes that are feasible to manufacture according to the ease of acquisition of materials and equipment; the structural design of the prototypes is made using 3D modeling software where each of the mechanical elements to be coupled is diagrammed, then the electrical drawings are generated where the selection of sensors, controllers, actuators according to design requirements, calculation of protections, cable selection, etc., is included. Finally, all the elements are coupled, and the programming algorithms are generated using the controllers' software, in order to generate functional tests that allow to verify the proper work of each element of the system.

The paper consists of 4 Sections including the Introduction. Section 2 describes the materials and methods for the development of the prototypes; Sect. 3 presents the results and discussions and finally the conclusions are presented in Sect. 4.

2   Methods

This section describes an evaluation of the DM used, the analysis of the supply found on the market, the construction of the new automated prototypes and the performance tests.

2.1   Preliminary Evaluation

Didactic Models. The didactic prototypes that were used within the institution were made with electrical and electronic elements that are assembled in a wooden structure; these could be used for simulation practices of: stamping, classification and sealing of wooden objects. Likewise, mechanical elements are used that help to couple different sensors and actuators as shown in Fig. 1. The distribution of the elements is adapted according to the type of process, which in some cases is distributed horizontally where they occupy a lot of space, for this reason the use of wood is considerable. Likewise, the electrical connections use banana plugs that are adapted in a space where there is no type of labels, which can cause some type of short circuit.

Simulation Software. In the Industrial Automation module, the application of programming simulators is necessary; in this context, the Tia Portal software was used with the PLC S7-1200 and CX-programer with the ZEN-20C3AR-A-V2 automaton. It should also be mentioned that the average number of students per course was 30, which made it difficult to use the DM in a personalized way, since there were only 3 processes as shown in Fig. 1. For this reason, the teacher was more focused on simulation activities in the programs, and in order to consolidate theory with practice, groups of 6 people were generated, so that they could interact with the different automation processes. Consequently, not all the members of the group were able to consolidate the skills in the handling of the elements of the automated systems, causing low grades and desertion of the module in the semester of study.

Condition of Practice Facilities. He area used for the Industrial Automation practices is 4 m long by 6 m wide, with a capacity of 20 students, who are distributed in the three DM, in addition to the existence of desks so that they can develop the activities. In addition, there are 3 electrical connection points distributed in the classroom, which

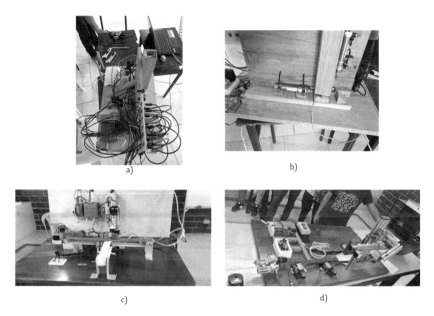

Fig. 1. a) Top view of the object stamping process, b) Front view of the object stamping process, c) Automated object sorting process, d) Object sealing process.

Fig. 2. Space used for Industrial Automation practices.

are insufficient for the number of students carrying their personal computers, as shown in Fig. 2.

2.2 Educational Models Available on the Market

With regard to the automation MD on the market, there are a variety of options from different manufacturers such as: Festo, Ariande, Lab-Volt, Phywe, among others, which have different processes with a variety of equipment. Taking the Festo brand as a reference, it is analyzed that it has stations with different operations that allow coupling elements to generate processes in production lines as shown in Fig. 3.

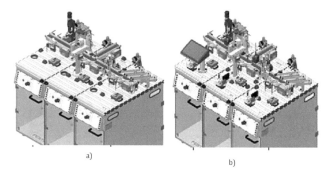

Fig. 3. a) Distribution, joining and sorting station, b) Composite manufacturing chain.

Figure 3a shows a system that performs the distribution, joining and classification of objects by means of three conveyor belts, pneumatic elements and electric actuators in order to form an industrial network; these are programmed using a PLC S7-1512. Likewise, Fig. 3b shows a manufacturing line composed of distribution, joining and classification stations, which, unlike the previous one, has a laptop computer. The analyzed stations have programming software and need an air pressure between 6 to 10 bars with an electrical supply of 110 V in alternating current at 60 Hz.

When quoting the two stations at Cole Didacticum the city of Quito, the following prices were determined: the distribution station, joining and classification of objects at a value of \$ 59650.97 dollars and the manufacturing chain at \$ 67978.54 dollars, which does not include VAT. It is necessary to emphasize that these stations are analyzed due to the processes and practices that can be performed and that are immersed in the study curriculum of the Industrial Automation course.

2.3 Construction of Low-Cost Educational Models

Study of Automation Processes. The following is an explanation of the automation processes proposed to generate the construction of the DM. The proposal is made taking as an example the composite manufacturing station made in the previous quotation, since it is related to the industrial areas that are handled within the province of Orellana, Sucumbíos and Napo.

i) Object distribution process: by means of the movement of a conveyor belt, a metal and plastic container-type specimen is moved towards a double-acting piston that distributes it to another conveyor belt. The materials to be used are detailed in Table 1, where the main element controlling the process is the PLC Logo V8.

ii) Object joining process: A conveyor belt moves the specimen from the previous process, and when it is detected by sensors, a pneumatic piston is activated to make the joint with a metallic cover, and then another piston is activated to seal it. The materials that make up the process are detailed in Table 2.

iii) Object classification process: according to the metal and plastic detection sensors, two double-acting pistons are activated to sort the specimen, which travels on a conveyor belt to a designated location. The list of materials is detailed in Table 3.

Table 1. List of materials for the object distribution process

Equipment	Quantify	Cost ($)
PLC Logo V8.2 12/24 VDC	1	182.40
Regulated power supply output 24 VDC 2.5 A	1	36.22
24 V DC motor	2	20.00
Solenoid valve 5/2 -24 VDC ¼ monostable	1	20.00
Photoelectric presence sensor 18-3A10NA	1	25.00
Double acting piston	1	10.00
20 A bipolar breaker	1	15.00
Pushbutton box	1	10.00
Pilot lights	3	10.5
Conveyor belt	2 m	80.00
Profiled aluminum	20 m	450.00
On-off pushbutton panel	2	10.00
Riel din	1	8.00
Electrical cable # 18	20 m	10.00
Fastening screws	20	10.00
Slotted trunking DXN10042 25X40mm	1	13.00
End terminals for wire end ferrules	200	8.00
6mm x 1/8" x 1/8" Range Type Flow Regulator	1	16.00
Pneumatic hose ¼	5	15.00
	TOTAL	**942.12**

Construction Process of Compact Manufacturing Stations. The process starts with the design of the stations in the CAD software, where an analysis of the aluminum is made and the mechanical and electrical elements of the DM are located; likewise, the electrical control circuit is created, whose main element is the PLC. Then, the aluminum profiles are coupled together with the fastening screws where the sensors, actuators, pneumatic elements, conveyor belts that allow the operation of the three processes are located. Finally, the programming algorithm is made using the software that manages the controller, where the following are declared: input and output variables, timing and counting function blocks, among others, which allow to generate the operation tests to verify if there are errors for subsequent correction as shown in Fig. 4.

3   Results and Discussions

In order to find the results of the present research work, we proceed to evaluate the students by means of 10 questions grouped in three parameters considering the 5-point Likert-type scale (1: totally disagree; 2: disagree; 3: neither agree nor disagree; 4: agree;

Table 2. Bill of materials for the object joining process

Equipment	Quantify	Cost ($)
PLC Logo V8.2 12/24 VDC	1	182.40
Regulated power supply output 24 VDC 2.5 A	1	36.22
24 V DC motor	1	10.00
Solenoid valve 5/2 -24 VDC ¼ monostable	3	60.00
Photoelectric presence sensor 18-3A10NA	2	50.00
Double acting piston	3	30.00
20 A bipolar breaker	1	15.00
Pushbutton box	1	10.00
Pilot lights	3	10.5
Conveyor belt	2 m	80.00
Profiled aluminum	20 m	450.00
On-off pushbutton panel	2	10.00
Riel din	1	8.00
Electrical cable # 18	20 m	10.00
Fastening screws	20	10.00
Slotted trunking DXN10042 25X40mm	1	13.00
End terminals for wire end ferrules	200	8.00
6mm x 1/8" x 1/8" Range Type Flow Regulator	1	16.00
Pneumatic hose ¼	5	15.00
	TOTAL	**1024.12**

5: totally agree). First, the activities developed during the framed practice are taken into account, which assesses the knowledge acquired by the student through the DM. The next parameter consists of the application of the methodology and finally an objective evaluation is made as shown in Table 4.

Table 3. Bill of materials of the object sorting process

Equipment	Quantify	Cost ($)
PLC Logo V8.2 12/24 VDC	1	182.40
Regulated power supply output 24 VDC 2.5 A	1	36.22
24 V DC motor	1	10.00
Solenoid valve 5/2 -24 VDC ¼ monostable	2	40.00
Photoelectric presence sensor	1	29.50
Inductive sensor	1	28.50
Capacitive Sensor	1	27.50
Double acting piston	2	40.00
20 A bipolar breaker	1	15.00
Pushbutton box	1	10.00
Pilot lights	3	10.5
Conveyor belt	1 m	40.00
Profiled aluminum	20 m	450.00
On-off pushbutton panel	2	10.00
Riel din	1	8.00
Electrical cable # 18	20 m	10.00
Fastening screws	20	10.00
Slotted trunking DXN10042 25X40mm	1	13.00
End terminals for wire end ferrules	200	8.00
6mm x 1/8" x 1/8" Range Type Flow Regulator	1	16.00
Pneumatic hose ¼	5	15.00
Signaling tower	1	28.50
	TOTAL	**1038.12**

The results of the application of the evaluation of learning achieved through low-cost DM take into consideration the technical, methodological and aptitudinal aspects applied to 80 students of the Electromechanics course. The results obtained are shown in Table (Table 5).

The results presented through the evaluation show that students found the activities motivating and geared to practical processes, therefore, the learning of practical activities is not only done through the use of simulators focused on automation processes.

Citing the work of Pinares-Mamani [17], who performs a DM with a cost of 100 dollars, to test advanced control algorithms, obtains as a result that students can make additional modifications only in the programming code, it is compared that with the DM executed in this research students can make modifications not only in the programming

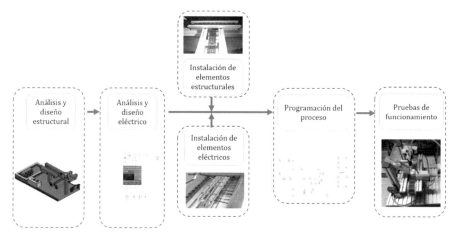

Fig. 4. Construction procedure for compact manufacturing stations

Table 4. Proposed student evaluation

N	Activities	1	2	3	4	5
1	Recognition and understanding of the complete manufacturing process					
2	Connection of the sensors and actuators on the didactic model					
3	Scheduling of the composite manufacturing process taking into account inputs, outputs and process					
	Methodology					
4	Integration of cooperative learning in the didactic methodology					
5	Adequacy of the use of software to simulate and program the composite manufacturing process					
6	Successful application of the experimental method					
	Evaluation					
7	Initial objective evaluation approach (Google forms)					
8	Evaluation approach according to teaching observation					
9	Evaluation approach according to delivery of activities (Practical report)					
10	Final objective evaluation approach (Moodle)					

part, but also in the part of the physical elements, since they can adapt them according to the needs of the automation process they are analyzing, in this way it is much more feasible to develop learning in students and expand the study related to industrial production environment.

Table 5. Results obtained from the evaluation

N	Result
Question 1	4.8
Question 2	4.6
Question 3	4.4
Question 4	5
Question 5	5
Question 6	5
Question 7	4.3
Question 8	4.9
Question 9	4.7
Question 10	4.8

4   Conclusions

The construction of a low-cost model designed to address industrial automation processes, with an approach focused on teaching-learning aimed at electromechanical students, stands as a very useful and effective tool. This model, when made available to students, gives them the opportunity to apply theoretical knowledge in a tangible and concrete environment, providing them with the ability to assimilate the concepts and principles inherent to industrial automation in a more pro-fund way, so that in the evaluations made in parameters of: Activities, Methodology and Evaluation the students obtain a result in the initial evaluation of 86% and in the evaluation of practical activity and final objective an average of 94% and 96% respectively.

Due to the low-cost nature of the model, it becomes an affordable option even for educational institutions facing resource constraints. This attribute greatly facilitates its incorporation into the educational program. Also, since it is set up with an interactive and didactic pedagogical approach, it promotes active and participatory learning among students. This approach motivates them to get involved in the exploration and experimentation of the various automation components and systems, preparing them for the challenges presented by contemporary labor dynamics, in which automation and the Industry 4.0 perspective play an increasingly important role.

References

1. Martinez, E.M., Ponce, P., Macias, I., Molina, A.: Automation pyramid as constructor for a complete digital twin, case study: a didactic manufacturing system. Sensors **21**(14), 14 (2021). https://doi.org/10.3390/s21144656
2. Pilco, M., Pilataxi, M.U. de las F.A.E.: Simulador 3d de un proceso de nivel de un tanque esférico para evaluar controladores, pp. 158–170 (2023)

3. Rivera, E.F., Pilco, M.V., Espinoza, P.S., Morales, E.E., Ortiz, J.S.: Training system for hybrid vehicles through augmented reality. In: 2020 15th Iberian Conference on Information Systems and Technologies (CISTI), pp. 1–6 (2020). https://doi.org/10.23919/CISTI49556.2020.914 1020

4. Rivera, J.: La gratuidad de la educación superior y sus efectos sobre el acceso: Caso Ecuador. Educ. Policy Anal. Arch. **27**, 29 (2019). https://doi.org/10.14507/epaa.27.3776

5. Sánchez, Ó.E.R., Cho, N.C., Acuña, H.G., Lázaro, J.G.M.: Proyecto integrador como estrategia pedagógica del curso de automatización industrial del programa de ingeniería mecatrónica de la Universidad Autónoma de Bucaramanga: caso de estudio. Encuentro Int. Educ. En Ing. (2022). https://doi.org/10.26507/paper.2534

6. Gomez, B.C.: La Maqueta como Recurso Didáctico para la Enseñanza de Matemática en Arquitectura. INGENIO **5**(2), (2022). https://doi.org/10.29166/ingenio.v5i2.4083

7. Basantes-Andrade, A., Cabezas-González, M., Casillas-Martín, S., Naranjo- Toro, M., Benavides-Piedra, A.: NANO-MOOCs to train university professors in digital competences. Heliyon **8**(6), e09456 (2022). https://doi.org/10.1016/j.heliyon.2022.e09456

8. Maltos, G.R., GonzÃ¡lez, J.R.G., GÃ³mez, E.J.M.: Propuesta de tablero de entrenamiento para automatizaciÃ³n y control/Training Board Proposal for Automation and Control. RECI Rev. Iberoam. Las Cienc. Comput. E InformÃ¡tica **8**(16), 16 (2019). https://doi.org/10.23913/reci.v8i16.94

9. Cabré, T.P., et al.: Didactic platform for DC motor speed and position control in Z-plane. ISA Transactions **118**, 116–132 (2021). https://doi.org/10.1016/j.isatra.2021.02.020

10. Proaño, Z.C., Andaluz, V.H.: Virtual training system of a horizontal three-phase separator. In: Mesquita, A., Abreu, A., Carvalho, J.V. (eds.) Perspectives and Trends in Education and Technology. Smart Innovation, Systems and Technologies, pp. 633–647. Springer, Singapore (2022). https://doi.org/10.1007/978-981-16-5063-5_52

11. Montenegro Rueda, M., Fernández Cerero, J.: Realidad aumentada en la educación superior posibilidades y desafíos. Augmented reality in higher education: possibilities and challenges (2022). https://doi.org/10.51302/tce.2022.858

12. Rivera, E.F., Morales, E.E., Florez, C.C., Toasa, R.M.: Development of an augmented reality system to support the teaching-learning process in automotive mechatronics. In: De Paolis, L.T., Arpaia, P., Bourdot, P. (eds.) Augmented Reality, Virtual Reality, and Computer Graphics, en Lecture Notes in Computer Science, pp. 451–461. Springer International Publishing, Cham (2021). https://doi.org/10.1007/978-3-030-87595-4_33

13. Garriazo, J.A.C., et al.: Aplicación del modelo didáctico 3D realidad aumentada en el aprendizaje colaborativo. Revisión sistemática. Horiz. Rev. Investig. En Cienc. Educ. **6**(22), 276–290 (2022). https://doi.org/10.33996/revistahorizontes.v6i22.335

14. Treviño-Elizondo, B.L., García-Reyes, H.: What does Industry 4.0 mean to Industrial Engineering Education?. Procedia Computer Science **217**, 876–885 (2023). https://doi.org/10.1016/j.procs.2022.12.284

15. Guerrero-Felix, J.G., Lopez-Miras, J., Rodriguez-Valverde, M.A., Moraila-Martinez, C.L., Fernandez-Rodriguez, M.A.: Automation of an atomic force microscope via Arduino. HardwareX **15**, e00447 (2023). https://doi.org/10.1016/j.ohx.2023.e00447

16. Oltean, S.-E.: Mobile robot platform with arduino uno and raspberry Pi for autonomous navigation. Procedia Manufacturing **32**, 572–577 (2019). https://doi.org/10.1016/j.promfg.2019.02.254

17. Pinares-Mamani, O.G.C., Cutipa-Luque, J.C.: A low-cost didactic module for testing advanced control algorithms. HardwareX **8**, e00148 (2020). https://doi.org/10.1016/j.ohx.2020.e00148

Organizational Commitment and Satisfaction in University Teaching Work: A Comparative Study

Aleixandre Brian Duche-Pérez[(✉)] , Cintya Yadira Vera-Revilla ,
Olger Albino Gutiérrez-Aguilar , Fanny Miyahira Paredes-Quispe ,
Milena Ketty Jaime-Zavala , Brizaida Guadalupe Andía-Gonzales ,
and Javier Roberto Ramírez-Borja

Universidad Católica de Santa María, Arequipa 04001, Perú
aduche@ucsm.edu.pe

Abstract. The exploration of working environments within educational settings has recently surged in significance, becoming paramount in the accreditation procedures centered on educational quality. Within this domain, organizational and management psychology bridge the gap between the professional milieu and broader institutional evolution. This study's primary aim is to discern the levels of job satisfaction and organizational dedication among undergraduate university educators and to probe any potential associations between these two factors. This investigation adopts a quantitative stance, with a descriptive-correlational, non-experimental, and cross-sectional framework. Data was garnered through the Multidimensional Scale of Teacher Job Satisfaction (EMSLD) as proposed by Barraza and Ortega (2009), in tandem with the Organizational Climate (OC) tool as presented by Barraza (2008). The SPSS statistical software (version 21.0) was employed to gauge the reliability of these tools, yielding Cronbach's Alpha coefficients of .89 and .91 for job satisfaction and organizational dedication respectively. The research involved 180 part-time assistant professors from three universities (one public and two private) situated in Arequipa. Notably, the findings indicate that correlations between job contentment and organizational loyalty can swing positive or negative based on the institution's legal structure and governance approach.

Keywords: Teacher · University · Job commitment · Job satisfaction

1 Introduction

The emphasis on university quality accreditation has heightened over recent years. Undeniably, global rankings, including the likes of QS World University Rankings and similar, have made a substantial impact on public perceptions, subsequently affecting university administrators and decision-makers. In this backdrop, the linkage between an advanced educated workforce and a nation's economic progress has shone a spotlight on the role

M. Z. Vizuete et al. (Eds.): CI3 2023, LNNS 1041, pp. 350–365, 2024.
https://doi.org/10.1007/978-3-031-63437-6_29

of universities. These institutions stand as pillars for societal change and the dissemination of knowledge. Hence, universities have the imperative to demonstrate the competence, productivity, and relevance of their faculty members in the pursuit of scholastic distinction.

Furthermore, trends like globalization and emerging economic landscapes are prompting universities to navigate various shifts and hurdles at an organizational scale. From this viewpoint, the educator has evolved as the foundational element of learning and the grooming of future generations who will eventually join the nation's workforce. Consequently, the educator's role is pivotal in realizing the institutional goals of universities [1].

The examination of living conditions and job contentment of university professors has emerged as a prevalent research topic in higher education studies. Authorities overseeing national-level universities are taking steps to uphold superior education quality standards. Nonetheless, Peruvian universities showcase variability in terms of academic standards, scale, faculty profiles, facilities, budgetary allocations, prevailing legislation, the nation's political climate, the institutional philosophy, and other elements, leading to a diverse higher education landscape. In this light, it's pivotal to highlight the plethora of factors within the university realm that influence faculty performance, subsequently affecting the student learning experience [2]. Two paramount variables among many are institutional allegiance and job contentment. These play a significant role as they can foreshadow the progression concerning the efficiency and durability of the concerned educational establishments.

Job satisfaction pertains to the collective sentiments towards one's profession, culminating in either positive or negative behavioral and psychological tendencies [3]. An individual's gratification with their job ranks prominently in matters related to organizational demeanor and superior workplace quality of life. In organizational and occupational psychology, job satisfaction is deemed an integral concept, given its intermediary role bridging workplace conditions and the resulting ramifications for both the institution and the individual [4].

In studies zeroing in on job contentment, one scholar articulates the understanding of scenarios or elements that lead a university educator to experience fulfillment or its absence, along with the stress emanating from certain conditions set by academic entities [5]. The significance of these insights cannot be overstated due to their direct implications on universities. Aspects causing discontent or occupational strain can impact educators' academic, administrative, and personal facets [6, 7]. On an academic front, this could manifest in areas like knowledge dissemination, teaching methodologies, evaluation techniques, knowledge stagnation, and lack of academic rejuvenation, to name a few [8]. Administratively, repercussions could span absenteeism, habitual lateness, disciplinary breaches, deceit, voluntary departures, and terminations. From a personal standpoint, it could lead to a drop in self-worth, depressive states, or other psychological and physical ailments.

Parallel research on job gratification and institutional loyalty, conducted by certain scholars [9], revealed that employees under stable employment contracts exhibit greater satisfaction and commitment, contingent upon contract specifics and work timings.

Secondly, the notion of organizational commitment, also referred to as employee loyalty, is depicted as the extent to which an individual aligns with an organization and harbors an aspiration to remain actively involved [10]. This sentiment is underscored by a profound trust in and endorsement of the institution's aims and standards, a drive to contribute notably for the organization's benefit, and an unwavering intent to remain affiliated with it [11, 12]. Essentially, it symbolizes the array of bonds that tether an individual to a specific entity, impacting their decision to either persevere with or depart from the organization [13].

In terms of the categorical breakdown, one scholar pinpoints three predominant facets of commitment: affective commitment, continuance commitment, and normative commitment [13]. Another researcher bifurcates commitment into three segments, rooted in the established definitions of organizational commitment: instrumental, affective, and behavioral. The behavioral dimension manifests in two forms: the inclination to retain membership within the organization, manifested in acts of association and additional contributions for its betterment [14].

In concurrence with this research trajectory, a different scholar underscores the fundamental role of studying sentiments towards collective endeavors in tandem with organizational commitment. Herein, job contentment emerges as an attitudinal factor potentially influencing commitment. This sentiment is intricately connected to other psychological elements, like psychological welfare and life contentment [15]. Hence, there exists an intimate bond between occupational satisfaction and overall life satisfaction. The latter is encapsulated as the comprehensive assessment an individual renders concerning their life [16]. Such an appraisal entails evaluating the tangible facets of life, both positives and negatives, and juxtaposing them against a benchmark or standard, culminating in a verdict about their contentment in life [17].

Various viewpoints from distinct studies underscore the disparity between occupational satisfaction and organizational commitment. Consequently, it's plausible to deduce that the latter embodies a more expansive construct, mirroring an overarching emotional response to the organization in its entirety. In contrast, job satisfaction pertains inherently to a designated occupational role and specific work-related nuances [18].

Extensive research has delved into the interplay between job contentment and organizational commitment, with many studies establishing a positive link between the two. For example, the sentiments of university employees concerning job satisfaction and organizational loyalty have been explored [19]. Additionally, a meta-analysis surveying the connections between these elements across various environments was conducted [20]. Job contentment has been recognized as a potent influencer of employee loyalty [21], and another study accentuated a robust positive association between job satisfaction and institutional allegiance [22].

Studies have further indicated that varying facets of commitment influence job satisfaction in diverse ways. Drawing from Meyer and Allen's categorization [13], affective commitment (AC) has been associated with positive emotions, fulfillment, workplace well-being, and satisfaction with peers, superiors, and advancement opportunities [23]. Continuance commitment (CC) has been correlated with contentment regarding compensation, superiors, and promotions, whereas normative commitment (NC) has been linked predominantly with salary satisfaction. A certain scholar illustrated that high degrees of affective and continuance commitment signify profound resonance with the organization [24].

Evidence has showcased pronounced affective commitment amongst educators, accompanied by propitious perceptions towards institutional allegiance [25]. While no marked correlation was discerned between organizational climate and continuance commitment (CO), a positive tie emerged between affective CO and educator efficacy [26]. Moderate CO levels were observed, along with diminished continuance commitment.

Our investigation narrows its scope to three universities situated in the Arequipa Region of Peru. The objective is to scrutinize the extents of job contentment and dedication among faculty, discern the strategies employed to foster commitment, and ascertain if this allegiance influences the anticipated outcomes of educational entities.

2 Methods

The research adopts a quantitative methodology and can be characterized as non-experimental, cross-sectional, and descriptive-correlational in nature. The survey method was employed to examine the research subject, utilizing a questionnaire as the primary tool.

For assessing job satisfaction, we used the Multidimensional Teacher Job Satisfaction Scale (in Spanish: Escala Multidimensional de Satisfacción Laboral Docente, EMSLD) [27]. This scale encompasses eight dimensions: interpersonal interactions, professional conduct, organizational elements, evaluation of tasks accomplished, engagement, work environment, physical surroundings, and leadership teams. To gauge institutional dedication, an instrument delineating seven dimensions was employed: organizational hierarchy, employee regulations, physical surroundings, material assets, motivation towards work and organizational commitment, communication practices, opportunities for professional growth, and work flow.

The research sample consisted of 180 undergraduate faculty members from three Arequipa-based universities: a state-run university (UP), a corporate-based private institution (UPS), and an associative private institution (UPA). Participation was on a voluntary basis. All the participants are employed as contracted faculty and fall under the auxiliary designation.

These educators hail from diverse educational backgrounds (engineering, biomedical, and humanities) and their roles and obligations fluctuate based on their specialized expertise, contractual details, and tenure in the teaching profession. Participant ages span from 20 to 59 years, with the majority of them falling within the 30–49 age bracket.

In terms of gender distribution, males constituted 60.00% of the participants, while females made up 40.00%. Every participant in the research is a Peruvian national residing in Arequipa.

Subsequent sections detail the distinct profiles of faculty across the different universities examined:

Before proceeding with the application of the measurement instruments, letters of authorization were sent to the rectors of each university, requesting their collaboration. The letters specified the characteristics of the research to be carried out and guaranteed absolute confidentiality in the manipulation of information.

Table 1. Characterization of the study subsamples.

Variable	Indicator	University		
		UPA	UPS	UP
Sex	Male	76.67	46.67	56.67
	Female	23.33	53.33	43.33
Age	20 to 29 years	6.67	0.00	6.67
	30 to 39 years	43.33	43.33	43.33
	40 to 49 years	36.67	30.00	36.67
	50 to 59 years	13.33	26.67	13.33
	60 to 69 years	0.00	0.00	0.00
	70 and over	0.00	0.00	0.00
Years of university teaching practice	Less than 03 years	13.33	0.00	23.33
	03 to 05 years	50.00	33.33	10.00
	06 to 10 years	13.33	33.33	33.33
	11 to 14 years	16.67	16.67	33.33
	15 years or more	6.67	16.67	0.00
Dedication regime	01 to 10 h	60.00	3.33	0.00
	11 to 19 h	23.33	60.00	0.00
	20 to 39 h	16.67	36.67	93.33
	40 h	0.00	0.00	6.67
Parallel teaching at other universities	Yeah	60.00	50.00	6.67
	No	40.00	50.00	93.33

After obtaining the agreement, the teachers were informed about the objective of the project and the importance of their collaboration and participation. The questionnaires were administered during working hours, and they were delivered personally to the teachers who voluntarily and randomly decided to participate.

3   Results

The following are the results of the questionnaire administered to the teaching staff of the universities under study. The primary aim of the research was to analyze the relationship between job satisfaction and institutional commitment among undergraduate assistant teachers. Information was gathered from each university's facilities, where three types of data were sought: identifying levels of job satisfaction, identifying factors of subjective well-being, and identifying existing relationships between these levels and factors.

3.1   Job Satisfaction

The results are presented in two sections: descriptive analysis and inferential analysis. In the descriptive analysis, job satisfaction is treated as a collective variable, and its analysis is conducted at three levels: empirical indicators, dimensions of job satisfaction, and job satisfaction itself. To interpret the results, a scale of intensity [27] expressed in percentage terms was used (Table 2).

Table 2. Scale of intensity.

%	Level
or to 20	Low
21 to 40	Medium-Low
41 to 60	Half
61 to 80	Medium-High
81 to 100	High

The inferential analysis was conducted using a group difference design without causal attribution, using five variables: sex, academic level, age, job seniority, and job category. Student's statistics and single-factor ANOVA were used in the analysis. In the second case, the Scheffe follow-up test was used to identify the specific groups where differences occurred. The decision rule to accept that a sociodemographic variable established a significant difference in job satisfaction among teachers in each university was $p < .05$. Both descriptive and inferential analyses were performed using SPSS Program version 21.0.

3.2   Empirical Indicators of Job Satisfaction

Regarding the specific empirical fields to which each of the EMSLD items refers, it was found, according to the university:

As shown, the empirical indicators that appear with the greatest intensity among teachers at the Public University are: "The interpersonal relationships I have with students of the institution" (91.11%), "The level of motivation I have towards my work at the moment" (87.78%), and "The teaching activities I perform at the moment" (86.67%), with the first belonging to the dimension of Interpersonal Relations and the other two located in the dimension of Professional Performance. Meanwhile, those with less intensity are: "The amount of work assigned to me" (61.11%), "The way in which I can move up in salary level" (50.00%), and "The economic remuneration I receive" (42.22%), all from the Labor Conditions dimension.

In the case of the Private University (Societaria), the empirical indicators that appear with greater intensity among teachers are: "The level of motivation I have towards my work at the moment" (91.11%), "How motivating my work is" (86.67%), and "The way students value my work" (86.67%), with the first two located in the Professional

Table 3. Level of intensity of each of the empirical indicators.

Empirical indicators of job satisfaction	University		
	UPA	UPS	UP
Interpersonal Relations dimension			
The interpersonal relationships I have with my fellow teachers	85.56	68.89	73.33
The interpersonal relationships I have with the students of the institution	87.78	82.22	91.11
The interpersonal relationships that I have with the management team of the institution	85.56	77.78	74.44
Professional Performance dimension			
The teaching activities that I carry out at the moment	88.89	85.56	86.67
The level of motivation I have at this moment towards my work	78.89	91.11	87.78
The autonomy I have to do my job	86.67	81.11	75.56
The freedom I have for the design and implementation of my work activities	84.44	80.00	82.22
How motivating my work is	90.00	86.67	76.67
Labor Conditions dimension			
The information provided by the institution in relation to my labor conditions or problems	62.22	67.78	63.33
The opportunities that the institution offers me to develop professionally	73.33	78.89	66.67
How can I promote myself to a management position?	54.44	72.22	64.44
The job level I have	84.44	85.56	83.33
The economic remuneration that I receive	58.89	76.67	42.22
The amount of work assigned to me	75.56	77.78	61.11
The way I can move up in salary level	100.00	58.89	50.00
Valuation of Work Developed dimension			
The way students value my work	83.33	86.67	80.00
The opinion that the Arequipa community has about the academics of this institution	81.11	75.56	73.33
The way in which the management team of the institution values my work	78.89	78.89	66.67
The way my fellow teachers value my work	82.22	73.33	71.11
Indicators of the Participation dimension			
My way of participating in decision making	74.44	83.33	74.44
The way they take my opinions into account	67.78	75.56	65.56

(continued)

Table 3. (*continued*)

Empirical indicators of job satisfaction	University		
	UPA	UPS	UP
Organizational factors			
The way in which institutional regulations are respected	76.67	82.22	68.89
The work environment in the institution	78.89	80.00	77.78
The way in which university statutes are respected	46.67	68.89	61.11
The information provided to me by the different instances of the institution to carry out my work	74.44	68.89	66.67
The way the institution is organized	71.11	78.89	63.33
Physical Environment dimension			
The ventilation of the physical space where I work	63.33	76.67	81.11
The lighting of the physical space where I work	75.56	84.44	83.33
The physical space where I do my work	74.44	73.33	77.78
Management Team dimension			
The fairness with which you are treated by the management team	75.56	82.22	70.00
The way in which the management team of this institution supervises my work	78.89	82.22	75.56
The way the management team treats me	83.33	82.22	70.00
The way in which the management team resolves conflicts that arise in the institution	77.78	71.11	61.11
The support and facilities provided by the management team to develop professionally	76.67	72.22	65.56

Performance dimension and the third in the Work Assessment dimension. The one with less intensity is "The information the institution provides me in relation to my labor conditions or problems" (67.78%) and "The way in which I can move up in salary level" (58.89%), both from the Labor Conditions dimension.

Regarding the Private (Associative) University, the empirical indicators that appear with the greatest intensity among teachers are: "The way I can move up in salary level" (100.00%), "How motivating my work is" (90.00%), and "The teaching activities I perform at the moment" (88.89%). The first corresponds to the dimension "Working Conditions," while the following two correspond to "Professional Performance." The ones with less intensity are: "The way in which I can promote myself to a management position" (54.44%) and "The way in which university statutes are respected" (46.67%), both from the dimension "Working Conditions".

3.3 Dimensions of Job Satisfaction

The results obtained in the descriptive analysis, conducted for each dimension of job satisfaction, are presented in Table 4. The average has been transformed into a percentage to facilitate interpretation, and a column has been added to show the result interpreted with the established scale.

Table 4. Job satisfaction, according to university and dimensions.

Dimensions of job satisfaction	University		
	UPA	UPS	UP
Relations interpersonal	79.63 (Medium-High)	76.30 (Medium-High)	86.30 (High)
Performance Professional	81.78 (High)	84.89 (High)	85.78 (High)
Conditions labor	61.59 (Medium-High)	73.97 (Medium-High)	72.70 (Medium-High)
Work Valuation Developed	72.78 (Medium-High)	78.61 (Medium-High)	81.39 (High)
Stake	70.00 (Medium-High)	79.44 (Medium-High)	71.11 (Medium-High)
Factors organizational	67.56 (Medium-High)	75.78 (Medium-High)	69.56 (Medium-High)
Atmosphere Physical	80.74 (Medium-High)	78.15 (Medium-High)	71.11 (Medium-High)
Management team	68.44 (Medium-High)	78.00 (Medium-High)	78.44 (Medium-High)
Relations interpersonal	79.63 (Medium-High)	76.30 (Medium-High)	86.30 (Medium-High)
Performance Professional	81.78 (Medium-High)	84.89 (Medium-High)	85.78 (Medium-High)

The level of satisfaction reported in each of the eight dimensions that make up the EMSLD is presented in Table 1. As can be seen, the surveyed teachers from the Public University and Private (Societary) University state that they are more satisfied with regard to "professional performance", while those of the Private (Associative) University report higher satisfaction in "interpersonal relations". However, the teachers of the UP and UPS, similarly, have a lower level of satisfaction in relation to "working conditions", while those of the UPA report lower satisfaction in relation to "organizational factors".

The professors of the three universities in the Arequipa Region who were surveyed report a medium-high level of job satisfaction: Public University (77.05%), Private (Societary) University (78.14%), and Private (Associative) University (72.81%).

3.4   Institutional Commitment

The results are presented in two parts: in the first, the descriptive analysis is carried out, and in the second, the inferential analysis. In the descriptive analysis, organizational commitment is taken as a collective variable, and its analysis is carried out at three levels: empirical indicators, dimensions of organizational commitment, and organizational commitment. In the interpretation of the results, the following scale of intensity Barraza, 2008 expressed in percentage terms was applied (Table 5):

Table 5. Scale of intensity.

Range	Category
or to 33	Weak
34 to 66	Moderate
67 to 100	Strong

The inferential analysis was carried out using a group difference design without causal attribution with five variables: sex, academic level, age, job seniority, and job category. The Student's statistics and ANOVA of a single factor were used in the analysis; in the second case, the Scheffe follow-up test was used when it was necessary to identify between which groups the difference occurred. The decision rule to accept that the sociodemographic variable establishes a significant difference in the organizational commitment of the professors of each university was $p < .05$. Both the descriptive and inferential analysis were performed using SPSS Program version 26.0.

3.5   Empirical Indicators of Organizational Commitment

The results obtained in the descriptive analysis carried out in each of the empirical indicators of organizational commitment are shown in Table 6, where the average has been transformed into a percentage to facilitate interpretation.

As can be seen, among teachers of the Public University, the empirical indicators that are presented with greater intensity are: "Working in this institution means a lot to me" (83.33%), "I would like to continue the rest of my professional career in this institution" (82.67%), and "I am proud to work in this institution" (82.00%), all belonging to the affective dimension. The ones with less intensity are "I think that if I left this institution I would not have many options to find another job" (48.00%) and "I work in this institution more because I need it than because I want to" (40.67%), both of which belong to the calculated dimension.

In the case of the Private University (Corporate), the empirical indicators that appear with greater intensity among teachers are: "If I continue in this institution it is because in another I would not have the same advantages and benefits that I receive here" (48.00%) and "One of the disadvantages of leaving this institution is that there are few possibilities of finding another job" (46.67%), both of the calculated dimension. The one with the

Table 6. Level of intensity of each of the empirical indicators.

Empirical indicators of organizational commitment	University		
	UPA	UPS	UP
Affective dimension			
I would like to continue the rest of my professional career in this institution	82.67	39.33	91.33
I really feel that any problem in this institution is also my problem	79.33	40.67	86.00
Working in this institution means a lot to me	83.33	40.67	92.00
In this institution I feel like family	70.67	43.33	82.00
I am proud to work in this institution	82.00	41.33	91.33
I feel like an integral part of this institution	74.67	44.00	88.00
Normative dimension			
I think it would not be right to leave this institution even if it benefits me in the change	70.00	43.33	80.67
I think I owe a lot to this institution	73.33	42.67	74.00
This institution deserves my loyalty	80.67	43.33	86.67
I don't think I could leave this institution because I have an obligation to its people	60.00	43.33	81.33
Calculated dimension			
If I continue in this institution it is because in another I would not have the same advantages and benefits that I receive here	62.00	48.00	65.33
Even if I wanted to, it would be very difficult right now to leave this job	62.00	43.33	68.00
One of the disadvantages of leaving this institution is that there is little chance of finding another job	58.00	46.67	50.00
If I decided to leave this institution now, many things in my personal life would be interrupted	56.00	41.33	60.00
At this time, leaving this institution would be at great cost to me	58.67	38.67	58.00
I think that if I left this institution, I would not have much chance of finding another job	48.00	40.00	45.33
I work in this institution more because I need it than because I want to	40.67	43.33	44.00
I could leave this job even if I don't have another one in sight	50.00	43.33	52.00

least intensity is "At this moment, leaving this institution would mean a great cost for me" (38.67%), also of the calculated dimension.

Regarding the Private (Associative) University, the empirical indicators that are presented with greater intensity among teachers are: "Working in this institution means a lot to me" (92%), "I am proud to work in this institution" (91.33%), and "I would like to continue the rest of my professional career in this institution" (91.33%), all of the affective dimension. The ones with less intensity are "I think that if I left this institution

I would not have many options to find another job" (45.33%) and "I work in this institution more because I need it than because I want to" (44.00%), both of the calculated dimension.

3.6 Dimensions of Organizational Commitment

The results obtained in the descriptive analysis, carried out in each of the dimensions of organizational commitment, are expressed in Table 4, where the average has been transformed into a percentage to facilitate its interpretation. A column has been added to interpret the result with the established scale (Table 7):

Table 7. Level of intensity with which each of the dimensions of organizational commitment is presented.

Dimensions of job satisfaction	University		
	UPA	UPS	UP
Affective	78.78 (Strong)	41.56 (Moderate)	88.44 (Strong)
Normative	71.00 (Strong)	43.17 (Moderate)	80.67 (Strong)
Calculated	54.42 (Moderate)	43.08 (Moderate)	55.33 (Moderate)

As can be seen, the dimension of organizational commitment that is more strongly manifested among teachers of the Public University and Private (Associative) University is the affective and normative one, while the one with less intensity is the calculated one. In the case of the Private (Corporate) University, the three dimensions are at the same level, with a focus on the calculated dimension.

The level of organizational commitment exhibited by the professors of the Public University and the Private (Associative) University is 68.07% and 74.81%, respectively. Interpreted with the already established scale, this suggests that both have a strong organizational commitment, with Private (Associative) University slightly higher. Meanwhile, the teachers of the Private University (Corporate), with a level of commitment of 42.60%, exhibit weak organizational commitment.

4 Discussion and Conclusions

The central objective of this study was to identify the level of job satisfaction and institutional commitment among undergraduate assistant university teachers hired from three universities located in the Arequipa Region, and to determine if there is a correlation between these variables and their demographic variables. In order to achieve the general objective, four specific objectives were formulated that, once the investigation was completed, were also considered achieved, as shown below.

Objective one: To establish the level of job satisfaction of the teachers of the UP, UPS, and UPA. The level of organizational commitment manifested by the teachers of the UPS is 78.14%, UP is 77.05%, and UPA is 72.81%, allowing us to affirm that the three case studies present a medium-high organizational commitment. Similar results were reported by other researchers [10, 17].

Objective two: To determine the dimensions of job satisfaction that present the greatest and least presence among the teachers of the UP, UPS, and UPA. The dimension of job satisfaction that is manifested with greater force in the teachers of the UP and UPS is that of professional performance, and in the case of the UPA, it is that of interpersonal relationships. The dimension of less intensity in the UP and UPS is that of working conditions, and in the UPA, it is that of organizational factors. Similar results were reported by other researchers [7, 19].

Objective three: To establish the level of organizational commitment of the teachers of the UP, UPS, and UPA. The level of organizational commitment manifested by the teachers of the UPS is 42.60.14%, UP is 68.07%, and UPA is 74.81%, allowing us to affirm that the three case studies present a medium-high organizational commitment. Similar results were reported by other researchers [14, 19, 23].

Objective four: To determine the dimensions of the organizational commitment that present a greater and lesser presence among the teachers of the UP, UPS, and UPA. The dimension of organizational commitment that is manifested with greater force in the teachers of the UP is the affective one, in the UPS it is the normative one, and in the case of the UPA, it is the affective one. The dimension of less intensity in the UP and UPA is the calculated one, and in the UPS, it is the affective one. Similar results were reported by other researchers [7, 20].

Objective Five: Determine the role of sociodemographic variables in job satisfaction and organizational commitment of teachers from UP, UPS, and UPA. The study found that sociodemographic variables do not significantly influence organizational commitment, although job seniority has a particular influence on the calculated dimension. Moreover, the academic level variable influences two indicators, and age and job seniority variables influence only three indicators each. Similar results were reported by other researchers [22, 24].

The teachers of the three universities report a medium-high level of job satisfaction, based on a scale of three values (0 to 33%, low; 34% to 66%, medium; and 67% to 100%, high). However, the result obtained is located at the upper limit of this level. Additionally, job satisfaction is higher in private universities than in UP. The study's findings are consistent with some authors, who report high job satisfaction among teachers in general [9, 13, 18]. However, the findings differ from others, who report a regular level of job satisfaction [8, 12, 16, 21].

The study found that the "professional performance" dimension is where the teachers surveyed are most satisfied, while the "working conditions" dimension has the greatest dissatisfaction. This result is consistent with two researches, who affirm that intrinsic factors are determinants of job satisfaction [11, 15, 26].

The study's findings have practical and theoretical implications. In practical terms, the EPD of the UJED is in a privileged position to achieve high development in the coming years, given the strong organizational commitment and postgraduate studies of its professors. Furthermore, the teachers have a deep psychological identification with their institution and low attachment of a material nature. Theoretically, the study confirms that

EMS teachers have a greater organizational commitment than basic education teachers, with the affective dimension being prevalent among them [25].

The results of the study suggest that university professors are more satisfied with their jobs when they have good interpersonal relationships, strong professional performance, positive organizational factors, fair evaluation of their work, participation in decision-making, good working conditions, an appropriate physical environment, and an effective management team. Therefore, steps can be taken to improve these aspects and increase the job satisfaction of university professors.

The study also found that the institutional commitment of university professors is influenced by hierarchy, labor regulations, the physical environment, material resources, job motivation, communication, opportunities for professional development, and work dynamics. Therefore, steps can be taken to foster the institutional commitment of university professors, such as providing opportunities for professional development, improving communication and work dynamics, and providing an appropriate physical environment and adequate material resources.

Finally, the study suggests that a culture of quality is an important factor in the job satisfaction and institutional commitment of university professors. Therefore, steps can be taken to promote a culture of quality in educational institutions, such as establishing clear quality standards, encouraging faculty participation in decision-making and continuous improvement, and recognizing and rewarding good performance.

References

1. Romi, M.V., Ahman, E., Disman, Suryadi, E., Riswanto, A.: Islamic work ethics-based organizational citizenship behavior to improve the job satisfaction and organizational commitment of higher education lecturers in Indonesia. Int. J. High. Educ. **9**(2), 78–84 (2020)
2. Mainous, A.G., Rahmanian, K.P., Ledford, C.J.W., Carek, P.J.: Professional identity, job satisfaction, and commitment of nonphysician faculty in academic family medicine. Fam. Med. **50**(10), 739–745 (2018)
3. Khoshhal, A., Keshtegar, A.: The effect of structural empowerment and organizational commitment on job satisfaction of the personnel at islamic azad university of birjand. Soci. Sci. (Pakistan) **11**(7), 1234–1241 (2016)
4. Durai, K., Sakthivel Rani, S., Sriram, V.P.: Moderating effect of organisational commitment on the relationship between organisational culture and job satisfaction in higher education institutions. Int. J. Emerg. Technol. **10**(3), 428–435 (2019)
5. Karya, D.F., Zahara, R., Anshori, M.Y., Herlambang, T.: Work-family conflict and organizational commitment of female lecturers of Nahdlatul Ulama university of Surabaya: an investigation of job satisfaction as a mediator using partial least square. IOP Conf. Ser.: Earth Environ. Sci. **747**(1) (2021)
6. French, K.A., Allen, T.D., Miller, M.H., Kim, E.S., Centeno, G.: Faculty time allocation in relation to work-family balance, job satisfaction, commitment, and turnover intentions. J. Vocat. Behav. **120** (2020)
7. Allahbakhshian-Farsani, L., Taghikhani, G., Latifi, M., Asadollahi-Nezhad, M., Maghami-Mehr, A.: The relationship between personality type and organizational commitment with job satisfaction among school librarians in Isfahan university of medical sciences, Iran. J. Isfahan Med. School **37**(351), 687–694 (2019)

8. Ashraf, M.A.: Demographic factors, compensation, job satisfaction and organizational commitment in private university: an analysis using SEM. J. Glob. Responsib. **11**(4), 407–436 (2020)

9. Mohamed, H.H.A., B., Mohamad, M.H., Basir, S.A., Mohaiyadin, N.M., Saudi, N.S.M., Loong, W.W.: The relationship between management's commitment with job satisfaction according to Shari'ah-based QMS (Ms1900:2014) at Malaysian higher education institution. Int. J. Eng. Adv. Technol. **8**(6 Special Issue 3), 1002–1006 (2019)

10. Janib, J., Rasdi, R.M., Omar, Z., Alias, S.N., Zaremohzzabieh, Z., Ahrari, S.: The relationship between workload and performance of research university academics in malaysia: the mediating effects of career commitment and job satisfaction. Asian J. Univ. Educ. **17**(2), 85–99 (2021)

11. Yadav, R., Khanna, A., Dasmohapatra, S.: The effects of quality of work life on organisational commitment and job satisfaction: a study of academic professionals in higher education sector. Int. J. Learn. Chang. **11**(2), 129–144 (2019)

12. Afif, A.H.: The relationship between perceived organizational supports with job satisfaction and organizational commitment at faculty members of universities. Sleep Hypn. **20**(4), 290–293 (2018)

13. Meyer, J.P., Allen, N.J.: A three component conceptualization. Hum. Res. Manage. Rev. **1**(1), 61–89 (1991)

14. Trivellas, P., Santouridis, I.: Job satisfaction as a mediator of the relationship between service quality and organisational commitment in higher education. an empirical study of faculty and administration staff. Total Q. Manage. Bus. Excell. **27**(1–2), 169–183 (2016)

15. Chukwusa, J.: Gender difference in organizational commitment, job satisfaction and job involvement: evidence from university library staff. Int. Inf. Libr. Rev. **52**(3), 193–201 (2020)

16. Otoum, R., Hassan, I.I., Ahmad, W.M.A.W., Al-Hussami, M., Nawi, M.N.M.: Mediating role of job satisfaction in the relationship between job performance and organizational commitment components: a study among nurses at one public university hospital in malaysia. Malays. J. Med. Health Sci. **17**(3), 197–204 (2021)

17. Cenas, P.V.: Individual performance commitment rating and work satisfaction of Pangasinan state university faculty with multi-designations: a basis for a mathematical model. WSEAS Trans. Bus. Econ. **19**, 1797–1803 (2022)

18. Bravo, G.A., Won, D., Chiu, W.: Psychological contract, job satisfaction, commitment, and turnover intention: exploring the moderating role of psychological contract breach in national collegiate athletic association coaches. Int. J. Sports Sci. Coach. **14**(3), 273–284 (2019)

19. Asrar-ul-Haq, M., Kuchinke, K.P., Iqbal, A.: The relationship between corporate social responsibility, job satisfaction, and organizational commitment: Case of pakistani higher education. J. Clean. Prod. **142**, 2352–2363 (2017)

20. Guzman, S.A., Foster, P.F., Grandon, E.E., Ramirez-Correa, P., Alfaro-Perez, J.: Exploring the relationship information systems and organizational performance, through job satisfaction and labor commitment in universities. In: Iberian Conference on Information Systems and Technologies, pp. 1–4 (2018)

21. Batugal, M.L.C., Tindowen, D.J.C.: Influence of organizational culture on teachers' organizational commitment and job satisfaction: the case of catholic higher education institutions in the philippines. Univ. J. Educ. Res. **7**(11), 2432–2443 (2019)

22. Mwesigwa, R., Tusiime, I., Ssekiziyivu, B.: Leadership styles, job satisfaction and organizational commitment among academic staff in public universities. J. Manage. Dev. **39**(2), 253–268 (2020)

23. Wang, P., et al.: Association between job stress and organizational commitment in three types of chinese university teachers: Mediating effects of job burnout and job satisfaction. Front. Psychol. **11** (2020)

24. Ngirande, H.: Occupational stress, uncertainty and organisational commitment in higher education: job satisfaction as a moderator. SA J. Hum. Resour. Manage. **19**, 1–11 (2021)
25. Al-Shutafat, M.S., Halim, B.B.A., Awang, H.Z.: Influence of job satisfaction on organizational commitment in public universities of jordan. International J. Innov. Creat. Change **5**(2), 138–156 (2019)
26. Jordan, G., Miglič, G., Todorović, I., Marič, M.: Psychological empowerment, job satisfaction and organizational commitment among lecturers in higher education: comparison of six CEE countries. Organizacija **50**(1), 17–32 (2017)
27. Barraza, M., Acosta, M.: Compromiso organizacional de los docentes de una institución de educación media superior. Revista Innovación Educativa **8**(45), 20–35 (2008)

Virtual Classroom Leveling Model Based on Self-regulation for the Higher Technological Institute "San Isidro"

Ángel Humberto Guapisaca Vargas[✉] [ID] and Verónica Gabriela Venegas Riera[ID]

Instituto Superior Tecnológico, Universitario San Isidro, Cuenca, Ecuador
aguapisaca@sanisidro.edu.ec

Abstract. Technology has been part of the educational system for years, but its application has not reached unprecedented levels of utilization compared to the present time, where the COVID-19 pandemic has led educational institutions to include virtual education, which has been applied in an emergent manner. The objective of this research was to design a virtual classroom model, based on self-regulation of learning and supported by information technologies, to meet the learning achievements in the subject of Mathematics of the first year students of Gastronomy and Financial Administration of the Instituto Superior Tecnológico Universitario San Isidro, through a projective study with documentary research of the institution's database and analysis with box and whisker diagrams for the different variables measured. The virtual classroom was designed according to Pintrich's self-regulated learning model and the ADDIE constructivist instructional design model, which are coupled to form an adequate methodology for virtuality. It was shown that students did not use self-regulation in their learning due to factors such as the distribution of time between work and study and, in some cases, family responsibilities, and it was also shown that the process of self-regulation should be acquired as a study habit.

Keywords: Technology · self-regulation · mathematics · pandemic · levelling

1 Introducción

The COVID-19 pandemic marked a turning point in the educational system at all levels; however, virtual education was adopted only as an emergency measure, with environments that were not necessarily academic, but rather informal, such as "videoconferencing platforms, among other technological proposals, because both students and teachers were not prepared for virtual education, as well as content and methodologies that were adapted without being adequate" [1]. Therefore, the teaching-learning process was affected by the quality and capacity of data transmission or adequate Internet connections, the accessibility of high-end mobile phones or insufficient computer equipment for the number of occupants in the families, the ability to use digital tools such as programs, including Teams and Zoom, which were tools to avoid missing classes and advance with the content [2]. On the other hand, self-regulated learning "explains the

M. Z. Vizuete et al. (Eds.): CI3 2023, LNNS 1041, pp. 366–377, 2024.
https://doi.org/10.1007/978-3-031-63437-6_30

way in which students take responsibility for their own learning process in a proactive way, initiating it personally and continuing it with perseverance and adaptability" [3]. In addition, self-regulation encompasses the cognitive, behavioral, and motivational aspects that students rely on to begin their learning process [4]. From a socio-cognitive perspective, self-regulation is conceived as "an active process in which the person selects the academic goals the student wishes to achieve and allows the person to regulate the cognitive, affective-motivational, contextual, and behavioral variables involved in learning in order to achieve them" [5]. The return to face-to-face teaching revealed the shortcomings in the learning outcomes that the teaching programs had had by not considering previous methods of ensuring parity in educational level achievement to ensure that the students had the necessary knowledge for the level they were studying. The diagnostic tests carried out to measure the level of knowledge at the beginning of each cycle only showed the cognitive gaps. The Covid-19 pandemic had a significant impact on both schools and universities because it immediately influenced the reorganization of all academic activities, especially by implementing creative educational options to avoid negative results in the teaching-learning process [6].

The Instituto Universitario San Isidro has been characterized by providing quality education, because it has the infrastructure, computer technology and it is appropriate to develop a virtual classroom leveling, with an educational model based on self-regulation, according to the reality of young people entering the professions of Gastronomy and Administration, because they are the professions that contain mathematics in their curriculum and is the subject in which the greatest difficulty in learning is contemplated by the abstract and numerical nature of it. The background presented proves that the problem within the teaching-learning process due to the limitation by the pandemic, was worldwide and the need to cover the academic gaps in students is a priority, for which several alternative solutions have been developed, among them virtual platforms converted into interactive classrooms. Therefore, this study is based on the analysis of the design of a virtual classroom model as a levelling tool based on the self-regulation of student learning, supported by the use of ICT, to improve the learning achievements of students of the Instituto Universitario San Isidro in the subject of Mathematics.

1.1 Theorical Framework

The relationship between higher education and self-regulation of learning lies in the existence of a greater degree of self-management independence on the part of students, which requires diversity in learning styles, time management, breadth in research, development of metacognitive skills, self-evaluation and feedback, these being the guidelines that will influence academic performance [7]. Learning self-regulation was developed by [8] The student must have the ability to organize his activities and environment in order to achieve the goals set in the academy. It is important to emphasize that during the planning of the lessons and the definition of the models for the virtual leveling classroom, it was considered to establish the mechanisms or strategies to motivate self-regulation, based on the self-regulated learning model of [5] which will be present from the beginning of the process. Therefore, students must directly assume their roles as autonomous learners in order to achieve the proposed objectives, with the realization of learning activities individually or in collaborative groups.

The strategies for working on students' self-regulation are defined as follows, according to the theory proposed by [9]: 1. Self-regulation of behavior, which includes the study environment, use of time, and collaborative work; 2. Self-regulation of motivation and affect, such as interests and beliefs, motivations and affect, values and moods; and 3. Self-regulation of cognition, referring to the metacognitive process. The proposed self-regulation strategy involves planning for the completion of the tasks and activities proposed in the study modules in order to achieve the learning objectives. For this reason, students set goals and objectives to be achieved in each module and activate cognitive resources to achieve them. [10]. Student compliance with respect to academic activities will depend on the planning of the assigned tasks and projects plus the exact combination of specific strategies according to the students' learning needs. Based on the previous antecedent, it is mentioned that self-regulation mentions that it is necessary to manage the students' time in relation to the assigned tasks, which is why it is advisable to monitor through a control mechanism known as progress bars for activities, which allows the control of the entries to the classroom and the time spent in it. This way the student will have control of his own development in the activities, reading time, or time spent in the classroom in order to improve his efficiency [11]. Therefore, the idea of activity progress bars will cause the learner to select and adjust strategies, increase or decrease effort against the task.

Similarly, at the beginning of the course, an Activity and Task Record Sheet is included, which the student can use to plan his or her own schedule for completion within a given time frame, depending on the learning outcomes [12]. In addition, the task control bar also serves as a motivational tool because students can feel a sense of satisfaction by visualizing the progress of the activities and resources presented in the course. The reflective component is important, the components of the module will be geared towards the learner making judgments and attributions related to the execution of the module, as well as evaluating the tasks and the learning environment, and measuring the achievement with an explanation of why the level was achieved. Therefore, the academic achievement will be reflected by the qualification that the student obtains in the course of the module, either through the completion of activities and evaluations based on a numerical scale [13]. There are several models for instructional design, such as Gange, Gagné and Briggs, the ASSURE model, Dick and Carey, Jonassen, and others [14], however, the ADDIE model was chosen because it makes it easy to set learning objectives, allows for dynamic activities, content structuring, and controlled student workload. On the other hand, instructional design is based on the ADDIE model, which has five phases, in sequential order: Analysis, Design, Development, Implementation, and Evaluation. [15], each of these phases addresses a specific aspect (Fig. 1). The intention of the ADDIE model is to provide a structure of diverse interactions governed by instructions, which have advantages such as: participation, evaluation and student learning by solving improved and current learning practices [16].

The analysis phase takes into account all the variables obtained in the data analysis for the design of the course, such as the students' characteristics, the students' previous knowledge, the resources available; this phase makes it possible to determine: the students' performance, their learning needs, to define the purpose of the teaching module, to identify the necessary resources, to specify the learning objectives, to deduce the

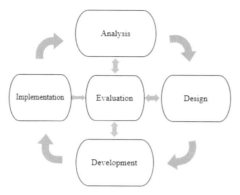

Fig. 1. ADDIE Model. Information retrieved from Branch [2010].

limitations of the course [17]. This phase is essential before starting a teaching-learning process through virtual media, because it allows to have a clear perspective of the situation of the potential participants to the educational project to be developed and above all it solves the organization of the resources and processes that will be applied in the educational module plan. The design phase includes the learning objectives, for which the program is developed in relation to certain didactic criteria according to how certain contents will be taught. Based on the learning objectives for the course, the materials, resources and activities are created and designed, based on the objectives of the module, the design of the evaluation process, systems and means of information transmission, didactic approach and activities focused on the development of the student's knowledge [18]. During the development phase, the contents are worked on in order to generate and validate the necessary didactic resources for the teaching-learning process that are required for the implementation of the module. Therefore, in this phase, the elaboration and testing of materials and resources are important because it will allow the adequate web programming that will include multimedia, development of guides, tutorials for teachers and students [18].

The implementation phase includes the installation of the course on the digital platform that will be used for the development of the module, so in this phase the learning environment will be created with the intention of involving students and training potential instructors to facilitate the teaching-learning strategies, in addition this process includes testing phases with the purpose of collecting information through feedback from those involved to establish present and future corrective actions [17]. The evaluation phase is carried out once the course has been completed and all its components have been evaluated, with appropriate feedback and subsequent adjustments, so that its application is in line with the objectives of self-regulation based leveling; this process makes it possible to evaluate the quality, but not only the products of the teaching-learning processes. Within the formulation of the evaluation, the formative evaluation should be considered, because it helps to propose a modification or a rethinking of one of the phases mentioned above [18].

Virtual classrooms have become a trend in education because they help students learn autonomously, which encourages the university student to take responsibility for learning, for this reason the close relationship with self-regulation is marked because teaching guidelines are established according to contextual and personal factors that favor the development of strategies directed within dimensions such as: cognitive, motivational, supportive and contextual [19]. The intention of educational development through a virtual platform or virtual classroom is to encourage students to develop analysis, reflection and appropriation. Currently, communication and information technologies have become important tools for the development of teaching and learning processes, because they allow the student to generate interaction and creative manipulation with the contents and materials for learning. For this reason, the virtual classroom is considered as a space where information can be shared, such as: search content, evaluations, documents, schedules, multimedia material, forums; then, by having a variety of materials and information, a virtual classroom becomes a necessary tool for didactic use for teaching-learning [20]. Finally, self-regulation and virtual classrooms, together with the ADDIE model, have a close relationship due to the interactivity of a self-directed and flexible nature of the learning environment. The complementarity generated between self-regulation and virtual classrooms is also due to the fact that the student seeks to develop the ability to control, monitor and adjust his own pace and the online learning process in a conscious and more effective way [17, 21, 16]. Therefore, self-regulation combined with virtual classrooms was critical for educational continuity in pandemic scenarios.

2 Methodology

The study developed has a projective model with a quantitative and qualitative approach (mixed), with a research design of experimental type, it also had a descriptive scope and because the existing relationship between the study variables was analyzed. The research carried out is of documentary type, for which the documents contained in the computer system of the Instituto Universitario San Isidro were analyzed, corresponding to the semesters developed in the years 2020, 2021 and 2022. In addition, the socio-demographic data of the students enrolled in the first levels of Financial Administration, Medical Emergency, Nursing and Gastronomy were requested from the Systems Department. In this way, two databases are available, the first with the averages for a quantitative analysis and a second database for a qualitative analysis of the socio-demographic situation. Based on these collected data, the study became retrospective in nature, as the collection of information was based on past events. The population considered was 946 enrolled and active students of the Instituto Universitario San Isidro, corresponding to the periods April 2020-August 2020, October 2020-March 2021, April 2021–August 2021, October 2021-March 2022 and April 2022-August 2022. As a sample, the first level students belonging to the University Gastronomy courses were chosen because their respective curricula include mathematics and it is the subject with the lowest performance, therefore the learning outcomes not achieved by the majority of students.

The technique used for data collection was documentary, for which we considered analyzing the averages at the general level of all subjects, then we proceeded to analyze the courses that have subjects related to numerical analysis, from which once compared the averages between subjects without numerical components and the subjects that handle numbers, we proceeded to evaluate the averages of students in the sample in mathematics, confirming that it is a subject where, in general, the averages are low. Then, the instructional design and the definition of the models for the design of the virtual classroom for the mathematics leveling were carried out, being the model of self-regulated learning of Pintrich, complemented with the instructional design of the ADDIE model. Similarly, the topics to be implemented in the classroom were defined in terms of mathematics, according to the basic requirements of the first level, so that students acquire numerical skills in elementary operations and, after completing the first level, the teacher becomes a bridge between knowledge and students, promoting self-learning [22]. For this reason, it was determined that the content to be worked on would be the law of signs, basic operations and application problems in a fixed time of three weeks with a total duration of 20 h. For the analysis of the data obtained in the socio-demographic data sheets, a linear regression model was performed to determine the variables that influence the students' grade and the relationship they have with it.

3 Result

In relation to the documented information, it was possible to obtain an analysis of the results according to the needs of the students, which allowed to determine the necessary elements that were considered for the development of the virtual classroom for the leveling of the mathematics course. The results are as follows.

3.1 Sociodemigraphic Data

Regarding the socio-demographic data of the students, it should be noted that, according to the completed forms, ethnicity and gender are not influential factors in the learning process. As for the disability situation, it can be a significant factor in the grades in cases where the disability is cognitive and of high impact, otherwise it is not a decisive factor. While the family and work aspects are factors that significantly influence the grades of students, due to the fact of working and studying at the same time together with the responsibility of having a family burden, added to absences due to work issues, leads to a gap between the learning process and the subject dictated. Therefore, according to [23] mentions that emotional, family, and work factors influence the educational performance of university students, generating different effects. On the other hand, the age of the students is an influencing factor, in this case it is positive because there is a relationship between the level of responsibility and the age of the student body, possibly due to an increase in commitment on the part of the person, therefore the analogy is that the older the student, the better the grades. The information presented can be seen in Table 1.

3.2 Mathematics Average in the Gastronomy Program

According to the hypothesis that the students after reintegration in the educational system after the pandemic lowered the educational performance, and before analyzing the

Table 1. Econometric Model related to the socio-demographic analysis

Dep.Variable:	Nota final	R-squared:	0,081		
Model:	OLS	Adj.R-squared:	0,08		
Method:	Least Squares	F-statistic:	56,93		
Date:	2022	Prob (F-statistic):	4,67E-154		
Time:	19:02:29	Log-Likelihood:	-4,21E+04		
N° de observations:	9057	AIC:	8,43E+04		
Df Residuals:	9042	BIC	8,44E+04		
Df Model:	14				
Covariance Type:	nonrobust				
	coef	std err	t	P>(t)	(0,025 | 0,975)
Intercept	34,771	6,879	5,055	0,000	21,287	48,255
C(Etniaest)(T,Blanco)	6,191	7,142	0,867	0,386	-7,810	20,191
C(Etniaest)(T, Indígena)	-2,142	7,042	-0,304	0,671	-15,946	11,662
C(Etniaest)(T, Mestizo)	9,102	6,789	1,341	0,180	-4,206	22,410
C(Etniaest)(T, Montubio)	4,474	8,322	0,538	0,591	-11,840	20,787
C(Etniaest)(T, Otro)	0,374	7,888	0,047	0,962	-15,088	15,835
C(Genero)(T,1)	-0,712	0,566	-1,258	0,209	-1,821	0,398
C(Genero)(T,2)	-2,607	7,658	-0,340	0,734	-17,619	12,404
C(Trabajo)[T, Trabaja y estudia]	-2,937	0,588	-4,990	0,000	-4,090	-1,783
C(Discapacidad)(T, si)	1,197	0,759	1,578	0,115	-0,290	2,864
Faltas	-1,162	0,062	-18,878	0,000	-1,283	-1,041
Créditos	2,990	0,193	15,486	0,000	2,611	3,368
Hijoestudiante	-2,016	0,380	-5,311	0,000	-2,760	-1,272
Edadestudiante	0,398	0,039	10,262	0,000	0,322	0,474
NumMatri	1,504	0,482	3,121	0,002	0,559	2,449
Omnibus:	583,232	Durbin-Watson:	0,464		
Prob(Omnibus):	0	Jarque-Bera (JB):	533,815		
Skew:	-0,532	Prob (JB):	1,21E-116		
Kurtosis:	2,468	Cond, N°:	1,46E+03		

averages of all the subjects by careers, taking into account the scores in internal tasks, external tasks, these two references the score is over 15 points and the lessons over 5 points, to this score is added the grade obtained in the exam of the first partial, which is also over 15 points, fulfilling in this way a total of 50 points. This process is repeated in the second semester until a total of 100 points is obtained. Students who achieve a grade of 70 points or more are promoted to the next level. Therefore, based on this premise, it is determined that the Gastronomy course has a significant deficiency in terms of the general average in the subject of mathematics, which is taught in the first level of the course. This result is obtained by comparing the general average of the grades for the years 2021 and 2022, which are presented with 69.40 and 71.09, that is, they are not significant averages that will cause problems for the students in the following levels to be studied. This reality has been detected worldwide, according to studies analyzed in various sources, for which several authors claim that the pandemic has influenced the low performance of students, undoubtedly this phenomenon affects students in their curricular progress [24]. Detailed information is shown in Fig. 2.

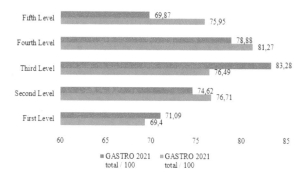

Fig. 2. Overall average of the gastronomy course by level of study.

3.3 Virtual Leveling Classroom

The formulation of the virtual classroom is based on the results obtained in the general analysis of averages, where the gastronomy career revealed that the biggest problem is the lack of knowledge in the subject of mathematics. Thus, the development of a leveling process is formulated to be carried out through a virtual classroom. For its success, it is proposed to follow the self-regulation model added to the ADDIE model, which strengthens a consolidated, organized and friendly structure with the students with the intention of managing their time and commitment to the execution of the leveling.

The training process to be carried out in the virtual classroom will be in the subject of Mathematics, with a total duration of 20 h and the objective of leveling students who are going to enter the first level, in the career of Gastronomy, where the numerical subject is also offered. The content has been distributed in three weeks, where the first and second week of content have seven hours of duration each and the third week with a duration of six hours, another option for the student is to consider the execution of the leveling classroom in a period of 20 days. Regarding the virtual classroom, it has been installed on the Moodle platform of the Instituto Universitario San Isidro, thus promoting a more active participation of the students in a dynamic process of interactive learning, based on the pedagogical philosophy of Moodle, that is, based on a social-constructivist pedagogy that guides the subjects to build their own knowledge through experiences, using active methodologies, continuous evaluation and feedback, and also using the method of cooperative work [25]. As a result of the analysis of the basic contents in which the students have more difficulties, it was determined that the contents of the levelling classroom should include the laws of signs, divisibility criteria, basic operations and application problems, topics which, according to the evaluations of the respective subject, have more difficulties of comprehension and application. The following virtual classroom structure is proposed for them. Table 2 shows this structure.

Table 2. Structure of the Virtual Leveling Classroom

Elements	Parámetros a considerar
Course Structure	Three content modules
Introductory video	Self-authored video presenting the module topic and objectives. The video has a maximum length of 5 min
Didactic resources	Main resources (study guides, instructional videos, main activities) Supplementary resources (web links, infographics, multimedia resources, reinforcement activities)
Interactive learning activities	Academic forums, assignments, databases, interactive quizzes, lessons, glossaries, games
Formulation of evaluative activities	Summative and linked to the content and learning objectives of the module. Each module will have one summative activity and at least one of the activities will have an assessment rubric
Meetings	Synchronous meeting in the first week for the presentation of the course. Asynchronous meetings for the modules

Source: Authors

4 Discussion

The pandemic had a significant impact on the academic performance of the students, causing a drop in grades in various subjects, which is the basis for an educational setback [26] that teaching and learning mathematics in the virtual modality imposed by the emergency may negatively affect student achievement. This assertion and the analysis of the results developed in this study confirm that there is a notorious influence with respect to low student grades in times of pandemic. However, in addition to the causes of low student performance, factors such as economic stability, work situation, Internet connectivity, emotional factors and family burden are added because in some cases it has a positive influence and in other cases it is a negative factor that generates constant conflict. [23].

With regard to the study, an analysis of additional factors that could influence the performance of students such as age, ethnicity and disability was planned, from which it was found that there is no causality with academic performance; but it was found that the older the student, the more responsibility he/she presents, despite the social factors mentioned above. Based on this construct, it can be inferred that age could be a factor of student responsibility, although it would not be causal. In order to remedy the students' learning deficiencies, especially in mathematics, a virtual classroom has been developed based on the self-regulation model, which helps with academic planning so that each student fulfills his or her learning commitment within the development time of the module to be taught. For this reason, the student must learn to manage time and resources during the learning process, but it is necessary that the teacher formulates a virtual classroom with clear content, didactic spaces, interactive material, evaluations, and it is also essential that there is follow-up and constant feedback [27]. As a solution to the lack of academic performance, especially in numerical subjects such as mathematics, it is essential to propose solution strategies that consider teaching the content to the student, but at the same time assume the responsibility of self-learning through a virtual platform developed with content appropriate to the needs of the student and the subject to be taught. Therefore, the adoption of self-regulation is considered, which aims to encourage the student to manage time, resources, evaluation processes and feedback, in addition to the fulfillment of activities. In addition to self-regulation, it is essential to add the ADDIE model, which consists of analysis, design, development, implementation and evaluation, considering that these elements would consolidate the structure of the teaching-learning process through virtual environments [28].

Finally, the teaching-learning process in cases of students with deficiencies in academic performance, must be considered under structural parameters through successful models such as the implementation of self-regulation plus trend models that allow organizing the necessary elements of knowledge, as is the case of ADDIE design, especially when building virtual classrooms aimed at the improvement and academic recovery of students who present difficulties due to various factors. The intention of generating innovative proposals in education is to achieve significant learning in higher education.

5 Conclusion

In conclusion, although virtual education is a current trend, it is necessary to mention that an adequate construction of virtual teaching proposals must be composed according to the self-taught component of self-regulation, although in a pilot test it was evident that students showed little commitment to their learning process because they no longer had the constant presence of the teacher. In addition, the model is designed for students entering the first level of professional training. For this reason, the process, with its respective components, could be chosen for high school levels, even at the school level, which would make it obligatory to work with data corresponding to each level of training.

It is important to mention that in the case of application to training in general, whether business or institutional, not necessarily educational, the model should be adapted to the objectives of each situation and to the knowledge needs of the potential beneficiaries. Once the adjustments have been made after the classroom evaluation, it could be used as a model and proposed for use with other subjects, with the same objective of leveling students who have knowledge deficiencies in any subject. In the future, a general virtual classroom model could be created for leveling according to the subjects included in the induction for the first cycle of each degree program. Finally, when presenting the model of virtual classroom for leveling based on self-regulation, we worked with students entering the first levels of the Institute's courses. Since it is a model, it is important to validate and even reconsider the statistical data collection process in terms of socio-demographic variables, because if it were to be implemented in other educational units other than the Instituto Universitario San Isidro, the situational reality of the students could vary and, therefore, other teaching and self-regulation models should be analyzed.

References

1. Gordillo, M.M.: Revista Iberoamericana de Docentes ¿Entornos digitales sin contornos educativos? Revista Iberoamericana de Docencia [Internet] (2021). [citado 20 de agosto de 2023], 1–9. Disponible en http://revistaiberoamericanadedocentes.com
2. Esteche-Cabaña, E., Gerhard-Wasmuth, Y.: Factores que inciden en la educación virtual en tiempos de pandemia (COVID-19) de los estudiantes universitarios de una universidad privada. Revista Iberoamericana de Docentes (2020)
3. Zimmerman Schunk, B.J.: Theories of self-regulated learning and academic achievement: an overview and analysis. Self-regulated learning and academic achievement: theoretical perspectives (2001)
4. Panadero, E., Alonso-Tapia, J.: How do students self-regulate? Review of Zimmerman's cyclical model of self-regulated learning. Anales de Psicologia 30(2) (2014)
5. Pintrich, P.R.: The role of goal orientation in self-regulated learning. Boekaerts & Niemivirta, pp. 451–502 (2000)
6. Gómez-García, G., Ramos-Navas-Parejo, M., de la Cruz-Campos, J.C., Rodríguez-Jiménez, C.: Impact of COVID-19 on university students: an analysis of its influence on psychological and academic factors. Int. J. Environ. Res. Public Health 19(16) (2022)
7. Hernández Barrios, A., Camargo Uribe, Á.: Autorregulación del aprendizaje en la educación superior en Iberoamérica: una revisión sistemática, vol. 49, Revista Latinoamericana de Psicologia (2017)
8. Zimmerman, B.J.: Attaining Self-regulation a Social a Cognitive Perspective. Academic Press, pp. 13–39 (2000)
9. Berridi Ramírez, R., Martínez Guerrero, J.I.: Estrategias de autorregulación en contextos virtuales de aprendizaje. Perfiles Educativos 39(156) (2017)
10. Barreto-Trujillo, F.J., Álvarez-Bermúdez, J.: Estrategias de autorregulación del aprendizaje y rendimiento académico en estudiantes de bachillerato. Revista de Estudios e Investigación en Psicología y Educación 7(2) (2020)
11. De Raadt, M., Dekeyser, S.: A simple time-management tool for students' online learning activities. En: ASCILITE 2009 - The Australasian Society for Computers in Learning in Tertiary Education (2009)
12. Dieser, M.P.: Estrategias de autorregulación del aprendizaje y rendimiento académico en escenarios educativos mediados por tecnologías de la información y la comunicación [Tesis doctoral]. [Buenos Aires]: Universidad Nacional de La Plata (2019)
13. García-Marcos, C.J., López-Vargas, O., Cabero-Almenara, J.: Self-regulated learning in online vocational education and training: effects of time management. Revista de Educación a Distancia 20(62) (2020)
14. Belloch, C.: Diseño Instruccional Consuelo. Chin. J. Microbiol. Immunol. 24(6) (2004)
15. Templos, L.: Modelo Instruccional ADDIE. Logos Boletín Científico de la Escuela Preparatoria No 2, 7(14) (2020)
16. Morales González, B.: Diseño instruccional según el modelo ADDIE en la formación inicial docente. Apertura 14(1) (2022)
17. Branch, R.M.: Instructional design: The ADDIE approach. Instructional Design: The ADDIE Approach (2010)
18. Morales, B., Edel, R., Aguirre, G.: Modelo ADDIE (análisis, diseño, desarrollo, implementación y evaluación): Su aplicación en ambientes educativos. En: Los Modelos Tecno-Educativos, revolucionando el aprendizaje del siglo XXI (2014)
19. Martínez-Sarmiento, L.F., Gaeta González, M.L.: Utilización de la plataforma virtual Moodle para el desarrollo del aprendizaje autorregulado en estudiantes universitarios. Educar 55(2) (2018)

20. Aguilar Ponce, L.D.J., Zambrano Montes, L.C.: Uso didáctico de las aulas virtuales en la enseñanza-aprendizaje. Revista Iberoamericana de Tecnología en Educación y Educación en Tecnología (32) (2022)
21. Miño-Puigcercós, R., Domingo-Coscollola, M., Sancho-Gil, J.M.: Transformar la cultura de enseñanza y aprendizaje en la educación superior desde una perspectiva diy. Educación XX1 **22**(1) (2019)
22. Mendoza, H.H., Burbano, V.M., Valdivieso, M.A.: El papel del docente de matemáticas en Educación superior a distancia y virtual: una mirada desde los métodos mixtos de investigación. Revista Espacios **40**(39) (2019)
23. Casiano Inga, D.A., Cueva Vega, E., Zumaeta Barrientos, M.R., Casiano Inga, C.A.: Impacto de la covid-19 en el desempeño académico universitario. Un análisis comparativo para la Universidad Nacional Toribio Rodríguez de Mendoza, en Amazonas (UNTRM-A). Actualidades Pedagógicas **1**(77) (2022)
24. Monroy-Varela, S.E., Gallego-Vega, L.E., Amórtegui-Gil, F.J., Vega-Herrera, J.M., Díaz-Morales, H.: Impact of the COVID 19 pandemic on the student's academic performance at the School of Engineering-Universidad Nacional de Colombia, Bogotá Campus. DYNA, Colombia, vol. 89 (2022)
25. Viteri Rade, L.Y., Valverde Alcívar, M., Torres Gangotena, M.W.: La plataforma Moodle como ambiente de aprendizaje de estudiantes universitarios. Revista Publicando **8**(31) (2021)
26. Sinche Delgado, A.V.: Incidencia del covid-19 en el rendimiento académico de la asignatura de matemáticas. Ciencia Latina Revista Científica Multidisciplinar **7**(1) (2023)
27. Molina Acuña, M.A., Fossi Becerra, L.F.: Autorregulación del aprendizaje: mediador en la adaptación, motivación y permanencia en la educación superior distancia tradicional. Perspectivas [Internet]. [citado 20 de agosto de 2023];7(22). Disponible en (2022). http://portal.amelica.org/ameli/journal/638/6383364007/html/
28. Saza Garzón, I.D., Mora Marín, D.P., Agudelo Franco, M.: El diseño instruccional ADDIE en la Facultad de Ingeniería de UNIMINUTO. HAMUT'AY **6**(3) (2019)

Author Index

A

Aguirre, Vicente Reinaldo Cango 216
Alejandra, Vallejo 131
Almeida, Patricia Lisbeth Esparza 82, 324
Álvarez-Sánchez, Ana 243
Álvarez-Tello, Jorge 301
Andía-Gonzales, Brizaida Guadalupe 350
Andrade, Paola 155
Arévalo Bonilla, Verónica Patricia 260
Aroca, Irlanda 3
Ávila-Ascanio, Luis Fernando 314
Avilés-Castillo, Fátima 144
Ayala, Karla 229
Ayala-Mendoza, Asdrúbal 301

B

Barona, Diego 30
Basantes, Carolina 275
Beltrán, Jorge 275
Buestán, Marco Guamán 190
Bullón-Solís, Omar 251

C

Cabrera, Diana Sánchez 122
Calero, Liliana M. 339
Carrera Guerrero, Sergio 260
Chabla, Anddy Sebastián Silva 287
Chasi, Carolina 203
Chasi, Consuelo 155
Cobos Lazo, Richard Santiago 15
Cóndor, Mariela 69
Córdova, M. 98
Crespo, Galo Patricio Hurtado 287
Cruz, Patricio 178
Cuenca Soto, María del Cisne 15
Cuzme-Rodríguez, Fabián 167

D

del Villar-Labastida, Alexis Suárez 243
Duche-Pérez, Aleixandre Brian 350
Duque-Romero, Marco 301

E

Espín, Francisco 155
Espinoza, Paul S. 216, 339
Estacio, Karen 110

F

Farinango, Mauricio 69
Farinango-Endara, Henry 167
Flores-Armas, Stefany 167

G

Galárraga, Juan Carlos Jaramillo 324
García, Mónica Monserrath Chorlango 82, 324
Gavilanez, I. 98
Geremy, Novoa 131
Gutiérrez-Aguilar, Olger Albino 350

H

Hoppe-Coronel, Sheylah 251

I

Iturra, Francisco 30

J

Jaime-Zavala, Milena Ketty 350
Jaramillo, Hugo Jonathan Narváez 82
Jaramillo, Karina Salom é Ayala 314
Josselyn, Solórzano 131
Juiña, Daniela 30

K

Kevin, Toapanta 131

L

Lara, Paulina 3
López, Pablo Crespo 190

M. Z. Vizuete et al. (Eds.): CI3 2023, LNNS 1041, pp. 379–380, 2024.
https://doi.org/10.1007/978-3-031-63437-6

M
Machay, Edwin 178
Máiquez, Diego 69
Manzano, Eduardo 155
Martínez-Gómez, Javier 203
Mayra, Quilumbaquin 131
Méndez, Marlon Fabricio Hidalgo 324
Michilena-Calderón, Jaime 167
Minango, Juan 229
Molina, Jaime 203
Mollocana Lara, Juan Gabriel 54
Montenegro, Byron Sebastián Trujillo 82, 324
Morales, Luis Shayan Maigua 324
Moreno, Gustavo 203
Morocho Ochoa, Pablo Gerónimo 15

N
Nolazco-Labajos, Fernando Alexis 251

O
Oyasa, Ana 69

P
Pabón, Diego 69
Paguay, Angélica 3
Paredes Obando, Johanna Beatriz 54
Paredes, Wladimir 229
Paredes-Quispe, Fanny Miyahira 350
Parra, Marco Antonio Gómez 122
Pazmiño-Guevara, Lizzie 301
Peñaranda, Nelson David Cárdenas 287
Peralta, Diana 203
Pilco, Marco V. 216, 339
Pozo-Carrillo, Cristian 167

Q
Quintana, Alain 260
Quinzo, Maritza 275

R
Ramírez-Borja, Javier Roberto 350
Riera, Verónica Gabriela Venegas 366
Rivera, E. Fabian 216, 339
Robalino, A. 98
Rodríguez Muñoz, Paúl 260
Rodríguez, Gonzalo 69
Rodríguez-Palomino, Roli David 251
Román-Bermeo, Cynthia L. 43
Romero, M. 98

S
Schaffry, Jerry Medina 339
Silva, Byron 30

T
Taruchain-Pozo, Luis Fernando 144
Tituaña, Darwin 178

U
Umaquinga-Criollo, Ana C. 287

V
Vaca, Kevin Andrés Rivera 82
Vallejo, Daniel Ochoa 3
Vargas, Ángel Humberto Guapisaca 366
Vásquez-Ayala, Carlos 167
Velásquez, Carlos 30, 155
Vera-Revilla, Cintya Yadira 350
Verdezoto Carrillo, Marco 260
Vilema-Escudero, Segundo F. 43
Villegas, Richard Antonio Martínez 122

Y
Yar, Edison Robinson Rodríguez 82
Yépez, Rafael Maldonado 122

Z
Zambrano, Marcelo 229

Printed in the United States
by Baker & Taylor Publisher Services